Fundamentals of the Theory of Mechanical Vibrations

Rubens Gonçalves Salsa Junior

Fundamentals of the Theory of Mechanical Vibrations

Springer

Rubens Gonçalves Salsa Junior
Federal University of Rio Grande do Norte
Natal, Brazil

ISBN 978-3-031-83994-8 ISBN 978-3-031-83995-5 (eBook)
https://doi.org/10.1007/978-3-031-83995-5

This Springer imprint is published by the registered company Springer Nature Switzerland AG
The registered company address is: Gewerbestrasse 11, 6330 Cham, Switzerland

To my daughter, Tutu

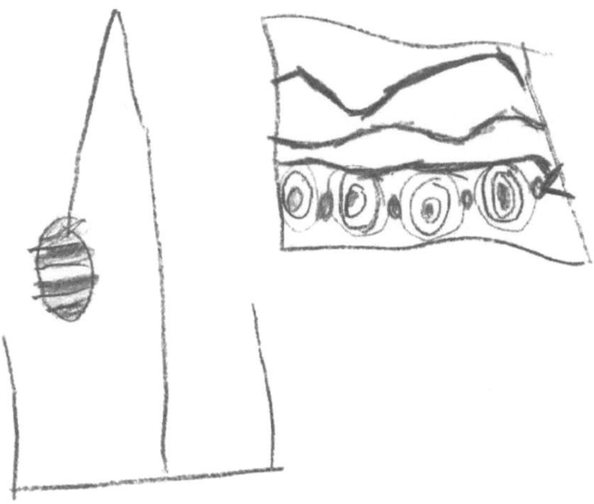

Preface

The *Vibrations of Mechanical Systems* course at the Federal University of Rio Grande do Norte (UFRN) aims to provide valuable training in the modeling and analysis of mechanical systems, establishing a solid and critical foundation for students aspiring to careers in various fields of mechanical engineering. As with any vibrations courses, it is essential that students have a good foundation in vector/matrix algebra, differential calculus, and differential equations. One of the greatest challenges reported by UFRN students is visualizing and understanding models applied to real-world situations.

To address this challenge, two educational projects were developed over the course of six semesters. The first project modernized the course by integrating classical theory with computational simulation, allowing the problems of interest to be more realistic and reducing the abstraction barrier of the course. Simulations also encouraged the development of critical thinking, as students need to actively participate in determining parameters, predicting outcomes, and investigating hypotheses. This approach also helped bridge the gap between classroom teaching and modern techniques used in professional design practice.

The second project aimed to further align theoretical classes with laboratory work, following the current trend in engineering education that values the balance between theory and practice. However, this aspect was compromised by nearly 3 years of remote instruction. To overcome this, activities were developed in which students built mechanical systems with simple everyday materials to validate qualitatively and quantitatively theoretical results. Similar to teaching geometry with compass and ruler, students performed practical experiments with common materials such as bottles, strings, and balls. These activities not only increased satisfaction with the course during remote instruction but also improved understanding of theoretical concepts. Due to their simplistic nature, these activities were obviously implemented for systems with one degree of freedom.

The goal of this book is to report the lecture notes that incorporated some of the activities from these two educational projects. The classical content is presented along with computational simulations whenever possible. Various MATLAB codes are provided throughout the text: some for integrating differential equations and others for generating illustrative plots. More complex codes are commented on so the parameters can be easily modified. MATLAB was chosen for its suitability for scientific applications, and an appendix with a brief review of differential equation integration using MATLAB has been included, assuming the reader has basic knowledge of vector operations and graph generation on this platform.

In addition, these notes include some of the practical experiments developed in the aforementioned teaching projects. These experiments are used to validate theoretical results and help students understand how equations represent real situations. All experiments are built with simple and accessible materials, making them replicable by readers. My hope is that these experiments can help students and instructors that either still use remote instruction or that do not have access to sophisticated laboratory facilities. None of the experiments presented are original. For example, the cylindrical buoy experiment was adapted from J. Casey's article titled "The Bobbing Buoy." Various references are listed in the bibliography but were not cited directly in the text, as the primary focus of this book is to meet the needs of undergraduate students.

The material is suitable for a semester-long undergraduate vibrations course aimed at mechanical engineering students who have completed *Dynamics*, as well as for graduate students who wish to review vibrations or even instructors seeking new teaching methods. An appendix with a brief review of key concepts from Dynamics has been included. The notation used follows that of O. M. O'Reilly in "Engineering Dynamics: A Primer", including the four steps for problem-solving described in his book.

The book is divided into two parts for didactic reasons. The first part addresses free and forced vibrations of single-degree-of-freedom systems. The second part focuses on systems with multiple degrees of freedom. I included two chapters in which I discuss the general form of the equations of motion. This topic is generally implicitly addressed in undergraduate courses, but I believe it helps students understand the limitations and applicability of the theory. Each chapter contains suggested problems that can be used for practice. For those in need, there are appendices reviewing vectors, matrices, and complex numbers as well.

Natal, RN, Brazil — Rubens Gonçalves Salsa Junior

Acknowledgments

This project could not have been accomplished without the help of many people. First and foremost, I would like to thank Springer and Executive Editor Michael Luby for believing in and endorsing this book. I am also grateful to Production Editor Dhivya Savariraj for the patience during the manuscript production.

The educational projects that I coordinated and laid the foundation of this book would not have been possible without the grants provided by the Federal University of Rio Grande do Norte. I extend my thanks to my Teaching Assistants V. L. Von Sohsten and L. T. A. Oliveira, whose assistance was fundamental in crafting some of the experiments.

I want to thank my colleagues R. A. Cardoso and R. C. S. Freire Jr. for reading an earlier version of this book. I specially grateful to T. P. Sales and M. E. Yıldızdag for carefully reviewing the first draft of this book and providing me with valuable feedback. I would also like to thank my wife and art historian, Luiza, for always being available to read the manuscript, even though it was outside her field.

Although I have done my utmost, it is inevitable that a project of this magnitude may contain errors. If you find any typographical errors in this book, please kindly contact me at rubens.salsa@ufrn.br.

Natal, RN, Brazil Rubens Gonçalves Salsa Junior

Contents

Commonly Used Symbols and Abbreviations

Latin letters:

A_j constant of integration

B_j constant of integration

c viscous damping coefficient

C center of mass

$\mathbf{C}$ $n \times n$ damping matrix

$\mathbf{D}$ $n \times n$ modal damping matrix

$\{\mathbf{e}_1, \mathbf{e}_2, , \mathbf{e}_3\}$ corotational basis vectors

$\{\mathbf{e}_r, \mathbf{e}_\theta\}$ unit vectors associated with polar coordinates

E total mechanical energy

F_0 force amplitude

$\mathbf{f}(t)$ n-dimensional forcing vector

g acceleration of gravity

G transmissibility of displacement

$\mathbf{g}(t)$ n-dimensional forcing vector of the decoupled system

H frequency response function

$\mathbf{H}$ $n \times n$ matrix of frequency response functions

i imaginary unit $i = \sqrt{-1}$

$\mathbf{i}$ unit vector of the Cartesian system

$\mathbf{I}$ identity matrix of order n

$\mathbf{j}$ unit vector of the Cartesian system

k linear spring stiffness

$\mathbf{k}$ unit vector of the Cartesian system

$\mathbf{K}$ $n \times n$ stiffness matrix

m mass

$\mathbf{M}$ $n \times n$ mass matrix

n system dimension ($n \geq 1$)

p_j j-th modal coordinate

$\mathbf{p}$ n-dimensional vector of modal coordinates

q_j j-th generalized coordinate

Q_j	constant amplitude
$\mathbf{q}$	n-dimensional column vector of generalized coordinates
$\mathbf{s}_j$	j-th mode of vibration
t	time
T	(natural) period of oscillation
T_d	damped period of oscillation
$\mathbf{u}$	n-dimensional mode shape vector
U	potential energy
$\mathbf{U}$	$n \times n$ modal matrix
$\mathbf{v}$	n-dimensional eigenvector associated with λ

Greek letters:

α	scalar constant
ϵ	change of length of spring
ζ	damping ratio
η	coefficient of proportional damping
θ	angle
λ	eigenvalue
ϕ	phase angle
ψ	holonomic constraint
ω_n	natural frequency
ω_d	damped frequency
ω_j	j-th natural frequency of a system with n degrees of freedom
ω	frequency of excitation
$\mathbf{\Omega}$	$n \times n$ spectral matrix

Special symbols:

$\mathbb{C}$	set of complex numbers
$\mathbb{R}$	set of real numbers
$\mathbf{0}$	zero matrix of order n or n-dimensional zero vector

Mathematical symbols:

$\overline{\lambda}$	complex conjugate of λ
$\mathbf{M}^T$	transpose of $\mathbf{M}$
$\dot{q}$	time differentiation of q

Abbreviations:

DOF(s)	degree(s) of freedom
FBD(s)	free body diagram(s)
FRF(s)	frequency response function(s)
IC(s)	initial condition(s)
MDOF	multi degrees of freedom
ODE(s)	ordinary differential equation(s)
SDOF	single degree of freedom

Part I
Vibration of Systems with One Degree of Freedom

Chapter 1
The Equation of Motion of Systems with One Degree of Freedom

This chapter explores the fundamental characteristics of vibratory systems and presents the general equation of motion for single-degree-of-freedom systems. The approach follows the engineering analysis cycle, beginning with the formulation of mathematical models, progressing to dynamic response analysis through computational simulation, and concluding with experimental validation. The activities outlined in this chapter can be easily reproducible by readers, offering visual aids that clarify abstract concepts like equilibrium points, stability, phase portraits, and the distinction between large and small vibrations. Four illustrative examples are provided to further demonstrate these concepts.

1.1 Why Study Vibrations?

Dynamics encompasses the broad area of mechanical engineering that focuses on the study of the motion of material objects and its causes. A specific subarea of interest within this domain is *vibrations*, which involves the analysis of oscillatory motion. Vibrations may manifest in a variety of ways in daily life. For instance, sound is transmitted by the longitudinal vibration of air particles. These waves enter the ear and set off vibrations in the eardrum and the trio of minuscule bones located within the middle ear, which engage in the propagation of auditory impulses. Another everyday example is in the vibrations of a washing machine due to the rotational motion of wet clothing during the spin cycle. In this case, the uneven distribution of cloths around the rotation axis results in a centrifugal force that makes the machine vibrate.

Supplementary Information The online version contains supplementary material available at (https://doi.org/10.1007/978-3-031-83995-5_1).

R. Gonçalves Salsa Junior, *Fundamentals of the Theory of Mechanical Vibrations*,
https://doi.org/10.1007/978-3-031-83995-5_1

The vibration of a washing machine described earlier is analogous to vibrations experienced by unbalanced rotors, which can lead to increased wear and tear on bearings and seals, potentially resulting in catastrophic failure. Through condition monitoring, engineers continuously monitor rotating machinery such as motors, pumps, and turbines to detect changes in vibration patterns that may indicate developing faults or degradation. These changes can include increases in amplitude, changes in frequency spectra, or the appearance of new vibration frequencies. This real-time monitoring belongs to the large class of techniques of predictive maintenance that uses data analysis to predict and prevent equipment failures.

The design of many machines and structures needs to properly consider how they vibrate under various conditions. Automotive engineers, for instance, employ vibration analysis to design many vehicle components. One of them is the suspension system, which is important for passengers to ride comfortably on a bumpy road. To reduce the amount of vibration transmitted to the passengers, the suspension needs to be more flexible. If it is too flexible, the vehicle might not handle well. Engineers need to consider these conflicting factors when fine-tuning the suspension system. The crankshaft is also a critical component in terms of vibration analysis. The cyclical nature of engine operation subjects it to torsional vibrations. This type of vibration is undesired, and it can be mitigated by fitting a harmonic damper to one end of the crankshaft. This type of mechanical appendage is named vibration absorber, and it is designed to bring oscillations to an acceptable level.

Vibration absorbers are a type of passive vibration control strategy that dissipate vibrational energy without requiring external power or control inputs. They function based on their inherent mechanical properties and do not actively adjust their response to changing vibration conditions. A common example of vibration absorbers can be found in electrical overhead transmission lines, where dumbbell-shaped devices called Stockbridge dampers are attached to cables to reduce vibrations caused by strong wind currents. Consideration of the vibrations caused by fluid-structure interplay is important in the design of many civil structures. One prominent example is Taipei 101, one of the tallest buildings in the world, which is particularly susceptible to lateral vibrations from strong winds. The pendulum installed at its top is designed to mitigate these wind-induced vibrations. One notorious case of uncontrolled vibrations is the Millennium Bridge in London. Shortly after it opened to the public, it experienced unexpected lateral swaying motion caused by the synchronization of the pedestrians' footsteps with the bridge's own natural frequency. The bridge was temporarily closed for modifications to dampen the swaying, and it has since been reopened. Given the prevalence of vibration in engineering applications, studying this field assumes significant importance.

The objective of this chapter is to introduce fundamental concepts of vibrations. The presentation follows the engineering analysis cycle, beginning with the formulation of mathematical models, progressing to dynamic response analysis through computational simulation, and concluding with experimental validation. The chapter is organized as follows. Section 1.2 explores the main characteristics of vibratory systems through a simple example, introducing key concepts such as kinematic constraints, degrees of freedom (DOF), damping, the equation of motion, and stability. In Sect. 1.3, the discussion advances in complexity with the introduction of the simple pendulum, a classical example that demonstrates the use of phase portraits to analyze nonlinear dynamics, numerical simulations, and linearization for small motions. These examples lay the foundation for Sect. 1.4, which discusses the general form of the equation of motion for single-degree-of-freedom (SDOF) systems. Section 1.5 provides two comprehensive illustrative examples, and the chapter concludes with a summary in Sect. 1.6.

1.2 The Classical Oscillator

Vibratory systems[1] possess five distinct characteristics: (i) an equilibrium position to oscillate about, (ii) some way to store kinetic energy, (iii) a mechanism that tries to restore the system back to equilibrium, (iv) a form to dissipate energy, and (v) stimuli that set the system into motion. These five traits can be identified through a straightforward demonstration outlined in the sequence.

1.2.1 A Demonstration

A simple vibratory system is shown in Fig. 1.1. It was constructed by attaching a plastic bottle filled with coins to a linear spring. This spring is a common one that can be removed from inside toys or bought in hardware stores. Once this *mass-spring system* is given an initial downward displacement and subsequently released from rest, the bottle oscillates vertically. If one waits long enough, the amplitude of oscillation starts to decrease, and eventually, the bottle comes to rest at the static *equilibrium position*, where the gravitational force is exactly balanced by the elastic force exerted by the stretched spring. Any other *stimuli*, such as nudging the container away from this position, will trigger renewed oscillations of the bottle around the static equilibrium position. While in this illustrative example, the initial stimulus was an imposed displacement, a vibratory system can be *perturbed* from equilibrium in various ways, including imposed velocity, force, torque, or a combination thereof.

Once the system is set into motion, oscillations arise in part because of the *elasticity* of the linear spring that is always trying to restore the bottle back to equilibrium. Intuitively, restoring forces are associated with linear springs, like the one shown in Fig. 1.1. Other common sources of restoring forces in vibratory systems are torsional springs, elastic structural elements such as rods, beams, shafts, or even the gravitational force. The mass, or *inertia*, stores kinetic energy and plays a crucial role in the exchange between kinetic and potential energies, contributing to

[1] A system is an assemblage of components acting as a whole.

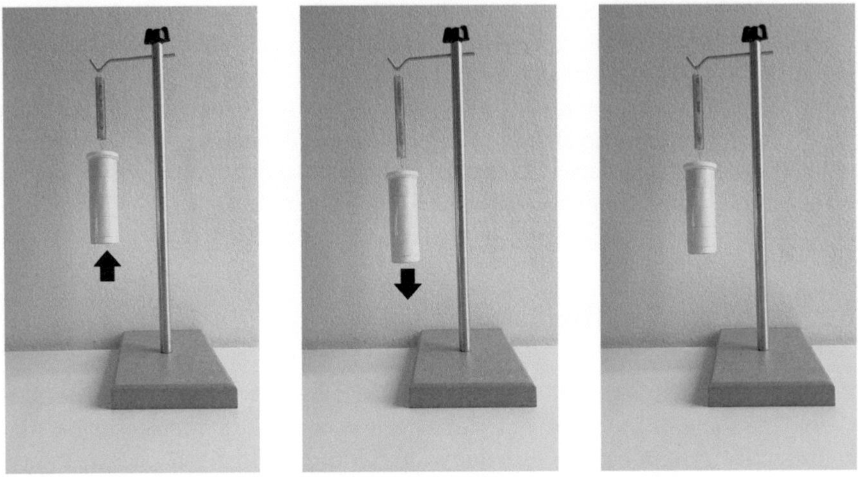

(a) Bottle moving upward and compressing the spring.

(b) Bottle moving downward and streching the spring.

(c) Bottle in static equilibrium.

Fig. 1.1 Snapshots of the oscillatory motion of a bottle filled with coins and attached to a spring

the occurrence of oscillations. Clearly, any physical system that possesses elasticity and inertia is capable of vibration. Finally, it was observed that oscillations do not persist indefinitely, but gradually decay over time. Energy is being dissipated somehow, and as a consequence, the bottle eventually settles in the static equilibrium position. This energy dissipation is referred to, in vibration terminology, as *damping*.

1.2.2 Building a Model

It is of fundamental importance to understand how the system in Fig. 1.1 vibrates due to different stimuli. Instead of physically testing the system for every possible scenario, engineers conceive mathematical representations for moving bodies and loads that act upon them, a process known as *modeling*. Through these mathematical models, the vibratory behavior can be investigated qualitatively and quantitatively using the various tools from calculus.

Developing a satisfactory model can often be challenging, as it requires striking a balance between simplifying the problem for mathematical tractability and preserving its essential physical features. Modeling is sort of an art form that requires some experience, but the present case is simple enough to serve as a first example.

For this spring-bottle setup, three simplifying assumptions can be made based on the observations: (i) the mass of the spring may be assumed negligible when compared to the mass of the bottle filled with coins, (ii) rotational motions of the

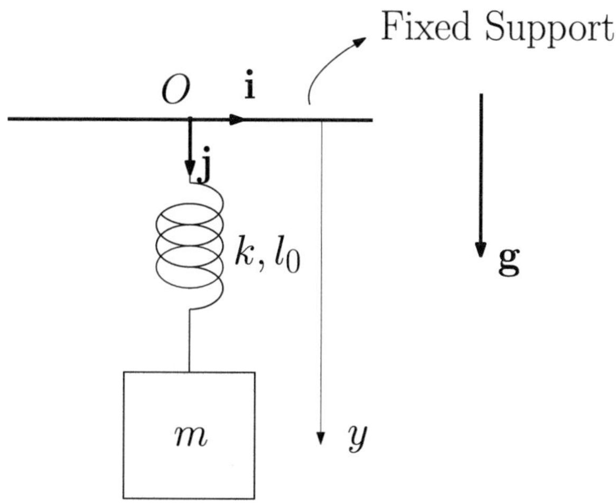

Fig. 1.2 A simplified model for the bottle-spring system from Fig. 1.1. The model consists of a particle with mass m connected to an ideal linear spring with stiffness k and unstretched length l_0

bottle are absent, so the motion of interest is primarily rectilinear along the vertical, and (iii) the spring is linear and follows Hooke's law. Consequently, a simplified model, depicted in Fig. 1.2, emerges, consisting of a particle with mass m attached to a rigid and stationary support via a linearly elastic spring characterized by stiffness k and unstretched length l_0.

The vibratory response predicted by this model can be solved using the Newton–Euler approach. This solution procedure, which is applicable to any other problem in vibrations, typically involves a systematic approach comprising four key steps. The first step is to perform *kinematic analysis*, where an appropriate coordinate system is chosen and expressions for position, velocity, acceleration are established. Next, it is necessary to construct a *free body diagram* (FBD) to identify all the forces acting upon the system. Afterward, Newton–Euler's *laws of motion* can be applied. The last step involves *analysis* of the equations gathered from the preceding step to solve for the vibratory behavior. Apart from actually solving the equations for the system vibrations, which will be covered in later chapters, these steps are systematically applied to the bottle-spring system in the following sections.

1.2.2.1 Kinematic Analysis

In the model shown in Fig. 1.2, a fixed origin is placed at O and a Cartesian coordinate system (x, y, z),[2] associated with the usual fixed orthonormal basis vectors $\{\mathbf{i}, \mathbf{j}, \mathbf{k}\}$, is used to describe the kinematics of the particle. In a general

[2] See Appendix D for a review of Newton–Euler dynamics.

setting, the position vector of a particle has the representation

$$\mathbf{r} = x\mathbf{i} + y\mathbf{j} + z\mathbf{k}. \tag{1.1}$$

However, the simplifying assumptions from Sect. 1.2.2 means not all of these coordinates are necessary. Because only the rectilinear motion along the vertical is of interest, coordinate y is free to vary while the other two Cartesian coordinates satisfy the imposed constraints

$$\psi_1 = x = 0, \quad \psi_2 = z = 0, \tag{1.2}$$

and the representation for the position vector becomes

$$\mathbf{r} = y\mathbf{j}. \tag{1.3}$$

Functions ψ_1 and ψ_2 in Eq. (1.2) are called *holonomic constraints* or coordinate constraints. Their presence means that the motion of the particle is completely characterized only by coordinate y, so this system is said do have only one DOF. Any set of independent variables that can be used to describe a mechanical system completely is a set of generalized coordinates for the system, the number of which is equal to the number of DOFs. In systems where only holonomic constraints are present, the number of DOFs may be computed according with

$$\begin{matrix}\text{Number of}\\ \text{DOFs}\end{matrix} = \begin{pmatrix}\text{Number of DOFs}\\ \text{if system is free}\\ \text{of constraints}\end{pmatrix} - \begin{pmatrix}\text{Number of holonomic}\\ \text{constraints}\end{pmatrix}. \tag{1.4}$$

To apply this to the aforementioned model, note that a particle has three DOFs if it is free to move in space, as shown in Fig. 1.3. The presence of the two coordinate constraints from Eq. (1.2) means that it actually has $3 - 2 = 1$ DOF, and its motion can be fully described by one generalized coordinate, i.e., coordinate y.

Once the bottle is displaced from equilibrium, coordinate y varies as time t progresses. Consequently, representation for velocity and acceleration vectors become

$$\mathbf{v} = \dot{y}\mathbf{j}, \quad \mathbf{a} = \ddot{y}\mathbf{j}, \tag{1.5}$$

respectively. Here, overdots are the usual notation to represent time differentiation:

$$\frac{dy}{dt} = \dot{y}, \quad \frac{d^2y}{dt^2} = \ddot{y}.$$

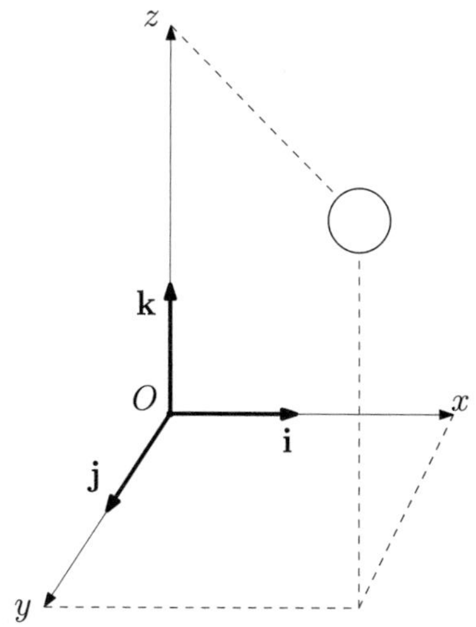

Fig. 1.3 A particle that is free to move in space has three DOFs, and its motion can be described by the Cartesian coordinates (x, y, z)

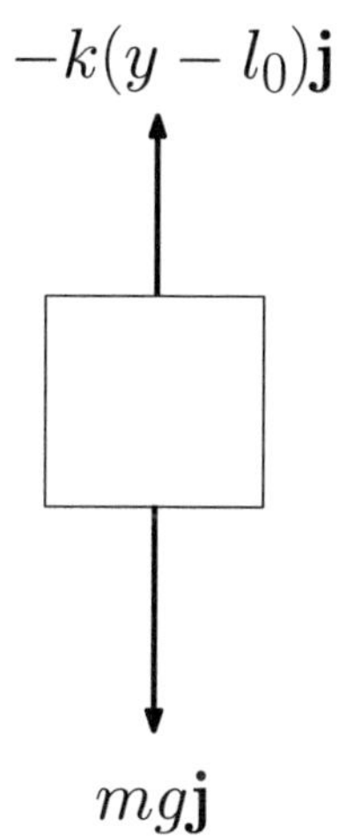

Fig. 1.4 FBD of the particle

1.2.2.2 Equation of Motion

The oscillatory motion of the bottle, denoted by $y(t)$, is the solution to the governing differential equation of motion, which is derived from the fundamental laws of dynamics. To obtain this equation using the Newtonian approach, it is necessary to construct the FBD shown in Fig. 1.4. Since $\mathbf{j}$ was chosen to point vertically downwards, the weight of the particle is $mg\mathbf{j}$. The spring is assumed to obey Hooke's law, so the elastic force exerted on the particle is dependent on the spring stiffness k and its change of length ϵ according with

$$\mathbf{F}_e = -k\epsilon\mathbf{j}, \tag{1.6}$$

where

$$\epsilon = \begin{pmatrix} \text{Current length} \\ \text{of spring} \end{pmatrix} - \begin{pmatrix} \text{Unstreched length} \\ \text{of spring} \end{pmatrix} = y - l_0. \tag{1.7}$$

Notice that $\epsilon = 0$ when $y = l_0$, i.e., when the spring is unstretched, it exerts no force on the particle. The spring is compressed if $\epsilon < 0$ or stretched when $\epsilon > 0$.

Application of Newton's second law yields

$$m\ddot{y}\mathbf{j} = mg\mathbf{j} - k(y - l_0)\mathbf{j}. \tag{1.8}$$

The vector notation may be dropped because the only direction involved is **j**, and the following differential equation is obtained:

$$m\ddot{y} + k(y - l_0) = mg. \tag{1.9}$$

The equation of motion is a second-order linear ordinary differential equation (ODE) that, together with the given the initial conditions (ICs),

$$\begin{cases} y(0) = y_0 \\ \dot{y}(0) = \dot{y}_0 \end{cases}, \tag{1.10}$$

can be solved for $y(t)$ using standard ODE theory. The ICs represent the stimuli exerted on the bottle at the instant $t = 0$: y_0 is the initial length of the spring, and $\dot{y}_0$ represents the initial velocity of the bottle. An initial velocity may arise if linear momentum is instantaneously transferred to the bottle, such as when it is abruptly struck.

1.2.3 Equilibrium Position

Being able to solve ODEs like Eq. (1.9) is the bread and butter of vibration analysis, a task that is left to later chapters. For now, it suffices to look for a special type of solution allowed by this equation. Specifically, look for solutions that correspond to equilibrium positions for the system:

$$y = y_e = \text{const.}$$

Substitute this into Eq. (1.9) to obtain the *equilibrium equation*

$$k(y_e - l_0) = mg, \tag{1.11}$$

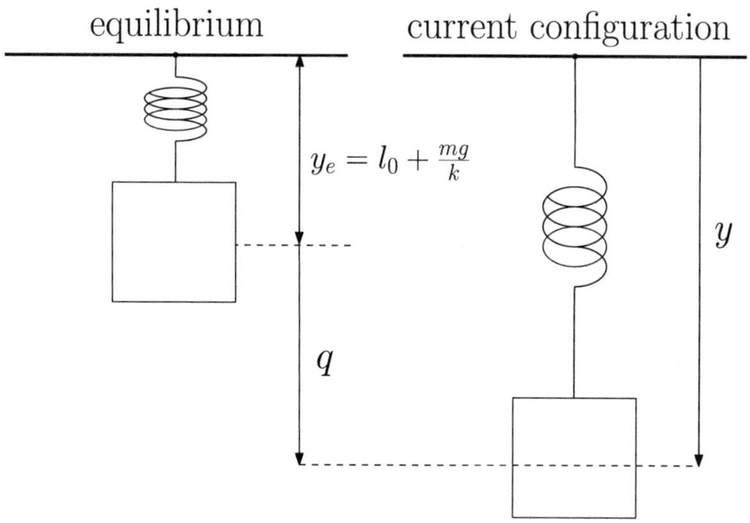

Fig. 1.5 Change of coordinates $y = y_e + q$. Coordinate y is measured from the fixed support while q starts from the equilibrium position

that has solution given by

$$y_e = \frac{mg}{k} + l_0. \tag{1.12}$$

Physically, y_e represents the position of the system when in static equilibrium, which is the sum of the unstretched length l_0 of the spring and the deflection due to weight $\delta_e = mg/k$.

The main objective in vibrations is to appraise how the mass-spring system moves when perturbed from equilibrium by any perturbation. To achieve that, it is convenient to introduce the change of coordinates $y = y_e + q$ shown in Fig. 1.5, where the new coordinate q measures the position of the particle relative to the static equilibrium position.

The meaning of this coordinate transformation is further elucidated by Fig. 1.6: coordinate y represents oscillations measured from the fixed support, while q captures only the fluctuations of y about the equilibrium position. In linear vibration analysis, it is irrelevant which one is used, but preference is generally given to q. One reason is that the equation of motion has fewer terms, as demonstrated in the following paragraph.

Since y_e is a constant, the derivatives of y and q are related by

$$\dot{y} = \dot{q}, \quad \ddot{y} = \ddot{q}. \tag{1.13}$$

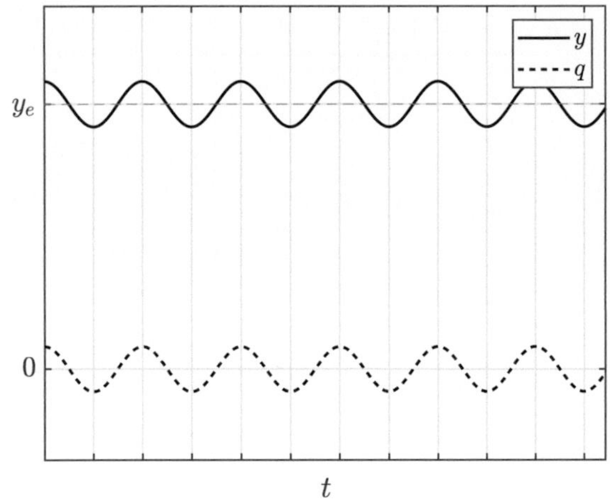

Fig. 1.6 The change of coordinates $y = y_e + q$. Coordinate y represents oscillations around y_e and q captures these fluctuations, representing oscillations around the origin

Substituting the coordinate transformation $y = y_e + q$ into (1.9) and considering Eqs. (1.12) and (1.13), the differential equation of motion can be rewritten in terms of q as

$$m\ddot{q} + kq = 0. \tag{1.14}$$

Even if it looks like the weight mg has disappeared from Eq. (1.14), in reality it is being balanced at every instant by the force $k\delta_e$. The conclusion here is that, **by defining the displacement variable to be zero at equilibrium, balancing forces associated with equilibrium can be ignored in the formulation**.

1.2.4 Energy Considerations

The total mechanical energy is the sum of kinetic and potential energies. For the mass-spring system, it is represented as

$$E = \frac{1}{2}m\dot{y}^2 + U, \tag{1.15}$$

where the potential energy is composed of elastic and gravitational parts:

$$U = \frac{1}{2}k\epsilon^2 - mgy = \frac{1}{2}k(y - l_0)^2 - mgy. \tag{1.16}$$

The total energy is expected to be conserved because the model developed so far has no mechanism that accounts for energy dissipation. To see this, differentiate E with respect to time to obtain

$$\dot{E} = (m\ddot{y} + k(y - l_0) - mg)\dot{y}. \tag{1.17}$$

Inspecting Eq. (1.9) reveals that $\dot{E} = 0$ during the motion. Therefore, E is constant or, in dynamics terminology, conserved.

An alternative viewpoint is useful. Integrating the Eq. (1.9) with respect to the displacement variable y gives

$$\frac{1}{2}m\dot{y}^2 + \frac{1}{2}k(y - l_0)^2 - mgy = \text{const.} \tag{1.18}$$

This integration leads to the expression of energy conservation. An expression like this is called a *first integral of motion* because it can be obtained by integrating the equation of motion once. A dynamical system might have other first integrals besides energy, and they all have the highest derivative one order lower than the highest derivative present in the equation of motion.

1.2.4.1 Potential Energy and Stability

The dynamics of the mass-spring system can be further elucidated by considering the associated energy. Because only conservative forces act on the particle (spring force and weight), the equation of motion (1.9) can be reformulated according with

$$m\ddot{y} = -\frac{dU}{dy}. \tag{1.19}$$

This equation illustrates how the particle's acceleration varies with y, because y ultimately affects what value dU/dy takes. Figure 1.7 demonstrates that the potential energy reaches its minimum value

$$U_e = \frac{1}{2}k(y_e - l_0)^2 - mgy_e \tag{1.20}$$

at the static equilibrium position y_e. This means that, if the bottle is displaced downward to a new position $y > y_e$ and released from rest, then the rate of change

$$\frac{dU}{dy} > 0 \tag{1.21}$$

because $U(y_e)$ is a minimum. Once released, the bottle accelerates upward ($m\ddot{y} < 0$), driven by the restorative force toward the equilibrium position. Since the total mechanical energy (1.15) is conserved, there is reciprocal exchange between kinetic

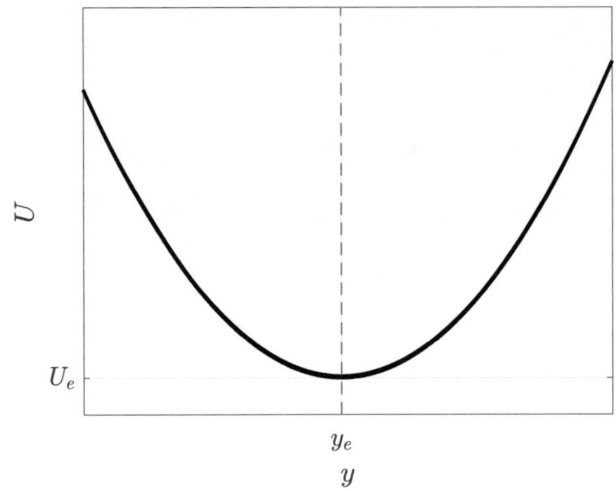

Fig. 1.7 System potential energy

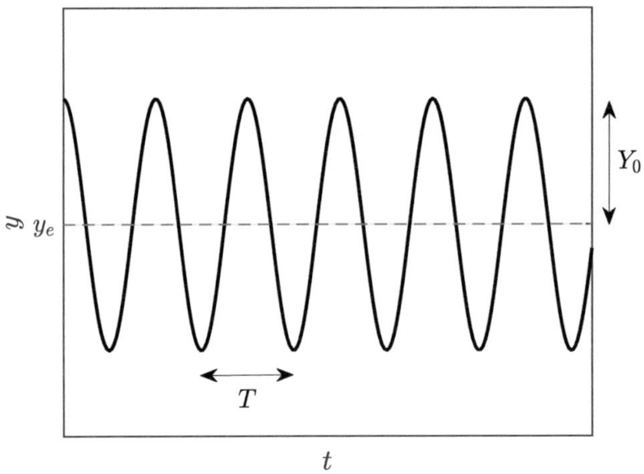

Fig. 1.8 Typical oscillatory behavior of the bottle-spring system. The amplitude of oscillation is Y_0, and the response is periodic with period T: $y(t+T) = y(t)$

and potential energies. Therefore, as the bottle is moving upward and finally reaches the equilibrium position, the potential energy is minimum while the kinetic energy becomes maximum. The bottle has to continue moving upward, reaching positions where $y < y_e$, and the slope of potential energy is $dU/dy < 0$. Now, $m\ddot{y} > 0$ and the conservative forces pull the bottle back to equilibrium. The bottle is decelerated until its velocity is instantaneously zero at the position of maximum potential energy, after which it starts to move back again. This pattern is repeated and results in the typical oscillatory motion depicted in Fig. 1.8: the bottle oscillates back and forth with period T and amplitude Y_0 around the equilibrium position y_e.

The root of the preceding analysis was that U is minimum at equilibrium: any *perturbation* would result in the bottle oscillating around this position. In this sense, the static equilibrium position y_e is referred to as a *stable equilibrium*. Conversely, if $U(y_e)$ were a maximum, similar analysis would reveal that any perturbation would make the bottle move away from equilibrium, and no oscillatory motion would be observed! In this case, y_e would be considered an *unstable equilibrium*. Since most systems are subject to "small" and unpredictable disturbances, stability is a crucial requirement to ensure that they remain close to the intended operating condition.

1.2.5 Damping

In formulating the mathematical model (1.14), no damping forces were accounted for, leading to a prediction of perpetual oscillations owing to the absence of energy dissipation. However, experimental observations reveal that oscillations gradually diminish over time due to damping. This disparity needs to be dealt with to enhance the model's predictive capability, through a process that is called model refinement.

Refining a model often involves incorporating additional factors or calibrating parameters based on experimental data. This iterative cycle of modeling, experimentation, and refinement, depicted in Fig. 1.9, serves as the cornerstone of engineering and scientific analysis. It facilitates the continuous enhancement of the accuracy and

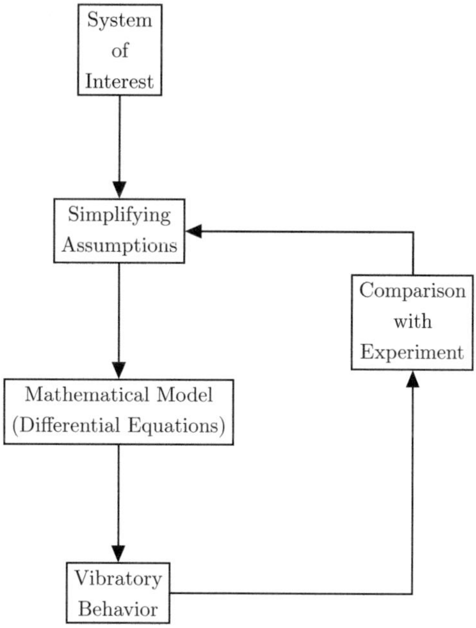

Fig. 1.9 Flowchart for engineering analysis

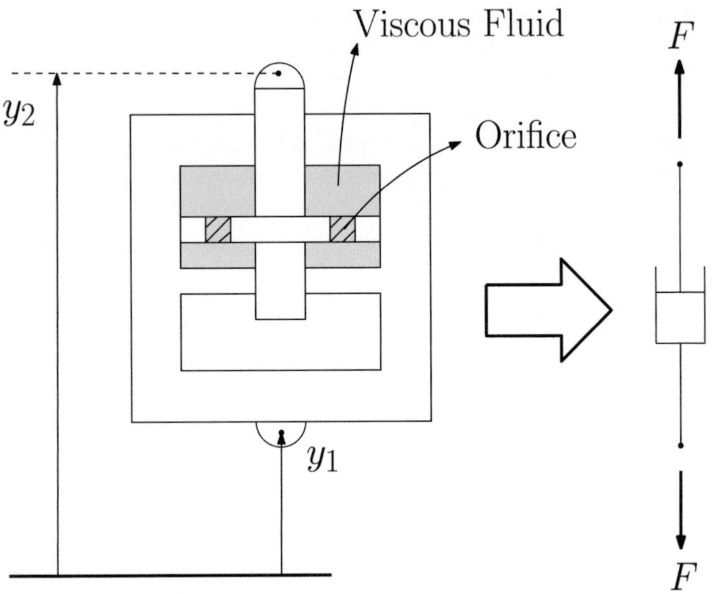

Fig. 1.10 Piston moving inside a cylinder. The internal fluid flows through the small orifices and generates a force $F = c(\dot{y}_2 - \dot{y}_1)$ that is linear in the velocity difference

reliability of mathematical models, ensuring that they reflect real-world behavior and thereby support informed decision-making in design.

In order to include energy dissipation in the model, firstly it is necessary to understand the mechanism of damping present in the system. In general, damping forces can arise from various forms, such as friction between the system and external agents (Coulomb damping), internal friction within the material (structural damping), opening and closing of cracks, or interaction between the system and viscous fluids. Given the complex nature of damping forces, their precise mathematical representation may not be straightforward. As a workaround, an idealized *viscous damping* force is often assumed. This terminology originates from the force observed when a piston moves inside a cylinder filled with a viscous fluid, as depicted in Fig. 1.10. As the piston of negligible mass moves, the internal viscous fluid flows through the orifices and a force $F = c(\dot{y}_2 - \dot{y}_1)$ develops in the damper. This damping force can be easily treated mathematically because it is linear in the velocity difference $(\dot{y}_2 - \dot{y}_1)$ through a proportionality constant $c \geq 0$ called *viscous damping coefficient*.

Once a damper is introduced in the model, as illustrated in Fig. 1.11, the following equation of motion is obtained:

$$m\ddot{q} + c\dot{q} + kq = 0, \tag{1.22}$$

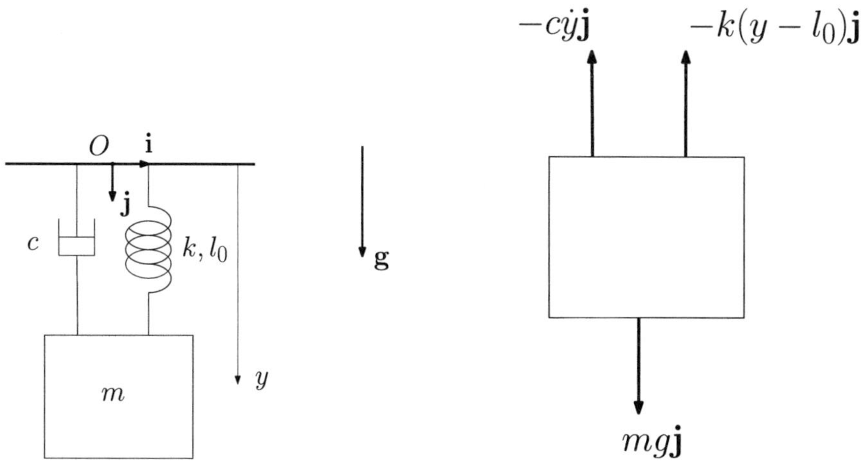

(a) A mass-spring-damper system.
(b) FBD of the damped oscillator.

Fig. 1.11 A mass-spring-damper system

where $y = y_e + q$. The presence of the damper means the system is *dissipative*, i.e., energy (1.15) is being dissipated according with

$$\dot{E} = -c\dot{q}^2 \leq 0. \tag{1.23}$$

Since the total mechanical energy is decreasing, it is expected that the amplitude of vibration will diminish over time.

1.3 The Simple Pendulum

The mass-spring system discussed in Sect. 1.2 served as a straightforward introduction to key concepts in the study of vibrations. Another classical example of oscillatory behavior is the motion of the simple pendulum, portrayed in Fig. 1.12. It consists of a light piece of string tied to a ball (with mass great enough so the string is not slack when it moves). If the free end of the string is fixed and the ball released from rest at a given angle, planar oscillations around the vertical are observed. The pendulum starts oscillating with constant angular amplitude, but the amplitude eventually diminishes until the pendulum comes to rest in the downward vertical position. The objective is to use the four steps described in Sect. 1.2.2 to develop a mathematical model that describes the observed behavior. As will become evident later on, this system is an excellent starting point for understanding complex vibration problems because the mathematical principles applied here find broader application in solving a wide range of examples.

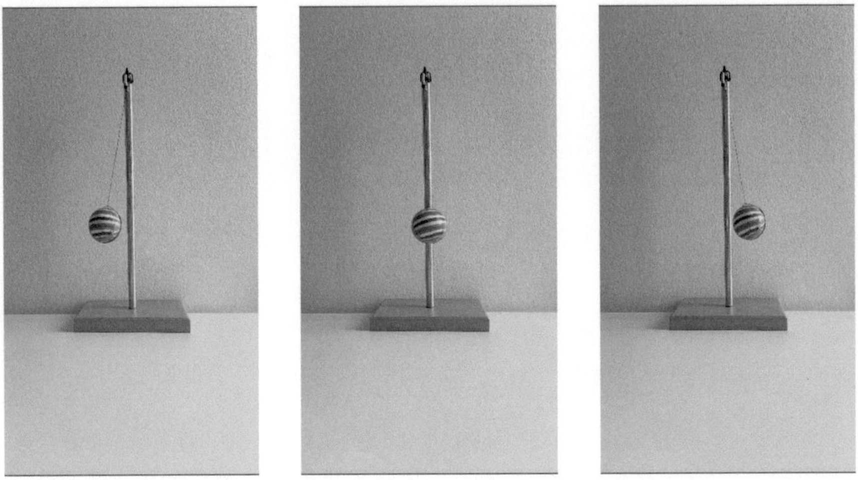

(a) Releasing the pendulum from rest. (b) Swinging through the vertical position. (c) Maximum amplitude reached.

Fig. 1.12 Snapshots of the motion of a simple pendulum constructed using 11 cm of string tied to a ball. The pendulum was released from rest

Fig. 1.13 Model of a simple pendulum and coordinate systems used in the analysis

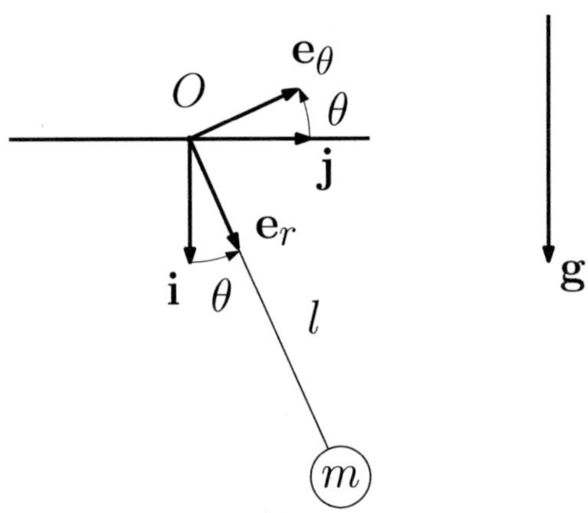

1.3.1 A Model for the Simple Pendulum

A model for the simple pendulum may be conceived according to Fig. 1.13. The pendulum bob is represented by a particle with mass m, which is attached to the fixed point O by an inextensible string of length l and negligible mass. Based on the observed motion, it is assumed that the pendulum oscillates on the plane spanned by $\{\mathbf{i}, \mathbf{j}\}$.

1.3.1.1 Kinematic Analysis

Using the cylindrical-polar coordinates (r, θ, z),[3] the position vector of a particle that is free to move may be expressed as

$$\mathbf{r} = r\mathbf{e}_r + z\mathbf{k}. \tag{1.24}$$

The simplifying assumptions imply that the pendulum moves under the kinematic constraints

$$\psi_1 = r - l = 0, \quad \psi_2 = z = 0. \tag{1.25}$$

The pendulum motion is completely characterized by the angular coordinate θ, and it has one DOF. The position vector can be expressed simply as

$$\mathbf{r} = l\mathbf{e}_r, \tag{1.26}$$

and the velocity and acceleration vectors have the representation

$$\mathbf{v} = l\dot{\theta}\mathbf{e}_\theta, \quad \mathbf{a} = l\ddot{\theta}\mathbf{e}_\theta - l\dot{\theta}^2\mathbf{e}_r, \tag{1.27}$$

respectively.

1.3.1.2 Equation of Motion

The forces acting on the particle can be identified from the FBD in Fig. 1.14. Because the string cannot resist compression, it exerts a traction force $-S\mathbf{e}_r$ on the particle, where S is the unknown magnitude of this force. By applying Newton's second law, the following equation is obtained:

$$mg\mathbf{i} - S\mathbf{e}_r = ml\ddot{\theta}\mathbf{e}_\theta - ml\dot{\theta}^2\mathbf{e}_r. \tag{1.28}$$

There are two unknowns: the traction force S and the motion θ. To obtain a differential equation that involves solely the free coordinate θ, take the projection of this vector equation along the $\mathbf{e}_\theta$ direction:[4]

$$\ddot{\theta} + \frac{g}{l}\sin\theta = 0. \tag{1.29}$$

[3] Appendix D contains a concise review of cylindrical-polar coordinates.

[4] A review of vector operations is provided in Appendix B.

Fig. 1.14 FBD of the particle. The traction force due to the string is $-S\mathbf{e}_r$

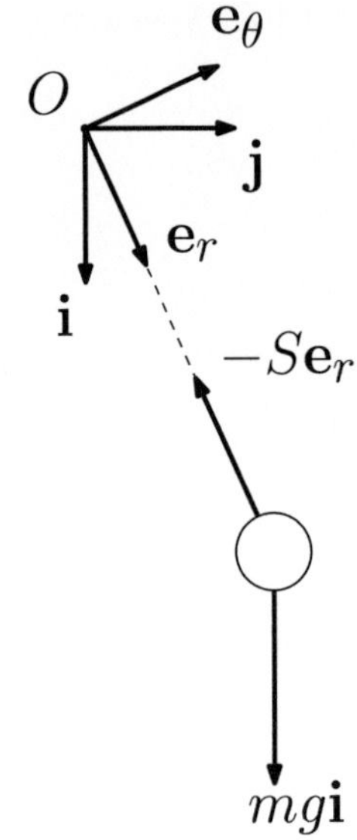

The presence of the term $\sin\theta$ makes this ODE *nonlinear*. While linear models, such as the one discussed in Sect. 1.2 for the mass-spring-damper system, are the main focus of these notes, there are many systems that are governed by second-order nonlinear ODEs because mechanical systems, in general, exhibit inherent nonlinearities attributable to various factors. One being material behavior: once a solid deforms too much and enters the plastic regime, it does not obey Hooke's law. In some applications, the external forces can have nonlinear nature, like drag forces, magnetic forces, Coulomb friction, or even impact forces. In the case of the simple pendulum, nonlinearity is due to the geometric characteristic of the system.

1.3.1.3 Energy Conservation

The total mechanical energy of a simple pendulum is given by

$$E = \frac{1}{2}ml^2\dot{\theta}^2 + U, \tag{1.30}$$

where the potential energy is $U = -mgl\cos\theta$. Computing the rate of change of E reveals that

$$\dot{E} = \left(\ddot{\theta} + \frac{g}{l}\sin\theta\right)ml^2\dot{\theta}. \tag{1.31}$$

The equation of motion (1.29) implies that the rate of change computed earlier is $\dot{E} = 0$. This indicates that the mechanical energy, as defined in Eq. (1.30), is conserved during the motion of the pendulum. This conservation represents a first integral of motion for the simple pendulum. Although the non-conservative force $-S\mathbf{e}_r$ acts on the system, it does no work ($-S\mathbf{e}_r \cdot \mathbf{v} = 0$), thereby not affecting the energy.

1.3.1.4 Equilibrium Position and Stability

The equilibrium positions of the pendulum are solutions of the form $\theta = \theta_e =$ const. Substitute this into Eq. (1.29) to obtain the equilibrium equation

$$\sin\theta_e = 0. \tag{1.32}$$

This is a transcendental equation with infinitely many solutions given by $\theta_e = 0, \pm\pi, \pm 2\pi, \ldots$, but only two are physically distinct, as shown in Fig. 1.15: $\theta_e = 0$ and $\theta_e = \pi$. The position $\theta_e = \pi$ is only physically possible if the string is replaced by a rigid rod of negligible mass because strings are not able to resist compression.

The assessment of equilibrium stability can also be approached through the lens of potential energy, akin to the analysis conducted in Sect. 1.2.4.1. In the case of the simple pendulum, where only the weight does work and energy is also conserved, the equation of motion (1.29) can be reformulated as

$$ml^2\ddot{\theta} = -\frac{dU}{d\theta}. \tag{1.33}$$

According to Fig. 1.16, the potential energy is minimum at $\theta_e = 0$. Thus, when perturbed from this position, the restorative force tries to bring the pendulum back to equilibrium ($\ddot{\theta} < 0$ if $\theta > 0$ or $\ddot{\theta} > 0$ if $\theta < 0$), resulting in oscillations around $\theta_e = 0$. This equilibrium is stable. Conversely, U is maximum at $\theta_e = \pi$, and any perturbation from this equilibrium leads the pendulum to diverge further away. Thus, the equilibrium point $\theta_e = \pi$ is classified as *unstable*.

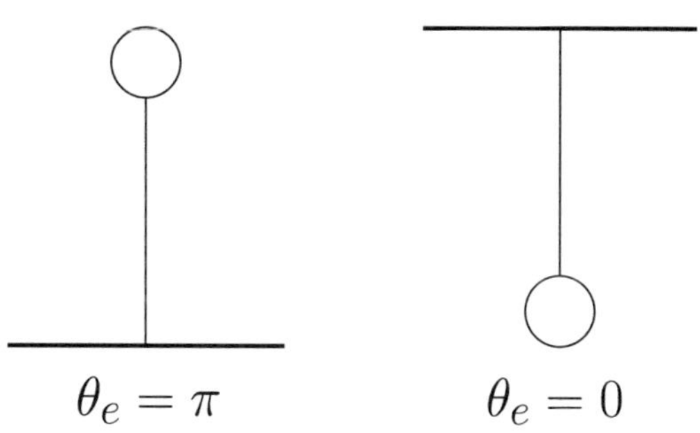

Fig. 1.15 The two equilibrium positions for the simple pendulum

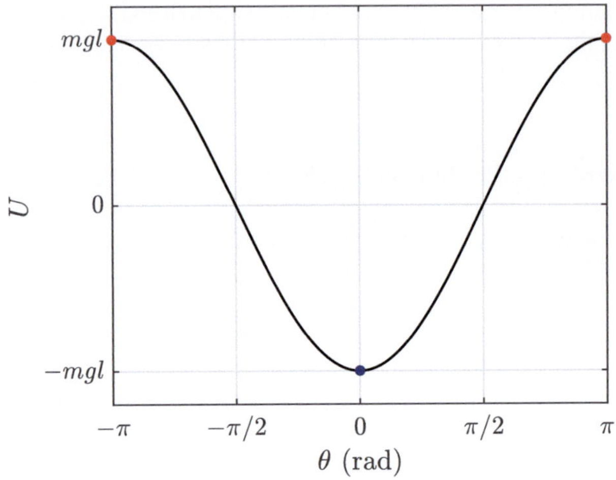

Fig. 1.16 Potential energy of the simple pendulum. The values corresponding to the stable (•) and unstable (•) equilibria are marked

1.3.2 *Simulating the Motion*

Exact solutions given in closed-form expressions are possible only for the relatively few systems whose nonlinear equations of motion have specific forms. When dealing with nonlinear systems, engineers rely on approximate and graphical solutions, frequently resorting to numerical methods. To numerically solve Eq. (1.29),[5] introduce the *state* variables $y_1 = \theta$ and $y_2 = \dot{\theta}$, which transforms Eq. (1.29) into a system of two first-order ODEs:

$$\frac{d}{dt}\begin{bmatrix} y_1 \\ y_2 \end{bmatrix} = \begin{bmatrix} y_2 \\ -\frac{g}{l}\sin y_1 \end{bmatrix}. \tag{1.34}$$

To replicate the experiment depicted in Fig. 1.12, consider a pendulum with $l = 0.11$ m and released with the following ICs:

$$\begin{cases} \theta(0) = y_1(0) = 10^\circ \\ \dot{\theta}(0) = y_2(0) = 0. \end{cases} \tag{1.35}$$

The MATLAB code `Chapt1_code1`, provided as *Electronic Supplementary Material*, integrates these equations from $t = 0$ to $t = 5$ s, generating the plot of $\theta(t)$ shown in Fig. 1.17. From this plot, one can deduce that the pendulum undergoes

[5] Introduction to numerical solution of ODEs using MATLAB is offered in Appendix E.

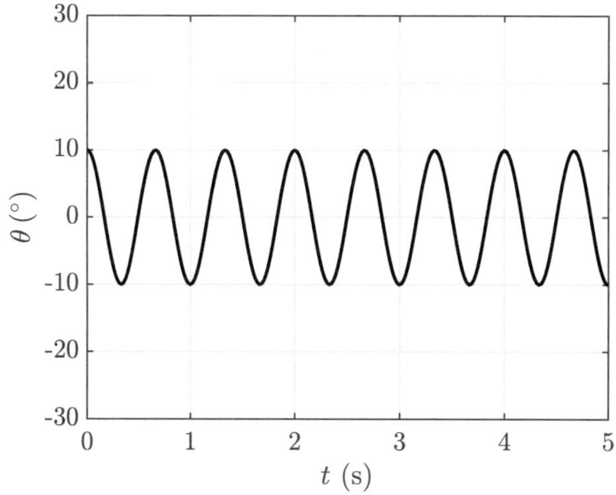

Fig. 1.17 Numerical solution of Eq. (1.29) with ICs $\theta(0) = 10°$ and $\dot{\theta}(0) = 0$

periodic oscillations with period $T \approx 0.67$ s. This is a good approximation for the early phase of motion when damping effects are not prevalent. The pendulum completes 10 oscillations in approximately 7 s, resulting in the measured period of $T^* = 0.7$ s. Considering the simplistic nature of the measurements used, the results exhibit surprisingly close agreement.

1.3.2.1 Phase Portrait

A complete overview of the dynamic behavior when the pendulum is perturbed from equilibrium can be obtained from the *phase portrait* depicted in Fig. 1.18. It is a plot in the *phase plane* $\{\theta, \dot{\theta}\}$ that consists of trajectories formed by solutions of Eq. (1.29) for various sets of ICs, each computed using the code provided earlier. Solutions belonging to trajectories in the upper half-plane move from left to right because θ increases with time when $\dot{\theta} > 0$. Likewise, trajectories in the lower half-plane are from right to left. The phase portrait of the simple pendulum repeats in intervals of 2π along the horizontal axis because θ and $\theta + 2\pi$ represent the same physical configuration of the pendulum.

A point $(\theta, \dot{\theta})$ represents the *state* of the system. The states $(\theta, \dot{\theta}) = (0, 0)$, $(\pm\pi, 0)$, $(\pm 2\pi, 0)$, ..., are the equilibrium points corresponding to the equilibrium positions. Different possible motions can be readily identified from this plot. Firstly, the closed trajectories around the origin represent periodic solutions when the pendulum oscillates back and forth about the vertical. The open trajectories at the top and bottom represent motions of the pendulum involving continuous rotation for which the angular coordinate θ is increasing (counterclockwise motion) or decreasing (clockwise motion). Continuous rotation and periodic oscillations are

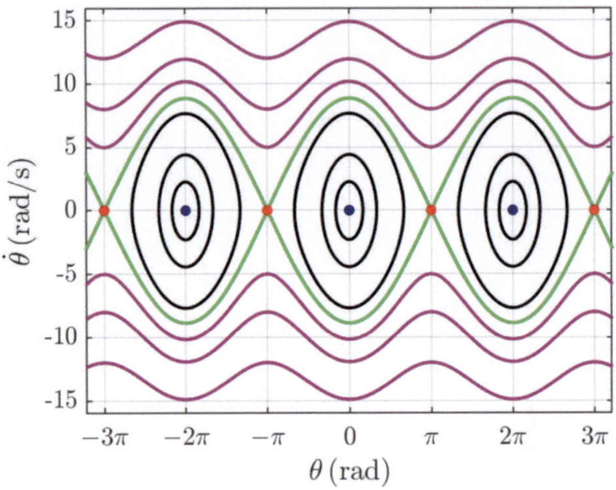

Fig. 1.18 Phase portrait of the simple pendulum with the stable (•) and unstable (•) equilibria shown. The trajectories (—) depict periodic motions around stable equilibria, while continuous rotational motions are shown as (—). The separatrices are represented by (—)

separated by the trajectory connecting $(-\pi, 0)$ directly to $(\pi, 0)$. This trajectory is called *separatrix* and corresponds to motions where the pendulum moves asymptotically, as $t \rightarrow \infty$, toward $(\pm\pi, 0)$. The interpretation is the same for trajectories connecting $(-3\pi, 0)$ to $(-\pi, 0)$, or $(\pi, 0)$ to $(3\pi, 0)$, and so on.

The phase portrait can also provide insights into the stability of equilibrium points. Consider a pendulum in equilibrium with $\theta_e = 0$. If this equilibrium state is slightly perturbed, the pendulum will end up on a nearby closed trajectory in the phase portrait, and the future states of the resulting motion will be confined to a small neighborhood around the origin $(0, 0)$, i.e., the pendulum will oscillate with a small angular amplitude around $\theta_e = 0$. This is the sense in which an equilibrium is deemed (Lyapunov) stable. If the pendulum is slightly perturbed from the unstable equilibrium $(\pi, 0)$, it will move away from this state, meaning the future states of its subsequent motion will not be contained within a small region around $(\pi, 0)$ in the phase plane.

1.3.3 Small Oscillation Approximation

In the experiment described in the beginning of this section, the pendulum was released from approximately 10°, a small initial angle. To examine the small oscillations of the pendulum around its stable equilibrium position, the nonlinear term in Eq. (1.29) can be expanded as a Taylor series around $\theta_e = 0$. This leads to

$$\ddot{\theta} + \frac{g}{l}\theta - \frac{g}{l}\frac{\theta^3}{3!} + \frac{g}{l}\frac{\theta^5}{5!} - \ldots = 0. \tag{1.36}$$

For small values of θ, the higher-order terms may be neglected, resulting in the equation of motion for small oscillations

$$\ddot{\theta} + \frac{g}{l}\theta = 0. \tag{1.37}$$

The nonlinear Eq. (1.29) was converted into the linear ODE (1.37), a process known as *linearization*. Working with linearized models offers several advantages, including the ability to obtain analytical solutions using well-established theory on ODEs. Additionally, an intriguing aspect is that this differential equation has the same form as the equation of motion for a mass-spring system, as given by Eq. (1.14). For "small oscillations" around $\theta_e = 0$, the simple pendulum and the mass-spring system exhibit similar dynamic behavior. Therefore, understanding the solution of Eq. (1.14) implies knowledge about the small oscillatory behavior of the pendulum described by Eq. (1.37).

The linearization process involves neglecting the higher-order terms and approximating $\sin\theta$ by θ. How good an approximation is this? To assess the accuracy of the small-angle assumption, let

$$e(\theta) = \left|\frac{\theta - \sin\theta}{\sin\theta}\right| \tag{1.38}$$

represent the error incurred when linearizing the term $\sin\theta$. As illustrated in Fig. 1.19, the error is $e < 1\%$ if $\theta < 14°$. The small-angle assumption, necessary for linearization, may not be overly restrictive.

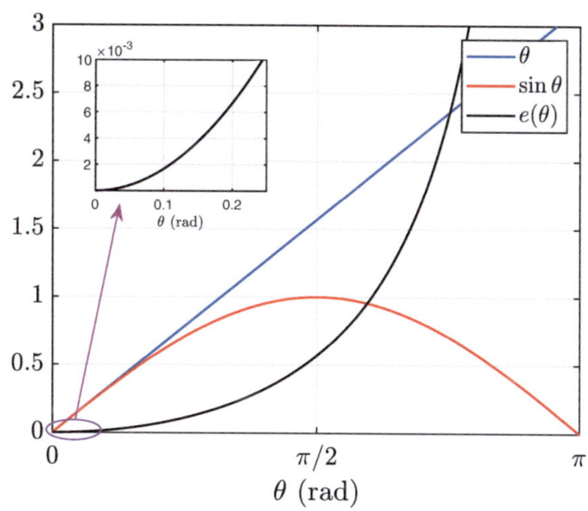

Fig. 1.19 Error committed when linearizing the equation of motion of a simple pendulum

1.3.3.1 Alternative Derivation of the Equation of Motion

In linear vibration analysis, it is customary to presume small motions around a stable equilibrium from the onset, which streamlines the derivation of the equation of motion. To demonstrate this approach with the simple pendulum, assume that it undergoes small oscillations around $\theta = 0$. A consequence is that $\cos\theta \approx 1$ and $\sin\theta \approx \theta$. The position of the pendulum bob can be approximated by

$$\mathbf{r} = l\mathbf{e}_r \approx l\mathbf{i} + l\theta\mathbf{j}. \tag{1.39}$$

This approximation implies that, for small angles, the particle predominantly moves horizontally. Upon differentiating this equation twice with respect to time, an expression for the linearized acceleration is obtained:

$$\mathbf{a} \approx l\ddot{\theta}\mathbf{j}. \tag{1.40}$$

Application of Newton's second law results in the vector equation

$$mg\mathbf{i} - S(\mathbf{i} + \theta\mathbf{j}) = ml\ddot{\theta}\mathbf{j}. \tag{1.41}$$

Equality of terms along the **i** direction reveals that $S = mg$. This result used in conjunction with the equality of terms along **j** leads to the same equation of motion obtained in (1.37).

1.3.3.2 Integral of Motion

It is straightforward to verify that a first integral for Eq. (1.37) is

$$E_{lin} = \frac{1}{2}ml^2\dot{\theta}^2 - mgl + \frac{1}{2}mgl\theta^2. \tag{1.42}$$

This expression represents the total mechanical energy of the linearized system. The linearized system does not capture the nonlinearity of the potential energy. As a result, the energy computed for the linearized system does not match the energy computed for the nonlinear system in Eq. (1.30).

Interestingly, the first integral for the linearized pendulum, E_{lin}, can be derived from the original energy expression in Eq. (1.30) by applying the small-motion approximation. To do this, expand the potential energy U as a Taylor series around $\theta = 0$ according with

$$U(\theta) \approx U(0) + \frac{dU}{d\theta}(0)\theta + \frac{1}{2}\frac{d^2U}{d\theta^2}(0)\theta^2. \tag{1.43}$$

Since $dU/d\theta = 0$ at $\theta = 0$ (the equilibrium position), the expansion simplifies to:

$$U(\theta) \approx -mgl + \frac{1}{2}mgl\theta^2. \tag{1.44}$$

Here, the term $U(0) = -mgl$ represents the potential energy at the equilibrium position, which is a constant. This constant term can be neglected because only the variation of potential energy is important. Thus, the energy of the linearized system E_{lin} closely approximates the energy of the original system E for small motions.

1.3.4 The Damped Pendulum

The model developed so far does not account for energy dissipation. Now, the dynamic behavior of the pendulum when damping is present will be investigated.

1.3.4.1 Equation of Motion

For simplicity, assume that the damping force acts on the pendulum bob and is opposite to the velocity, as shown in the FDB from Fig. 1.20. Application of Newton's second law gives

$$mg\mathbf{i} - S\mathbf{e}_r - cl\dot{\theta}\mathbf{e}_\theta = ml\ddot{\theta}\mathbf{e}_\theta - ml\dot{\theta}^2\mathbf{e}_r. \tag{1.45}$$

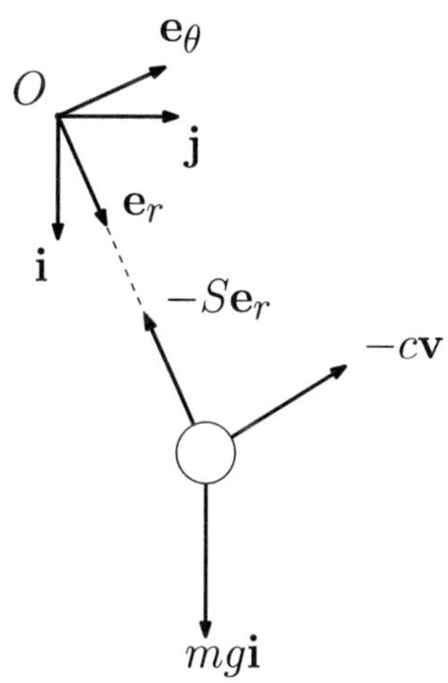

Fig. 1.20 FBD of the damped pendulum

Taking the projection of this equation results in the differential equation

$$\ddot{\theta} + \frac{c}{m}\dot{\theta} + \frac{g}{l}\sin\theta = 0. \tag{1.46}$$

If $c = 0$, this ODE reduces to that of the simple pendulum given by Eq. (1.29).

1.3.4.2 Numerical Integration

To understand the effect of damping on the response, take the arbitrary parameter values

$$l = 9.81 \text{ m}, \quad c = 0.15 \text{ Ns/m}, \quad m = 1 \text{ kg}, \tag{1.47}$$

and use the MATLAB code `Chapt1_code2` from the *Electronic Supplementary Material* to produce the plot shown in Fig. 1.21. Because of damping, the pendulum swings back and forth around the vertical with decaying amplitude.

1.3.4.3 Energy Considerations

As illustrated in Fig. 1.22, the presence of damping makes the total mechanical energy (1.30) decay with rate given by

$$\dot{E} = -cl^2\dot{\theta}^2 \leq 0. \tag{1.48}$$

The rate of change is $E = 0$ only at the instants when $\dot{\theta} = 0$.

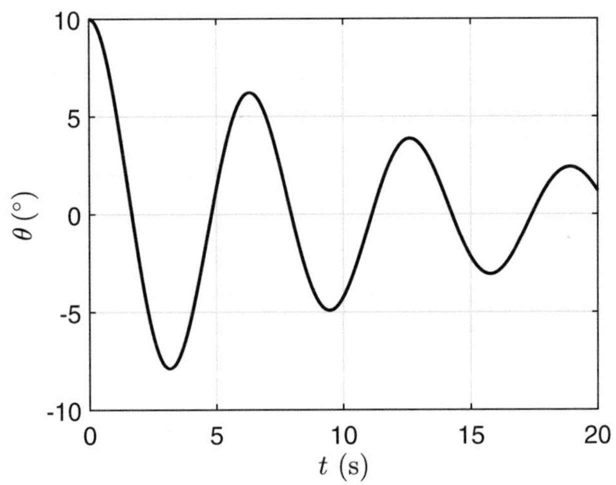

Fig. 1.21 Response of damped pendulum for the ICs $\theta(0) = 10°$ and $\dot{\theta} = 0$

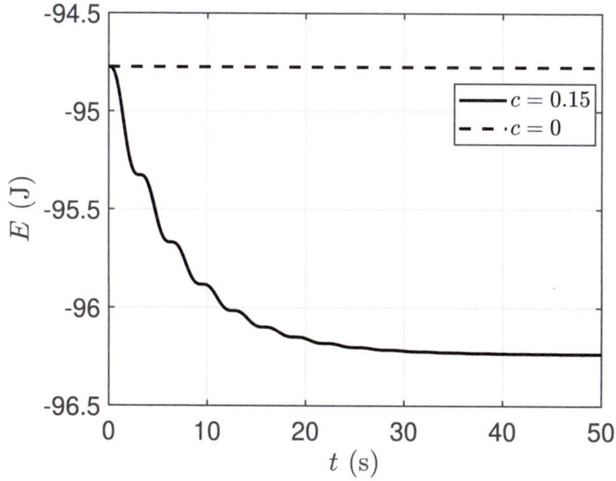

Fig. 1.22 Evolution of the total mechanical energy E of the simple pendulum and damped pendulum. The energy was computed with Eq. (1.30) using the numerical solution of the equation of motion for the cases when $c = 0$ and $c = 0.15$ Ns/m, under the same ICs $\theta(0) = 10°$ and $\dot{\theta}(0) = 0$

Fig. 1.23 Phase portrait of the damped pendulum with the asymptotically stable (•) and unstable (•) equilibria shown

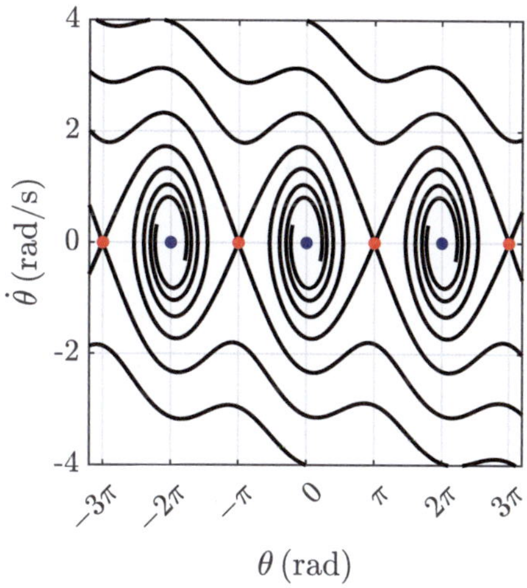

1.3.4.4 Phase Portrait

The numerical solutions can be used to construct the phase portrait shown in Fig. 1.23. Because of damping, energy is dissipated and the trajectories move inward. This gradual decay in amplitude of oscillation makes all trajectories ultimately converge to a state of rest. Trajectories characterized by high angular velocities undergo several complete oscillations before eventually settling into an equilibrium position. In contrast to the undamped pendulum's elliptical trajectories around the stable equilibrium (0, 0), the damped pendulum exhibits spiraling trajectories that converge toward this equilibrium. A slight perturbation from the state (0, 0) will result in a trajectory in its vicinity that moves toward it as $t \to \infty$. For this reason, this equilibrium is now termed *asymptotically stable*.

1.3.4.5 Small Oscillations

The equation of motion for small oscillations around $(\theta, \dot{\theta}) = (0, 0)$ is

$$\ddot{\theta} + \frac{c}{m}\dot{\theta} + \frac{g}{l}\theta = 0. \tag{1.49}$$

This equation has the same form as that of the mass-spring-damper in Eq. (1.22).

1.4 Form of the Equation of Motion

The fact that the equation of motion for small oscillations of the pendulum shares the same form as that of a mass-spring-damper system is not unique to the pendulum itself. This fascinating feature repeats, under broad conditions, in SDOF mechanical systems. To understand why this happens, suppose the generalized coordinate x of a SDOF system composed of interconnected particles and rigid bodies satisfies the equation of motion

$$\ddot{x} + g(x, \dot{x}, t) = F(t), \tag{1.50}$$

where $F(t)$ is an externally applied load that depends on time only and $g(x, \dot{x}, t)$ is a nonlinear function depending on x, $\dot{x}$, and t. Equation (1.50) represents a general form of the equation of motion of SDOF systems because the motion is completely specified by the position x and velocity $\dot{x}$ at any time t or, in other words, by the state $(x, \dot{x})$. The impressed load $F(t) - g(x, \dot{x}, t)$ can only depend on the state $(x, \dot{x})$ and time t; it cannot depend on $\ddot{x}$, which is generated as a consequence. In the absence of any loads, the system continues in its state of rest or uniform motion. When acted upon by an impressed load, an acceleration $\ddot{x}$ is generated.

Since the effect of $F(t)$ is to disturb the system from equilibrium, the equilibria may be obtained by ignoring $F(t)$ and substituting $x = x_e =$ const. into (1.50):

$$g(x_e, 0, t) = 0. \tag{1.51}$$

This equation is generally nonlinear and requires the use of proper numerical methods to be solved, e.g., Newton–Raphson method. The number of mathematical solutions will depend on the nature of the equilibrium equation. For example, there is at most N real solutions if (1.51) is a polynomial of degree N or infinitely many solutions in the case of a transcendental equation. However, only a subset of them will be physically distinct.

When analyzing small vibrations around a stable equilibrium point $(x, \dot{x}) = (x_e, 0)$, one can linearize Eq. (1.50) using a multivariate Taylor expansion of the nonlinear term $g(x, \dot{x}, t)$ according with

$$g(x, \dot{x}, t) \approx g(x_e, 0, t) + \frac{\partial g}{\partial \dot{x}}(x_e, 0, t)(\dot{x} - 0) + \frac{\partial g}{\partial x}(x_e, 0, t)(x - x_e). \quad (1.52)$$

Since $g(x_e, 0, t) = 0$ from Eq. (1.51), and denoting

$$\frac{\partial g}{\partial \dot{x}}(x_e, 0, t) = a(t), \quad \frac{\partial g}{\partial x}(x_e, 0, t) = b(t), \quad (1.53)$$

the equation of motion (1.50) can be rewritten as

$$\ddot{x} + a(t)\dot{x} + b(t)(x - x_e) = F(t). \quad (1.54)$$

Introducing the change of coordinates $q = x - x_e$, it becomes

$$\ddot{q} + a(t)\dot{q} + b(t)q = F(t). \quad (1.55)$$

The equation of motion for "small oscillations" is a second-order linear ODE. The advantage of approximating the nonlinear equation of motion (1.50) by its linearized counterpart (1.55) lies in the ability to simplify the mathematics involved. This simplification enables the utilization of well-established analytical methods from linear systems theory, facilitating the investigation of dynamic behavior, stability, and control. Most importantly, cause and effect are now related linearly, i.e., if the external excitation $F(t)$ is regarded as input to the system and the resulting response $q(t)$ as output, the output is doubled when the input is doubled. This characteristic is called the **principle of superposition**[6] and is depicted in Fig. 1.24.

A system is time-invariant if a delay in the excitation produces an identical delay in the response, as depicted in Fig. 1.25. A linear system is time-invariant if and only if it possesses constant coefficients.[7] When there is a time-dependent coefficient, the system is time-varying. Without a doubt, most of the literature on vibrations deals almost exclusively with linear time-invariant systems described by second-order linear ODEs with constant coefficients:

$$\ddot{q} + a\dot{q} + bq = F(t). \quad (1.56)$$

As will be seen in Chap. 2, the values assigned to the coefficients a and b significantly alter the nature of the dynamic behavior, affecting aspects such as sta-

[6] See Problem 1.5.

[7] See Problem 1.6.

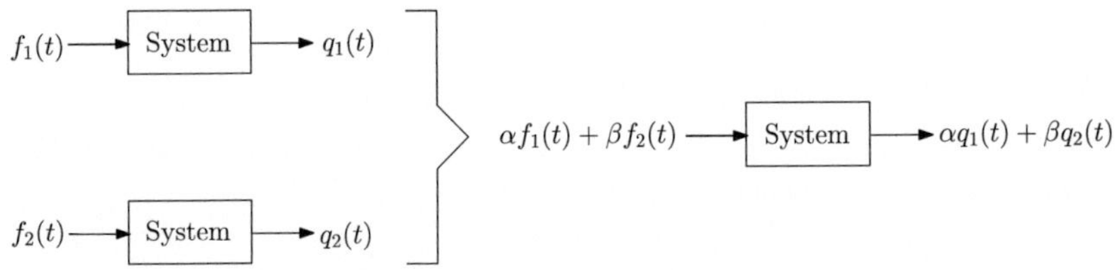

Fig. 1.24 Principle of superposition, where "System" means the differential operator $L[x(t)] = \frac{d^2x}{dt^2} + a(t)\frac{dx}{dt} + b(t)x$ that stems from the differential equation (1.55). If $L[q_1(t)] = f_1(t)$ and $L[q_2(t)] = f_2(t)$ under the same ICs, then $L[\alpha q_1(t)+\beta q_2(t)] = \alpha f_1(t)+\beta f_2(t)$ for any constants α and β

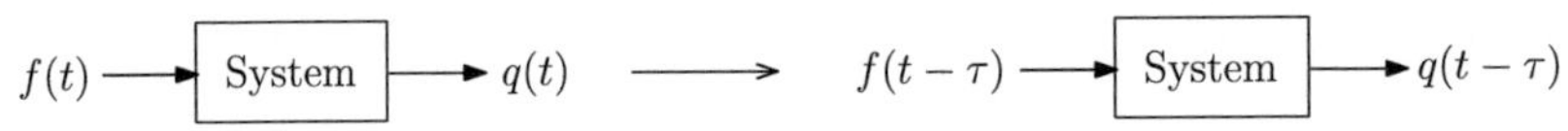

Fig. 1.25 Time-invariant system satisfies $L[q(t-\tau)] = f(t-\tau)$ for an arbitrary constant τ

bility, oscillation, or exponential growth/decay. Engineering analysis predominantly focuses on stable systems, characterized by non-negative coefficients given by

$$a = \frac{c}{m} \geq 0, \quad b = \frac{k}{m} \geq 0.$$

Equation (1.56) can be reframed as

$$m\ddot{q} + c\dot{q} + kq = f(t), \tag{1.57}$$

where $f(t) = mF(t)$. In conclusion, the "small" vibrations of SDOF systems are described by differential equations similar to that of a mass-spring damper from Eq. (1.22). For this reason, it is considered the **prototype system** for vibration analysis: solving Eq. (1.22) gives information on the small vibrations of other systems.

Linear analysis serves to explain much of the vibratory behavior of many systems, helping to make predictions and to gain a deeper understanding of the underlying dynamics. However, there are many phenomena that cannot be predicted or explained by linear theory. While nonlinear analysis is out of the scope of these notes, one can point that a striking difference between linear and nonlinear systems is that the relationship between cause and effect is no longer proportional. For example, the axial compression of a column may be proportional to the applied force for small loads, but at a certain critical load, it will buckle and snap over to a large displacement. Another important difference is that all forces, including the static forces associated with equilibrium, should be included in the analysis of nonlinear systems.

1.5 Illustrative Examples

Two comprehensive examples are provided to illustrate the derivation of the equation of motion for small vibrations of mechanical systems with one DOF.

1.5.1 A Rolling Cylinder

Consider the system shown in Fig. 1.26. It consists of a cylinder rolling without slipping on a horizontal surface, with a linear spring connecting points B on the cylinder and point A on the fixed vertical wall.

1.5.1.1 Kinematic Analysis

The cylinder is modeled as a rigid body undergoing planar motion. Typically, a motion of this nature is associated with three DOFs because general planar rigid motions can be described by two coordinates of any point on the body, along with an angle of orientation, as depicted in Fig. 1.27. However, two holonomic constraints govern the movement. The first constraint arises from the rectilinear motion of the cylinder's center of mass:

$$\mathbf{x}_C = x\mathbf{i}, \tag{1.58}$$

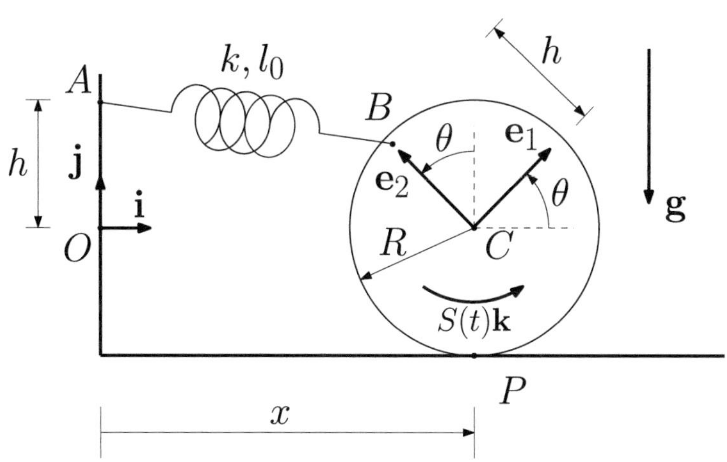

Fig. 1.26 A cylinder with radius R, mass m, and moment of inertia I_{zz} relative to the axis $\mathbf{e}_3 = \mathbf{k}$ passing through its mass center C. It is assumed that the spring is relaxed when $\theta = 0$, i.e., $x(\theta = 0) = l_0$. An external moment $S(t)\mathbf{k}$ acts on the cylinder

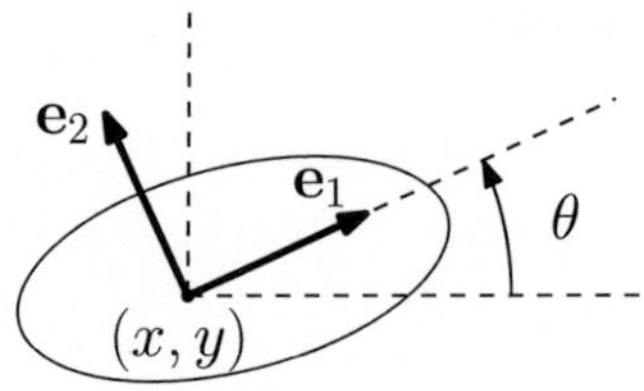

Fig. 1.27 A rigid body in planar motion. Three coordinates are needed to specify its motion, with one possible set being the Cartesian coordinates (x, y) of a point of the body and the orientation θ of the corotational basis $\{\mathbf{e}_1, \mathbf{e}_2\}$ with respect to the Cartesian basis $\{\mathbf{i}, \mathbf{j}\}$

or equivalently

$$\psi_1 = \mathbf{x}_C \cdot \mathbf{j} = 0. \tag{1.59}$$

The second constraint emerges from the rolling assumption, whereby the velocity of the instantaneous point of contact P with the ground satisfies $\mathbf{v}_P = \mathbf{0}$. This leads to the additional velocity constraint

$$\dot{x} + R\dot{\theta} = 0. \tag{1.60}$$

Integrating this equation with respect to time reveals the holonomic constraint

$$\psi_2 = x - x_0 + R(\theta - \theta_0) = 0, \tag{1.61}$$

where $x_0 = x(t = 0)$ and $\theta_0 = \theta(t = 0)$ are constants. For the sake of simplicity, these parameters are taken to be $\theta_0 = 0$ and $x_0 = l_0$, i.e., the spring is unstretched when $\theta = 0$. As a result, the system is left with only one DOF. The constraint ψ_2 indicates that either x or θ can be used to parameterize the motion.

1.5.1.2 Equation of Motion

The FBD of the system is shown in Fig. 1.28. The force exerted by the spring on the cylinder is governed by Hooke's law, and it acts along the direction of the imaginary line connecting points A and B:

$$\mathbf{F}_e = -k\epsilon \frac{\mathbf{x}_B - \mathbf{r}_A}{\|\mathbf{x}_B - \mathbf{r}_A\|}. \tag{1.62}$$

The vector $\mathbf{x}_B - \mathbf{r}_A/\|\mathbf{x}_B - \mathbf{r}_A\|$ is a unit vector of the direction of the spring force, where $\mathbf{x}_B$ is the position vector of point B

$$\mathbf{x}_B = \mathbf{x}_C + h\mathbf{e}_2 \tag{1.63}$$

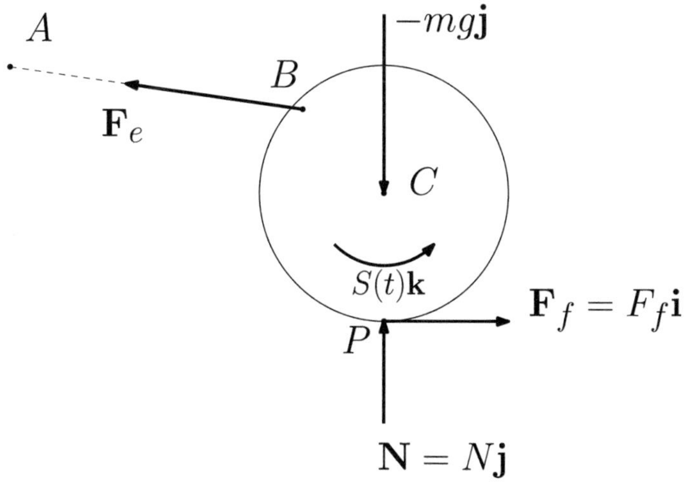

Fig. 1.28 FBD of the rolling cylinder

and $\mathbf{r}_A$ is that of point A

$$\mathbf{r}_A = h\mathbf{j}. \tag{1.64}$$

The change in length of the spring, ϵ, is given by

$$\epsilon = \|\mathbf{x}_B - \mathbf{r}_A\| - l_0. \tag{1.65}$$

The normal force $\mathbf{N}$ and the static friction force $\mathbf{F}_f$ act on the instantaneous point of contact P. Static friction is necessary to maintain rolling without slipping. The weight of the body acts on the center of mass C.

The equation of motion can be derived once all the forces and moments have been identified. Application of Euler's first law yields

$$\mathbf{F}_e + F_f\mathbf{i} + (N - mg)\mathbf{j} = m\ddot{x}\mathbf{i}. \tag{1.66}$$

From this vector equation, one can extract the following two scalar equations:

$$F_f = m\ddot{x} - \mathbf{F}_e \cdot \mathbf{i}, \tag{1.67}$$

$$N = mg - \mathbf{F}_e \cdot \mathbf{j}. \tag{1.68}$$

The inner products are not expanded for compactness. Similarly, application of Euler's second law with respect to the mass center gives

$$(h\mathbf{e}_2 \times \mathbf{F}_e) \cdot \mathbf{k} + \left(-R\mathbf{j} \times F_f\mathbf{i}\right) \cdot \mathbf{k} + S(t) = I_{zz}\ddot{\theta}. \tag{1.69}$$

Using the shifting property of vector triple products,[8] Eq. (1.69) simplifies to

$$I_{zz}\ddot{\theta} = RF_f - h\mathbf{e}_1 \cdot \mathbf{F}_e + S(t). \tag{1.70}$$

Expanding the expression for $\mathbf{F}_e$ and substituting Eq. (1.67) and the rolling constraint (1.60) into this equation, one arrives at

$$\begin{aligned} &\left(I_{zz} + mR^2\right)\ddot{\theta} \\ &- \frac{k\epsilon}{\epsilon + l_0}\left\{R\left[x(\theta) - h\sin\theta\right] + h\left[x(\theta)\cos\theta - h\sin\theta\right]\right\} = S(t), \end{aligned} \tag{1.71}$$

where $\epsilon = \sqrt{[x(\theta) - h\sin\theta]^2 + h^2(\cos\theta - 1)^2} - l_0$. The equation of motion is a nonlinear second-order ODE in terms of the coordinate θ since x is related to θ through Eq. (1.61).

1.5.1.3 Numerical Integration

To observe the dynamic behavior of the system, take the following arbitrary parameter values

$$\begin{aligned} &S = 0, \quad l_0 = 5\text{ m}, \quad R = 1\text{ m}, \\ &h = 0.5\text{ m}, \quad k = 10\text{ N/m}, \quad m = 1\text{ kg}. \end{aligned} \tag{1.72}$$

In the remainder, it is assumed the cylinder is released from rest with $\theta(0) = 15°$. The MATLAB code `Chapt1_code3` provided in the *Electronic Supplementary Material* can be used to numerically integrate Eq. (1.71) and generate Fig. 1.29, which shows the cylinder exhibits periodic oscillations with approximate amplitude of 15° and period $T = 1.64$ s.

1.5.1.4 Equilibrium Positions

The equilibrium positions are solutions $\theta = \theta_e =$ const. (or $x = x_e =$ const.). The applied moment $S(t)$ is an external agent that perturbs the system from equilibrium, so it may be ignored for now. The constant solution may be substituted into the equation of motion to yield the equilibrium equation

$$\frac{k\epsilon}{\epsilon + l_0}\left\{R\left[x_e - h\sin\theta_e\right] + h\left[x_e\cos\theta_e - h\sin\theta_e\right]\right\} = 0, \tag{1.73}$$

[8] Refer to Appendix B.

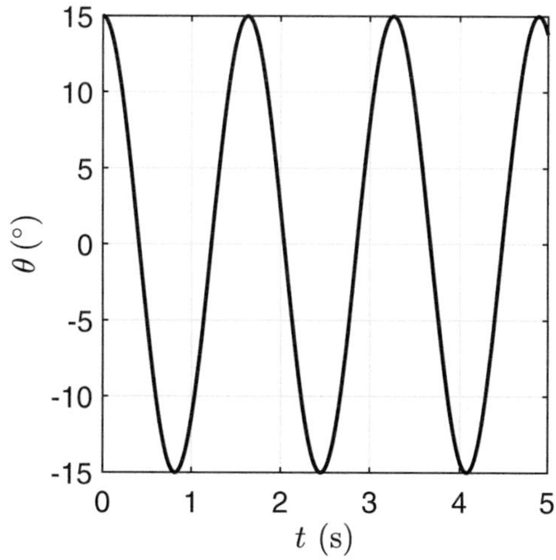

Fig. 1.29 Response of the cylinder for the data in Eq. (1.72)

where $\epsilon = \sqrt{[x_e - h\sin\theta_e]^2 + h^2(1+\cos\theta_e)^2} - l_0$ and $x_e = l_0 + R\theta_e$. A simple substitution reveals that one solution of this equation is $\theta_e = 0$ (or $x_e = l_0 = 5$ m). The other solution, namely $x_e = -5.3259$ m, can be obtained with MATLAB's nonlinear equation solver. This solver is implemented in the *Electronic Supplementary Material* `Chapt1_code4`. This second solution is not physically possible because the cylinder cannot move through the left wall where the spring is attached.

Alternatively, these solutions may be visualized by plotting the nonlinear function in the left side of Eq. (1.73) for different values of x_e, as shown in Fig. 1.30. It is worthwhile to note that the slope of this curve at $x_e = -5.3259$ is zero, a condition termed *zero stiffness*. If this configuration was physically attainable, it would represent a situation where the system offers little to no resistance to deformation or displacement, and it appears highly flexible. This behavior can be observed in certain mechanical systems designed for vibration isolation, shock absorption, or precise positioning.

Numerical solutions can be used to construct the phase portrait shown in Fig. 1.31. The closed trajectories represent motions where the cylinder oscillates periodically around the equilibrium. The trajectories closer to the equilibrium state $(x_e, 0)$ are motions where the cylinder travels roughly the same distance when it moves to and from the wall. In the outer trajectories, the cylinder oscillates with greater amplitude as it moves closer to the wall than when it moves away from it.

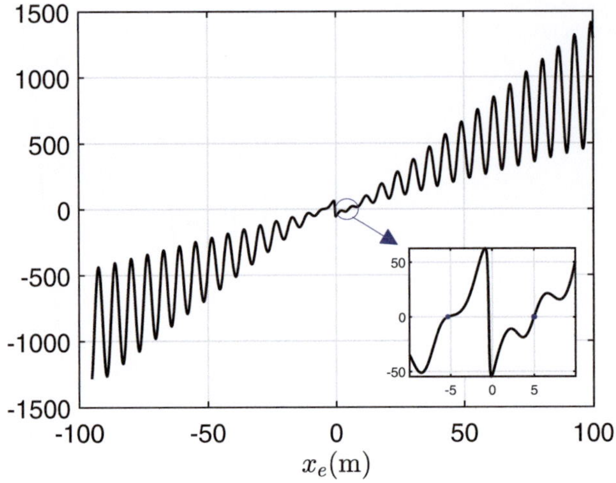

Fig. 1.30 A plot of the function in the left side of equilibrium equation (1.73) for different values of x_e. The equilibrium points are indicated by (•)

Fig. 1.31 Phase portrait of the rolling wheel with $S = 0$. Trajectories (—) represent oscillatory motion around the stable equilibrium (•), while trajectories (—) are motions where the wheel collides with the left wall with $\dot{x} \neq 0$

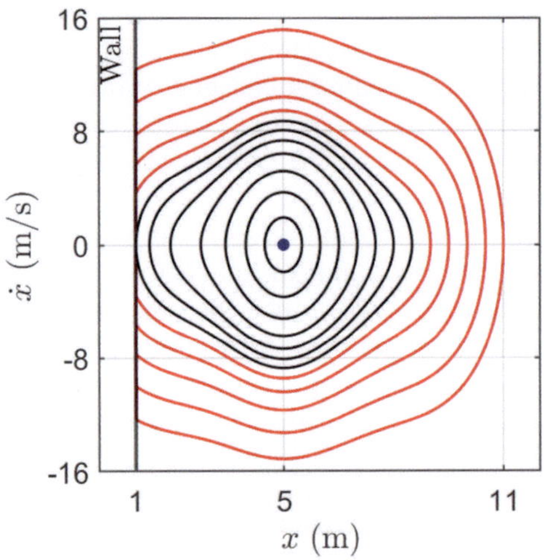

The phase portrait indicates the equilibrium point $x_e = 5$ m is stable because any minor perturbation on the equilibrium state $(x_e,\ 0)$ will result in states belonging to a closed trajectory in the vicinity of the equilibrium, which represents small oscillations around x_e. This is aligned with the plot of potential energy

$$U = \frac{1}{2}k\epsilon^2 \tag{1.74}$$

shown in Fig. 1.32, demonstrating U is minimum at this equilibrium position.

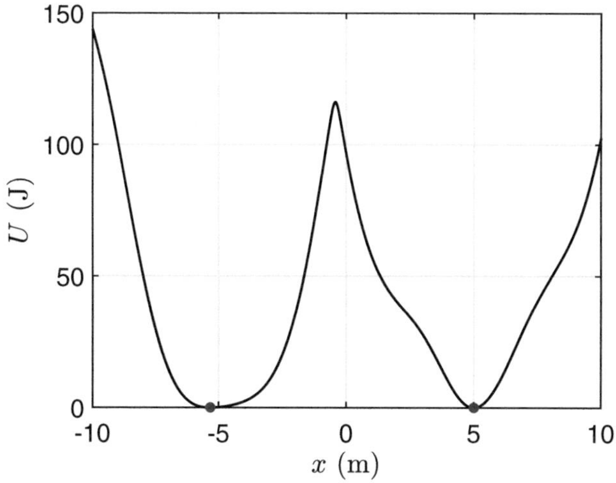

Fig. 1.32 Potential energy with equilibrium points indicated by (•)

1.5.1.5 Small Oscillations About the Equilibrium

The dynamic behavior of small vibrations about the stable equilibrium (0, 0) can be investigate once the equation of motion (1.71) is linearized. However, expanding the nonlinear term

$$g(\theta) = -\frac{k\epsilon}{\epsilon + l_0} \{R\,[x(\theta) - h\sin\theta] + h\,[x(\theta)\cos\theta - h\sin\theta]\} \tag{1.75}$$

as a Taylor series is a laborious task. To simplify the process, the MATLAB code `Chapt1_code5` from the *Electronic Supplementary Material* utilizes the symbolic toolbox to perform a Taylor expansion of this term according with:

$$g(\theta) \approx g(0) + \frac{dg}{d\theta}(0)\theta = k(R+h)^2\theta. \tag{1.76}$$

The equation of motion for small vibrations is

$$\ddot{\theta} + \frac{k(R+h)^2}{\left(I_{zz} + mR^2\right)}\theta = \frac{S(t)}{\left(I_{zz} + mR^2\right)}. \tag{1.77}$$

As expected, this equation has the same form as the equation of motion (1.14) of a mass-spring system. Figure 1.33 shows a plot of the error

$$e(\theta) = \left|\frac{k(R+h)^2\theta - g(\theta)}{g(\theta)}\right| \tag{1.78}$$

Fig. 1.33 Error due to linearization

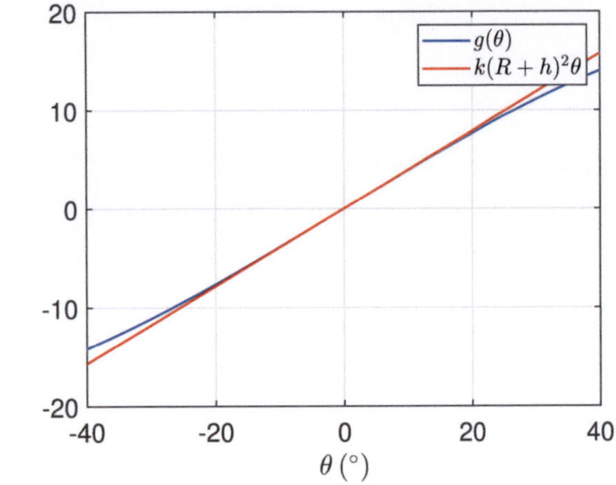

(a) Comparison between $g(\theta)$ and $k(R+h)^2\theta$.

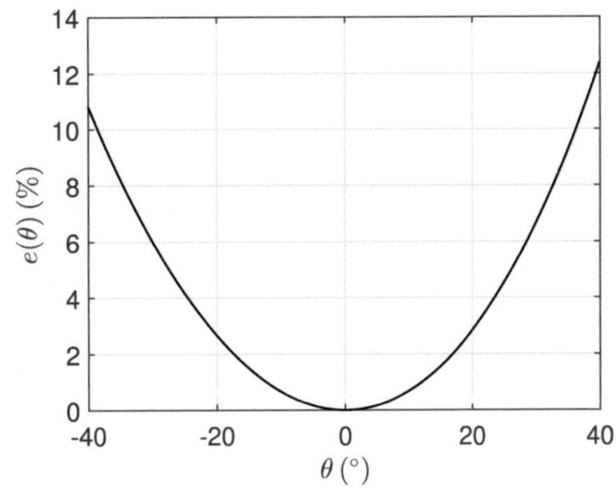

(b) Relative error computed according with Eq. (1.78).

that is committed because of linearization. The approximation is not that restrictive since the error is $e(\theta) < 12\%$ for a swing of $|\theta| < 40°$.

1.5.1.6 Energy Approximation

The total mechanical energy of the wheel is

$$E = \frac{1}{2}\left(I_{zz} + mR^2\right)\dot{\theta}^2 + \frac{1}{2}k\epsilon^2. \tag{1.79}$$

Using the small motion assumption, the potential energy can be approximated by a second-order Taylor expansion around $\theta = 0$, leading to

$$E \approx \frac{1}{2}\left(I_{zz} + mR^2\right)\dot{\theta}^2 + \frac{1}{2}k(R+h)^2\theta^2. \tag{1.80}$$

Equation (1.77) can be used to verify that this approximation satisfies

$$\dot{E} = S(t)\dot{\theta}. \tag{1.81}$$

Therefore, E is a first integral of motion of Eq. (1.77) if S is not acting on the wheel.

1.5.1.7 Alternative Derivation of the Equation of Motion

If the only objective is to obtain Eq. (1.77), it is faster to assume small motions from the outset. For small oscillations around $(\theta, \dot{\theta}) = (0, 0)$, introduce the first-order approximations $\cos\theta \approx 1$ and $\sin\theta \approx \theta$. The corotational basis becomes

$$\mathbf{e}_1 \approx \mathbf{i} + \theta\mathbf{j}, \quad \mathbf{e}_2 \approx -\theta\mathbf{i} + \mathbf{j}. \tag{1.82}$$

In turn, the position vector of point B can be approximated by

$$\mathbf{x}_B = \mathbf{x}_C + h\mathbf{e}_2 \approx (x - h\theta)\mathbf{i} + h\mathbf{j} = [l_0 - (R+h)\theta]\mathbf{i} + h\mathbf{j}, \tag{1.83}$$

where the rolling condition (1.61) has been used. Since point B moves roughly along the horizontal direction, the change of length of the spring is

$$\epsilon \approx -(R+h)\theta. \tag{1.84}$$

Based on the FBD from Fig. 1.34, Euler's first law gives

$$F_f = -mR\ddot{\theta} - k(R+h)\theta, \tag{1.85}$$

$$N = mg. \tag{1.86}$$

Apply Euler's second law with respect to the mass center to obtain

$$I_{zz}\ddot{\theta} = (-h\theta\mathbf{i} + h\mathbf{j}) \times [k(R+h)\theta\mathbf{i}] \cdot \mathbf{k} + (-R\mathbf{j}) \times \mathbf{F}_f \cdot \mathbf{k} + S(t). \tag{1.87}$$

Substitute (1.85) into this equation to show that θ satisfies

$$(I_{zz} + mR^2)\ddot{\theta} + k(R+h)^2\theta = S(t). \tag{1.88}$$

This equation is equal to Eq. (1.77), which was obtained by Taylor expanding the nonlinear term in (1.71) with MATLAB.

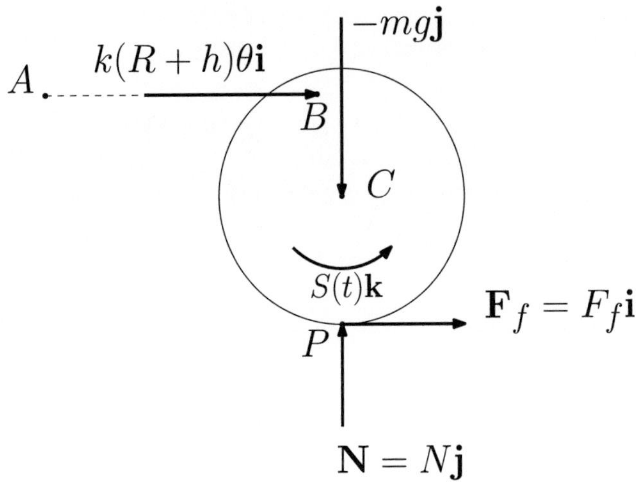

Fig. 1.34 FBD of the rolling cylinder under the assumption of small ocillations

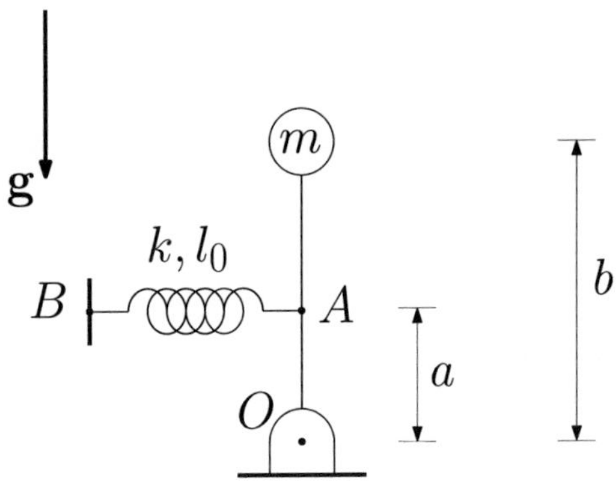

Fig. 1.35 A metronome modeled as a rigid massless rod attached to a particle with mass m. The spring is unstretched in the configuration shown

1.5.2 Motion of a Metronome

The objective of this example is to derive the equation of motion of a metronome. This is a mechanical device consisting of an inverted pendulum that swings back and forth at a set interval. It produces an audible tick sound with each swing to help musicians maintain a consistent rhythm while playing or practicing. The model for a metronome shown in Fig. 1.35 consists of a rigid massless rod attached at its end to a particle with mass m. The rod is also connected to a linear spring with stiffness k and unstretched length l_0. It is assumed friction in the pin joint generates a moment that opposes the motion.

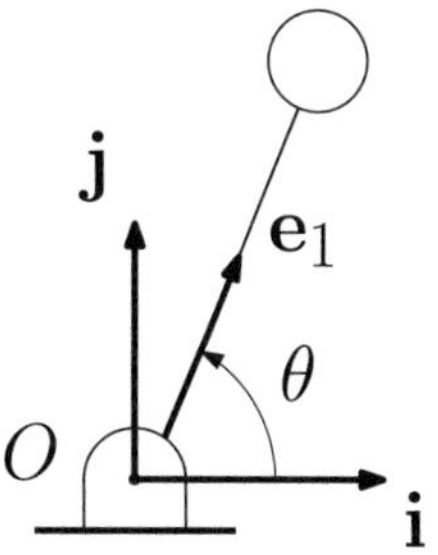

Fig. 1.36 Reference system. Vector $\mathbf{e}_1$ is chosen to lie along the rod

1.5.2.1 Kinematic Analysis

The corotational basis vectors are chosen such that $\mathbf{e}_1$ lies along the rod, as depicted in Fig. 1.36. The rigid body is moving under the holonomic constraints

$$\psi_1 = \mathbf{x}_O \cdot \mathbf{i} = 0, \quad \psi_2 = \mathbf{x}_O \cdot \mathbf{j} = 0. \tag{1.89}$$

Thus, it has only one DOF and the angular coordinate θ is free to vary. The position vector of the concentrated mass is

$$\mathbf{x} = b\mathbf{e}_1. \tag{1.90}$$

As a consequence,

$$\mathbf{v} = b\dot{\theta}\mathbf{e}_2. \tag{1.91}$$

The angular momentum with respect to O is

$$\mathbf{H}^O = \mathbf{x} \times m\mathbf{v} = mb^2\dot{\theta}\mathbf{k}. \tag{1.92}$$

1.5.2.2 Equation of Motion

The spring force is given by

$$\mathbf{F}_e = -k\epsilon \frac{\mathbf{x}_\mathbf{A} - \mathbf{r}_B}{\|\mathbf{x}_\mathbf{A} - \mathbf{r}_B\|}. \tag{1.93}$$

The position vectors of points A and B are

$$\mathbf{x}_A = a\mathbf{e}_1, \quad \mathbf{r}_B = -l_0\mathbf{i} + a\mathbf{j}, \tag{1.94}$$

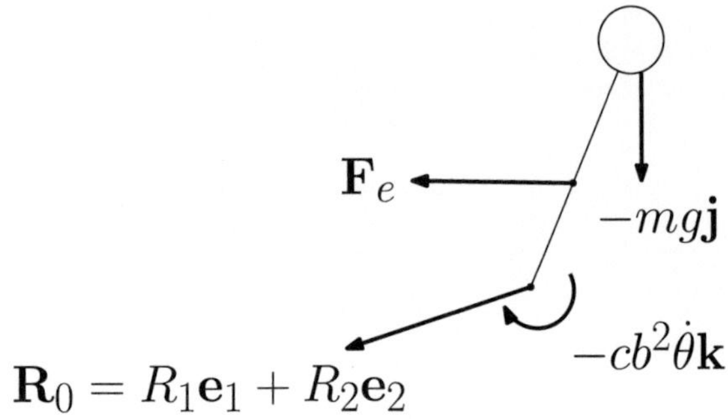

Fig. 1.37 FBD of metronome

respectively. They can be used to compute the change in length of the spring:

$$\epsilon = \|\mathbf{x_A} - \mathbf{r}_B\| - l_0 = \sqrt{(l_0 + a\cos\theta)^2 + a^2(\sin\theta - 1)^2} - l_0. \tag{1.95}$$

The remaining forces are the weight $-mg\mathbf{j}$ acting on the concentrated mass and the reaction force $\mathbf{R}_O$ acting at O due to the pinned connection. Assume that friction on the pin joint generates the moment $-cb^2\dot{\theta}\mathbf{k}$ that opposes the angular velocity of the metronome, where $c \geq 0$ is a constant. These forces are displayed in the FBD shown in Fig. 1.37.

Application of Euler's second law with respect to the fixed point O results in the second-order nonlinear ODE

$$mb^2\ddot{\theta} + cb^2\dot{\theta} - a\mathbf{e}_2 \cdot \mathbf{F}_e + mgb\cos\theta = 0. \tag{1.96}$$

For compactness, the spring force is not expanded.

1.5.2.3 Simulating the Motion

Assume the arbitrary values

$$l_0 = 0.25 \text{ m}, \quad b = 0.5 \text{ m}, \quad a = 0.25 \text{ m}, \quad m = 1 \text{ kg}, \tag{1.97}$$

and the ICs

$$\theta(0) = \frac{\pi}{2} \text{ rad}, \quad \dot{\theta}(0) = 1 \text{ rad/s}. \tag{1.98}$$

The MATLAB code `Chapt1_code6` available in the *Electronic Supplementary Material* can be used to numerically integrate Eq. (1.96) for $c = 0$ and $k = 200$ N/m, but it can be easily modified to incorporate any other values for these parameters. It also produces the plots shown in Fig. 1.38. These plots illustrate different types of motions permitted for the metronome. In the first one, shown in Fig. 1.38a, the metronome oscillates about the vertical ($\theta = \pi/2$) with constant amplitude and period $T \approx 1.15$ s. The other motion, from Fig. 1.38b, shows that θ increases (or

Fig. 1.38 Motion of the undamped metronome

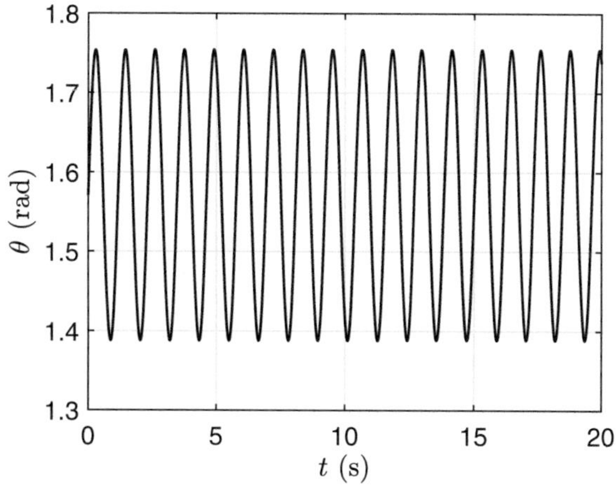

(a) Motion with $k = 200$ N/m.

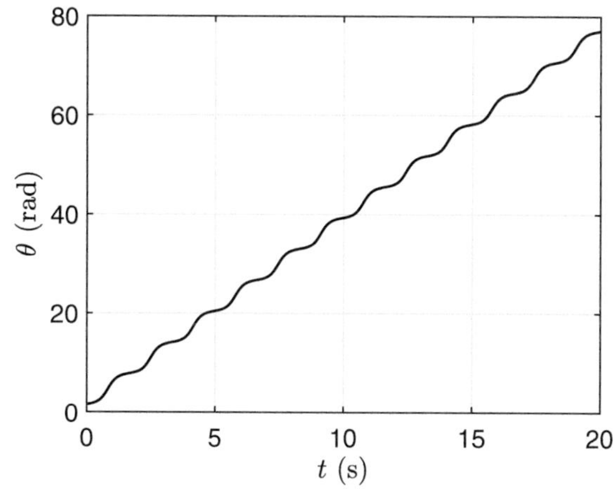

(b) Motion with $k = 10$ N/m.

decreases) due to the continuous oscillation of the metronome. This happens if the spring is too flexible to resist the force of gravity.

When $c \neq 0$, the total mechanical energy

$$E = \frac{1}{2}mb^2\dot{\theta}^2 + mgb\sin\theta + \frac{1}{2}k\epsilon^2 \tag{1.99}$$

decreases with time according with

$$\dot{E} = -cb^2\dot{\theta}^2. \tag{1.100}$$

As a result, the amplitude of oscillation decreases. This can be observed in Fig. 1.39.

Fig. 1.39 Motion of the damped metronome with $k = 200$ N/m and $c = 0.1$ Ns/m

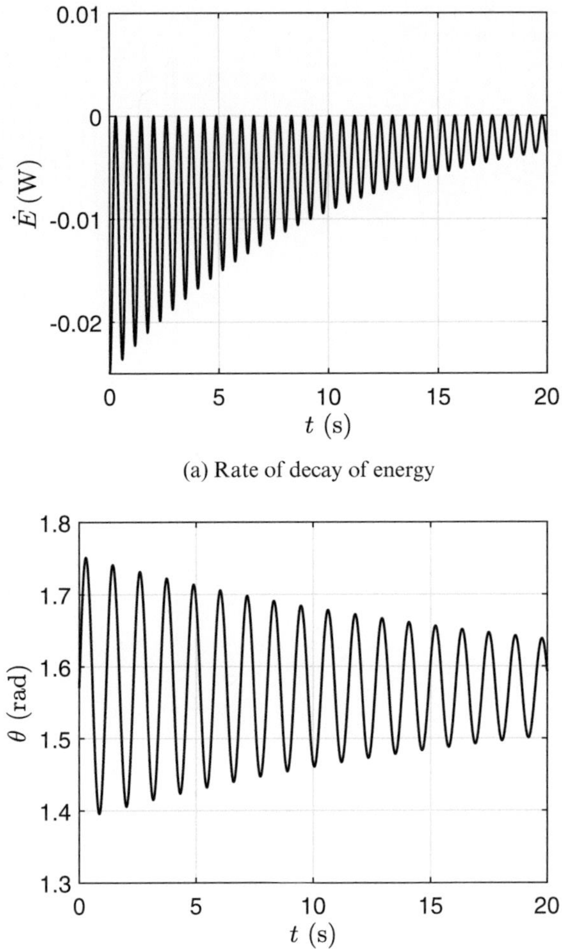

(a) Rate of decay of energy

(b) Oscillatory damped motion.

1.5.2.4 Equilibrium Positions

The equilibrium positions $\theta_e = \text{const.}$ are solutions of

$$-a\mathbf{e}_2(\theta_e) \cdot \mathbf{F}_e(\theta_e) + mgb\cos\theta_e = 0. \tag{1.101}$$

A direct substitution reveals that $\theta_e = \pi/2$ is an equilibrium position. Figure 1.40 shows a portion of the phase portrait in the neighborhood of this equilibrium. When $c = 0$, the trajectories form closed curves representing oscillatory motion around $(\theta, \dot{\theta}) = (\pi/2, 0)$: this equilibrium is stable. It is interesting to note that, in the outer trajectories, the metronome does not swing symmetric arcs around $\pi/2$. A

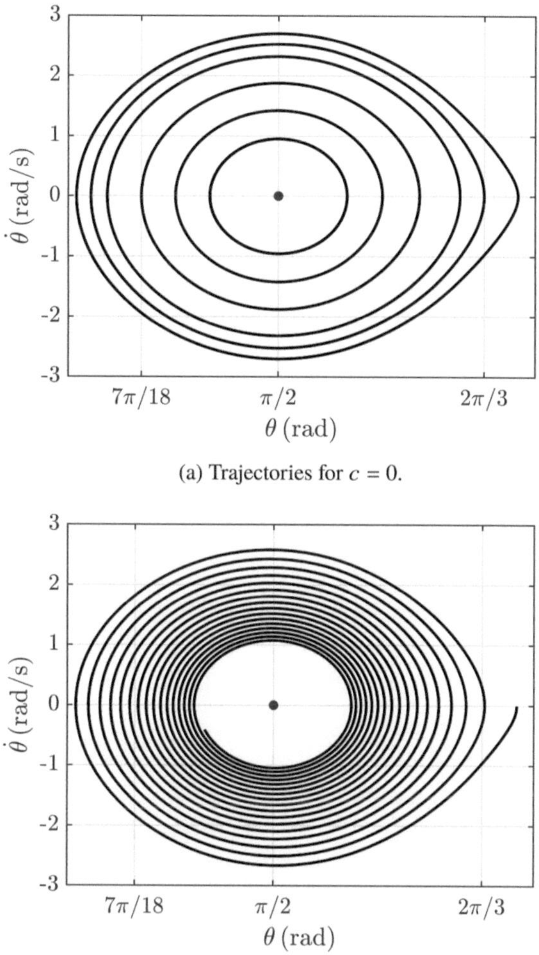

Fig. 1.40 Trajectories in the neighborhood of $(\theta, \dot{\theta}), = (\pi/2, 0)$ with the equilibrium state (•) shown. The value of $k = 200$ N/m is used in the simulations

(a) Trajectories for $c = 0$.

(b) A typical trajectory for $c = 0.1$ Ns/m.

typical trajectory spirals toward the equilibrium when $c > 0$ because the amplitude of oscillation is decreasing. In this case, the equilibrium is asymptotically stable.

1.5.2.5 Linearization

The nonlinear term in Eq. (1.96) can be expanded in a Taylor series about $\theta_e = \pi/2$ to reveal the equation governing the small oscillations of the metronome. The MATLAB code `Chapt1_code7` from the *Electronic Supplementary Material* can be

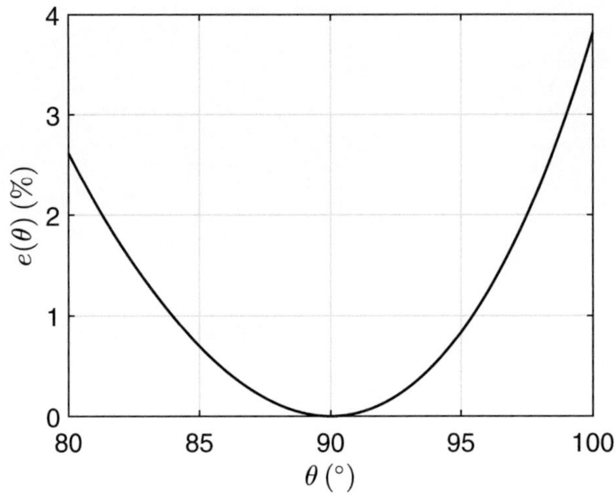

Fig. 1.41 Error due to linearization. Plot constructed using $k = 200$ N/m

used to linearize the function $g(\theta) = -a\mathbf{e}_2 \cdot \mathbf{F}_e + mgb\cos\theta$ according with

$$g(\theta) \approx g\left(\frac{\pi}{2}\right) + \frac{dg}{d\theta}\left(\frac{\pi}{2}\right)\left(\theta - \frac{\pi}{2}\right) = (ka^2 - mgb)\left(\theta - \frac{\pi}{2}\right) \tag{1.102}$$

Figure 1.41 shows that the error due to linearization

$$e = \left|\frac{(ka^2 - mgb)\left(\theta - \frac{\pi}{2}\right) - g(\theta)}{g(\theta)}\right| \tag{1.103}$$

is less than 4% for amplitudes less than 10°.

Finally, the linearized equation of motion becomes

$$mb^2\ddot{\theta} + cb^2\dot{\theta} + (ka^2 - mgb)\left(\theta - \frac{\pi}{2}\right) = 0. \tag{1.104}$$

Defining a new coordinate

$$\gamma = \theta - \frac{\pi}{2}, \tag{1.105}$$

to express the equation of motion as

$$mb^2\ddot{\gamma} + cb^2\dot{\gamma} + (ka^2 - mgb)\gamma = 0. \tag{1.106}$$

1.5.2.6 Energy Approximation

The total mechanical energy of the metronome is given by Eq. (1.99). Using up to the second-order term, the potential energy can be expanded as a Taylor series around $\theta = \pi/2$ to yield an approximation for E:

$$E \approx \frac{1}{2}b^2\dot{\gamma}^2 + \frac{1}{2}(ka^2 - mgb)\gamma. \tag{1.107}$$

This approximation is a first integral of motion for the linearized equation (1.106) if $c = 0$.

If $ka^2 < mgb$, the approximation for U indicates that it is a concave down parabola in the neighborhood of $\theta = \pi/2$. This means $U(\pi/2)$ is a local maximum, and the equilibrium is unstable. Conversely, U is concave up if $ka^2 > mgb$, so $\theta = \pi/2$ is a stable equilibrium. These behaviors were observed in Fig. 1.38.

1.5.2.7 Alternative Derivation of the Equation of Motion

Assuming that the metronome executes small oscillations around $\theta = \pi/2$, the first-order approximations

$$\cos\theta \approx -\left(\theta - \frac{\pi}{2}\right) = -\gamma, \quad \sin\theta \approx 1, \tag{1.108}$$

are introduced to express the position of the particle as

$$\mathbf{x} = b\mathbf{e}_1 \approx -b\gamma\mathbf{i} + b\mathbf{j}. \tag{1.109}$$

The angular momentum with respect to O is

$$\mathbf{H}^O = \mathbf{x} \times m\mathbf{v} \approx (-b\gamma\mathbf{i} + b\mathbf{j}) \times (-mb\dot{\gamma}\mathbf{i}) = mb^2\dot{\gamma}\mathbf{k}. \tag{1.110}$$

The position vector of point A is approximated by

$$\mathbf{x}_A = a\mathbf{e}_1 \approx -a\gamma\mathbf{i} + a\mathbf{j}. \tag{1.111}$$

Then, the change of length of the spring is simply

$$\epsilon = \|\mathbf{x_A} - \mathbf{r}_B\| - l_0 \approx -a\gamma, \tag{1.112}$$

and the spring force

$$\mathbf{F}_e \approx ka\gamma\mathbf{i}. \tag{1.113}$$

Using Euler's second law,

$$mb^2\ddot{\gamma} = -cb^2\dot{\gamma} + (-a\gamma\mathbf{i} + a\mathbf{j}) \times (ka\gamma\mathbf{i}) \cdot \mathbf{k} + (-b\gamma\mathbf{i} + b\mathbf{j}) \times (-mg\mathbf{j}) \cdot \mathbf{k}. \tag{1.114}$$

Expansion of the vector triple product results in the equation of motion (1.106).

1.6 Summary

Vibration is the study of oscillatory motion. Vibratory systems are characterized by (i) an equilibrium position for the system to oscillate about, (ii) inertia, (iii) elasticity, (iv) damping, and (v) stimuli that set the system into motion. The motion of SDOF systems can be completely characterized by a single generalized coordinate x that, in general, satisfies a second-order ODE of the form

$$\ddot{x} + g(x, \dot{x}, t) = f(t),$$

obtained from the fundamental laws of motion. Here, $f(t)$ is an externally applied load, and $g(x,\ \dot{x},\ t)$ is a nonlinear function of the variables x, $\dot{x}$, and t. Solution of the equilibrium equation

$$g(x_e,\ 0,\ t) = 0$$

gives the equilibrium points for the system. By introducing a change of coordinates $q = x - x_e$, the small oscillatory behavior about a stable equilibrium can then be obtained by analysis of the linearized equation of motion of the form

$$\ddot{q} + a(t)\dot{q} + b(t)q = f(t).$$

Most of engineering literature is concerned with stable systems whose coefficients are constant and non-negative. Under these conditions, the equation of motion is

$$m\ddot{q} + c\dot{q} + kq = f(t),$$

where $m > 0$, $c \geq 0$, and $k > 0$. Hence, any SDOF system can be represented by the mass-spring-damper equation.

1.7 Problems

1.1 Check that the mechanical energy (1.15) can be expressed equivalently as

$$E = \frac{1}{2}m\dot{q}^2 + \frac{1}{2}q^2 + U_e.$$

1.2 Determine the equivalent stiffness for the parallel association of two springs with unstretched length l_0 and stiffness k_1 and k_2, as shown in Fig. 1.42.

1.3 Determine the equivalent stiffness for the series association of two springs with unstretched length l_0 and stiffness k_1 and k_2, as shown in Fig. 1.43.

1.4 Use appropriate coordinate systems to express the kinematic constraints and determine the number of DOFs of:

(a) Two particles connected by a rigid rod, but otherwise free to move in the plane, as shown in Fig. 1.44a.
(b) Two particles connected by a spring, but otherwise free to move in the plane, as shown in Fig. 1.44b.
(c) A rigid body rotating around a fixed axis, as shown in Fig. 1.44c.
(d) A rigid body in translation, as shown in Fig. 1.44d.

Fig. 1.42 Parallel association of two springs

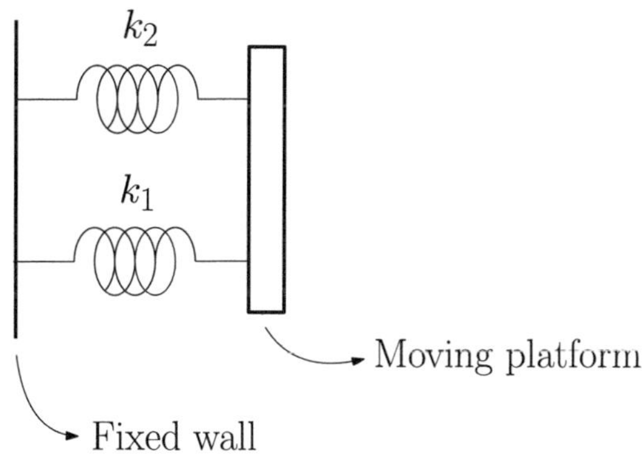

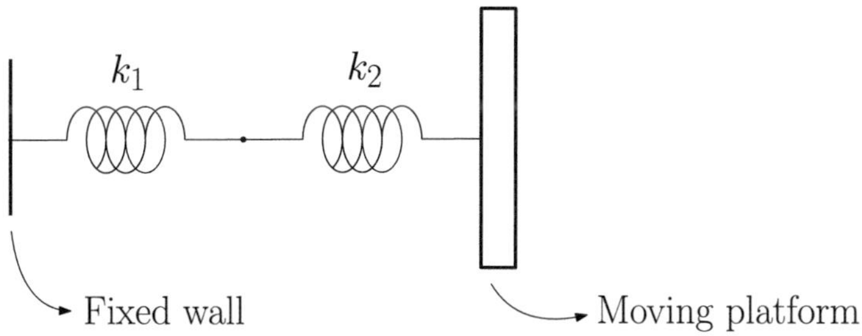

Fig. 1.43 Series association of two springs

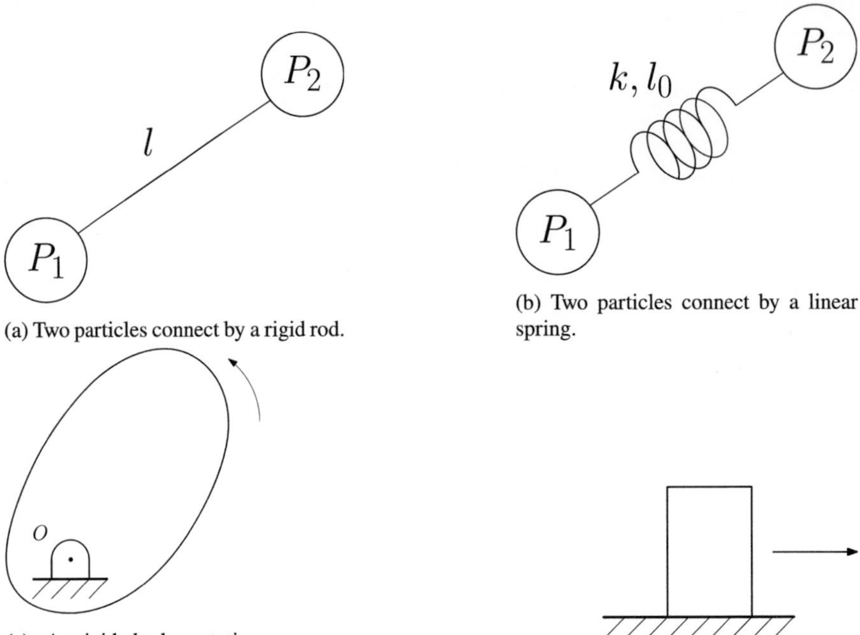

(a) Two particles connect by a rigid rod.

(b) Two particles connect by a linear spring.

(c) A rigid body rotating around the fixed axis at O.

(d) A rigid body translating.

Fig. 1.44 Figure for problem 1.4

1.5 Check that the linear system (1.55) satisfies the principle of superposition shown in Fig. 1.24.

1.6 Prove the linear system (1.55) is time-invariant if and only if it possesses constant coefficients.

1.7 Consider a mechanical system described by the equation of motion

$$m\ddot{x} + c\dot{x} + k(x + \epsilon x^3) = f(t). \tag{1.115}$$

Check that the principle of superposition does not apply to this system.

1.8 A mass-spring system is free to oscillate on a smooth horizontal surface. As depicted in Fig. 1.45, high-stiffness barriers that are not connected to the mass are introduced to limit the vibration amplitudes. These barriers are known as buffer springs. When $x = 0$, the springs connected to the particle are unstretched.

Use Newton's second law to show that the equation of motion for this system is given by

$$m\ddot{x} + f(x) = 0. \tag{1.116}$$

Is this system linear?

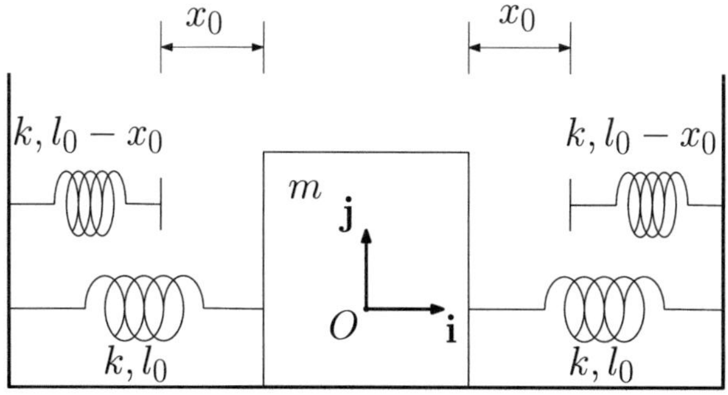

Fig. 1.45 A particle with mass m that is free to move on a smooth horizontal surface

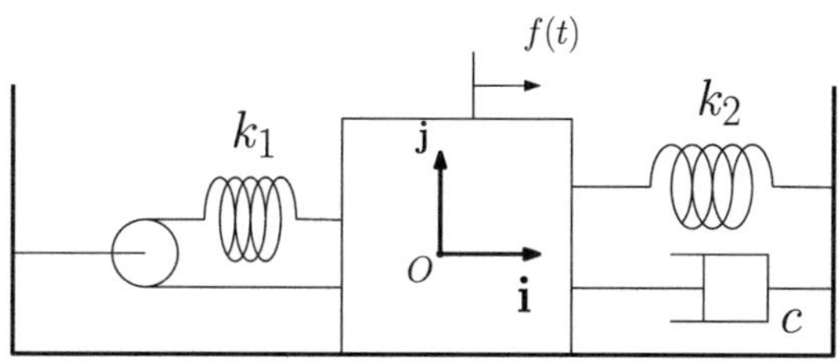

Fig. 1.46 A particle with mass m free to move on a smooth horizontal surface. The spring with stiffness k_1 is connected to a pulley

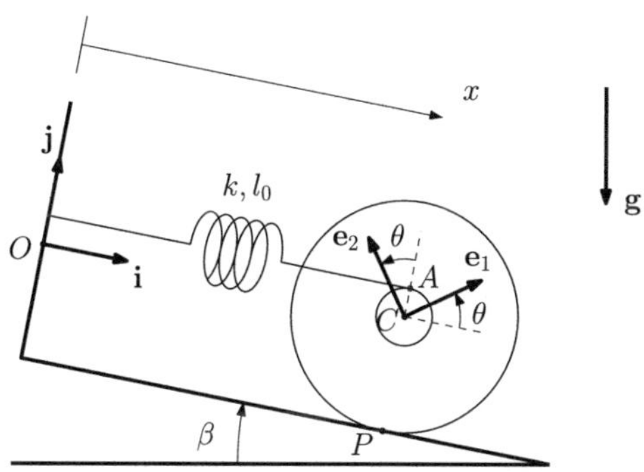

Fig. 1.47 Uniform disk rolling without slipping on an inclined plane

1.9 The mechanical system shown in Fig. 1.46 consists of a block with mass m excited by a force $f(t)$. The block is connected to a spring with stiffness k_2 and a damper with viscous damping coefficient c. An elastic cable with stiffness k_1 is connected to the block through a fixed pulley of negligible mass. In the configuration shown in the figure, the springs are relaxed. Find the equation of motion of this system.

1.10 A uniform disk of radius R rolls without slipping on an inclined plane. Concentric with the disk, and fixed to it, is a drum of radius r around which a spring with stiffness k and unstretched length l_0 is wound, as shown in Fig. 1.47. The total mass of the disk+drum assembly is m, and the moment of inertia of the assembly about the axis $\mathbf{e}_3 = \mathbf{e}_1 \times \mathbf{e}_2$ passing through the center of mass C is I_{zz}. Determine the equation of motion of this system using a generalized coordinate measured from the equilibrium position.

Chapter 2
Free Vibration of Systems with One Degree of Freedom

The objective of this chapter is to investigate the free vibratory response of systems with one degree of freedom. The presentation starts with undamped systems, and key concepts are presented using a practical demonstration that can be easily replicated. Subsequently, the analysis extends to cases where damping is present and the effect of different levels of damping on the system response is examined. Several examples are provided to illustrate the applicability of the theory. Some of them involve modeling, analysis, and validation of the theory by means of simple experiments that readers can carry out on their own.

2.1 What Is Free Vibration?

This chapter is concerned with free vibration of SDOF systems. Free vibration occurs when a mechanical system is set into motion and then allowed to vibrate without any external force applied after the initial disturbance. This type of analysis is important because it is used to determine natural frequencies. These frequencies are intrinsic properties of mechanical systems and dictate when violent vibrations (resonance) may occur in the presence of external forces. Additionally, free vibration tests are used to evaluate the dynamic properties of materials and structures, such as damping.

This problem can be tackled by analyzing the motion of the prototype system shown in Fig. 2.1. Specifically, the objective is to solve the equation of motion

$$m\ddot{q} + c\dot{q} + kq = 0 \tag{2.1}$$

Supplementary Information The online version contains supplementary material available at (https://doi.org/10.1007/978-3-031-83995-5_2).

R. Gonçalves Salsa Junior, *Fundamentals of the Theory of Mechanical Vibrations*,
https://doi.org/10.1007/978-3-031-83995-5_2

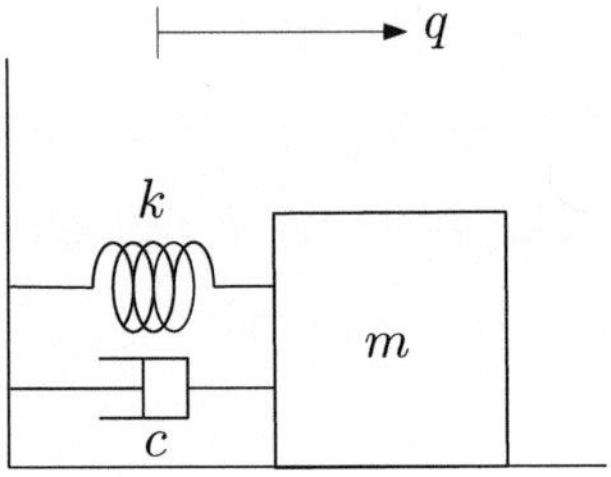

Fig. 2.1 Prototype SDOF system in free vibration. The system consists of a mass m, damping c, stiffness k, with no external force acting on it. The motion is parameterized by the generalized coordinate q, measured from the equilibrium position

for arbitrary ICs. The chapter starts with the analysis of undamped systems in Sect. 2.2. The theory of damped free vibration is presented in Sect. 2.3. Nine examples are provided to illustrate the application of the theory. Some examples also involve hands-on experiments that can be replicated by readers to reinforce the concepts. A summary of the main results is presented in Sect. 2.4.

2.2 Free Vibration of Undamped Systems

To understand free vibration, it is instructive to consider first the case without damping. While no real system operates without some mechanism of energy dissipation, this analysis offers useful results when damping is small.

2.2.1 *The Harmonic Oscillator*

Consider the demonstration from Sect. 1.2.1, where the plastic bottle oscillates up and down. After the bottle is pushed and released from rest, the amplitude of oscillation remains practically constant for some time, giving the impression that damping is absent. Focus on this stage of the motion and repeat the experiment as follows. If different initial displacements are imposed, it can be observed that the period of oscillation remains virtually unaltered. This indicates the frequency of oscillation is independent of ICs. However, the number of coins affects the period. If more coins are used, then the time required to complete one oscillation increases. Conversely, removing some coins hastens this process, illustrating the dependence of the period on the mass of the system. The spring stiffness also impacts the dynamic behavior. If two identical springs are connected to the bottle in parallel, the system becomes stiffer,[1] and the bottle oscillates more rapidly. It is now pertinent to verify whether the models established in Sect. 1.2 can accurately replicate these observations.

[1] See Problem 1.3.

2.2.1.1 Computing the Response

The oscillatory behavior around the static equilibrium position is described by Eq. (1.55). When damping is absent ($c = 0$) and no external forces are acting upon the system ($f = 0$), the equation of motion takes the form

$$m\ddot{q} + kq = 0. \tag{2.2}$$

This represents a second-order linear homogeneous ODE with constant coefficients. It can be solved using established ODE theory by considering the ICs $q(0) = q_0$ and $\dot{q}(0) = \dot{q}_0$, which correspond to the potential and kinetic energies of the system at $t = 0$. For convenience, it is advantageous to rewrite Eq. (2.2) in the *standard* form:

$$\ddot{q} + \omega_n^2 q = 0, \tag{2.3}$$

where $\omega_n = \sqrt{k/m}$ has units of rad/s and is called *natural frequency*. This parameter carries significant physical importance, which is elucidated further.

The general solution of Eq. (2.3) has the form

$$q(t) = Ae^{\lambda t}, \tag{2.4}$$

where $A, \lambda \in \mathbb{C}$ are constants to be determined. Substitute (2.4) and its second-order derivative

$$\ddot{q}(t) = A\lambda^2 e^{\lambda t} \tag{2.5}$$

into Eq. (2.3) to obtain

$$A(\lambda^2 + \omega_n^2)e^{\lambda t} = 0. \tag{2.6}$$

Since $e^{\lambda t} \neq 0$, this equation becomes

$$A(\lambda^2 + \omega_n^2) = 0. \tag{2.7}$$

It is necessary to impose $A \neq 0$ to avoid the trivial solution. Then, one arrives at the *characteristic polynomial*

$$\lambda^2 + \omega_n^2 = 0. \tag{2.8}$$

This is a quadratic polynomial in λ, with roots given by

$$\lambda_{1,2} = \pm i\omega_n, \tag{2.9}$$

where $i = \sqrt{-1}$ is the imaginary unit.[2] Because of linearity, the solution $q(t)$ is

$$q(t) = Ae^{i\omega_n t} + Be^{-i\omega_n t}. \tag{2.10}$$

Using Euler's identity

$$e^{i\omega_n t} = \cos(\omega_n t) + i \sin(\omega_n t), \tag{2.11}$$

the response can be further simplified to

$$q(t) = a \cos(\omega_n t) + b \sin(\omega_n t), \tag{2.12}$$

where $a,\ b \in \mathbb{R}$ are constants of integration determined from

$$q(0) = a = q_0, \tag{2.13}$$

$$\dot{q}(0) = b\omega_n = \dot{q}_0. \tag{2.14}$$

Solving for a and b reveals that

$$q(t) = q_0 \cos(\omega_n t) + \frac{\dot{q}_0}{\omega_n} \sin(\omega_n t). \tag{2.15}$$

This is the response of a mass-spring system in free vibration under general ICs.

2.2.1.2 Analysis of the Response

Response (2.15) is also commonly expressed in the following form:

$$q(t) = Q \cos(\omega_n t - \phi), \tag{2.16}$$

where Q and ϕ are constants that can be determined by comparing the expansion

$$Q \cos(\omega_n t - \phi) = Q \cos\phi \cos(\omega_n t) + Q \sin\phi \sin(\omega_n t) \tag{2.17}$$

with Eq. (2.15). By doing so, one obtains the system of equations

$$Q \cos\phi = q_0, \quad Q \sin\phi = \frac{\dot{q}_0}{\omega_n}. \tag{2.18}$$

[2] Appendix A reviews essential properties of complex numbers.

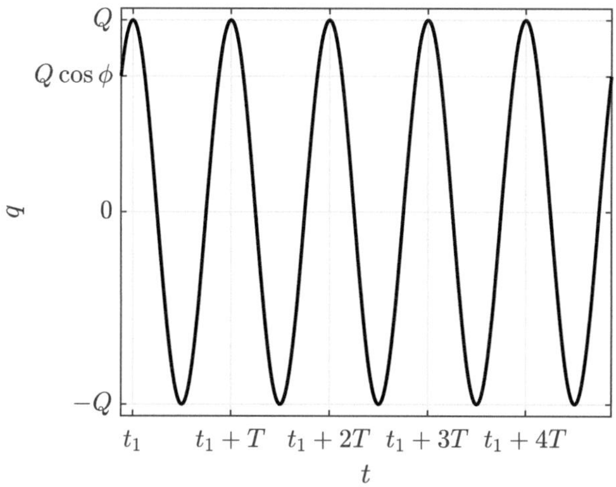

Fig. 2.2 Free vibration of undamped SDOF system (2.2). It oscillates harmonically with natural period $T = 2\pi/\omega_n$, where $\omega_n = \sqrt{k/m}$ is the natural frequency

These equations can be solved for Q and ϕ, and the following alternative expression for $q(t)$ is obtained:

$$q(t) = \sqrt{q_0^2 + \frac{\dot{q}_0^2}{\omega_n^2}} \cos\left(\omega_n t - \arctan\frac{\dot{q}_0}{q_0\omega_n}\right), \tag{2.19}$$

Several key observations can be readily made from Eq. (2.19), whose characteristic behavior is shown in Fig. 2.2. When the system is set into motion, it oscillates harmonically with frequency $\omega_n = \sqrt{k/m}$. For this reason, this system is called harmonic oscillator. The natural period of oscillation $T = 2\pi/\omega_n$ is determined solely by the system's physical properties, specifically the inertia m and stiffness k. This is in accordance with the experimental observations that (i) the period does not depend on the ICs, (ii) it increases if m is increased, and (iii) it decreases if k is increased. In addition, the amplitude of oscillation

$$Q = \sqrt{q_0^2 + \frac{\dot{q}_0^2}{\omega_n^2}} \tag{2.20}$$

remains constant over time. This behavior is expected in the absence of damping, when the system experiences sustained oscillations without any reduction in amplitude. The phase angle

$$\phi = \arctan\frac{\dot{q}_0}{q_0\omega_n} \tag{2.21}$$

Fig. 2.3 Phase portrait of the harmonic oscillator with the stable equilibrium (•) shown

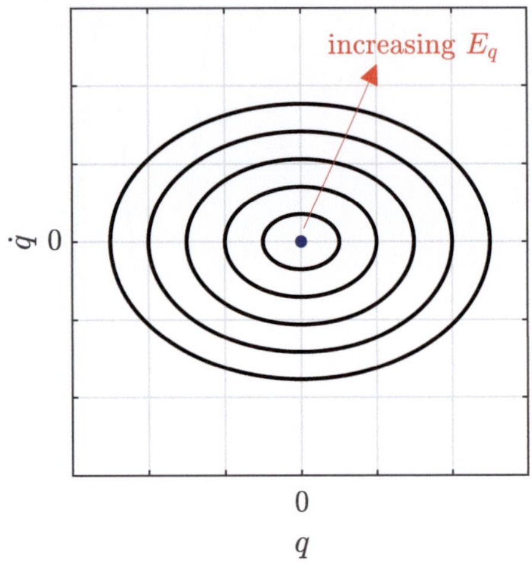

determines where the curve intersects the vertical axis, only influencing the starting position of vibration. In contrast to the frequency of oscillation ω_n, both Q and ϕ depend on external factors, i.e., the ICs.

According to Eq. (2.19), any perturbation on the equilibrium state $(q, \dot{q}) = (0, 0)$ causes the system to oscillate harmonically. This oscillatory motion is periodic, so a typical trajectory in the state space is a closed curve. Figure 2.3 illustrates the phase portrait of the harmonic oscillator. The trajectories shown in the phase portrait are not arbitrary closed curves, but ellipses. To understand why this occurs, recall that the mass-spring system conserves energy. This implies that

$$E_q = \frac{1}{2}m\dot{q}^2 + \frac{1}{2}kq^2 = \text{const.} \tag{2.22}$$

This equation can be rewritten as

$$1 = \frac{q^2}{\left(\frac{2E_q}{k}\right)} + \frac{\dot{q}^2}{\left(\frac{2E_q}{m}\right)}, \tag{2.23}$$

which represents the equation of an ellipse with semi-major/minor axes $2E_q/k$ and $2E_q/m$. Each trajectory corresponds to a level curve of constant energy, where the associated energy level increases as the trajectories are farther away from the origin.

2.2.2 General Case of Undamped Free Vibration

In Sect. 1.4, it was explained that the equation of motion governing small vibrations of undamped systems has the specific form Eq. (2.2). This means the key findings regarding the motion of the harmonic oscillator applies to other undamped SDOF systems in free vibration. An important consequence is that the natural frequency of vibration can be found by examining the equation of motion when it is in the standard form (2.3). Additionally, the trajectories in a small neighborhood around the stable equilibrium should resemble those from Fig. 2.3.

2.2.3 Illustrative Examples

This section demonstrates the application of the theory of undamped free vibration. It revisits three examples from Chap. 1, while also introducing three new examples that offer valuable opportunities for experimentation and firsthand validation of the theory.

2.2.3.1 Small Oscillations of a Simple Pendulum

In Sect. 1.3, it was shown that small oscillations of a simple pendulum satisfy the differential equation

$$\ddot{\theta} + \frac{g}{l}\theta = 0. \tag{2.24}$$

This equation shares the same form as Eq. (2.3) so the same results from Sect. 2.2.1.1 are applicable here. For example, if the pendulum is released from rest with initial angle $\theta(0) = \theta_0$, it will undergo harmonic oscillations given by

$$\theta = \theta_0 \cos(\omega_n t), \tag{2.25}$$

where $\omega_n = \sqrt{g/l}$ is its natural frequency. The natural frequency depends on the pendulum size but not its mass. For a pendulum of length $l = 11$ cm, the natural period of oscillation is

$$T = 2\pi\sqrt{\frac{l}{g}} = 0.6653\,\text{s}. \tag{2.26}$$

This is close to the period observed in the experiment from Sect. 1.3 or the period estimated when integrating the nonlinear ODE (1.29).

It is fascinating that the frequency of oscillation of a pendulum is constant. This phenomenon was first observed by Galileo, reportedly during a Mass at the cathedral in Pisa. Using his own heartbeat as a time reference, he noted that the swinging period of a chandelier was not affected by the pendulum's amplitude. Another interesting characteristic is that the period of oscillation is influenced by the acceleration due to gravity. This principle was used by Huygens, who invented a pendulum clock that can measure gravity's acceleration across various geographical locations.

2.2.3.2 The Rolling Cylinder

Consider the problem of the rolling cylinder discussed in Sect. 1.5.1. The equation of motion for small oscillations is

$$\ddot{\theta} + \frac{k(R+h)^2}{\left(I_{zz} + mR^2\right)}\theta = 0. \tag{2.27}$$

Inspection of this equation reveals that the natural frequency of vibration is given by

$$\omega_n = \sqrt{\frac{k(R+h)^2}{\left(I_{zz} + mR^2\right)}}. \tag{2.28}$$

Given that $I_{zz} = mR^2/2$ for a homogeneous cylinder, this expression simplifies to

$$\omega_n = \frac{(R+h)}{R}\sqrt{\frac{2k}{3m}}. \tag{2.29}$$

Therefore, the natural frequency increases linearly with h, i.e., it is greater when connected at the outer rim of the cylinder, as shown in Fig. 2.4.

Using the data from Sect. 1.5.1, the small motion assumption predicts the cylinder oscillates with period

$$T = \frac{2\pi}{\omega_n} = 1.6223\,\text{s}. \tag{2.30}$$

This result closely aligns with the period obtained from Fig. 1.29, which was constructed from the numerical solution of the nonlinear equation of motion.

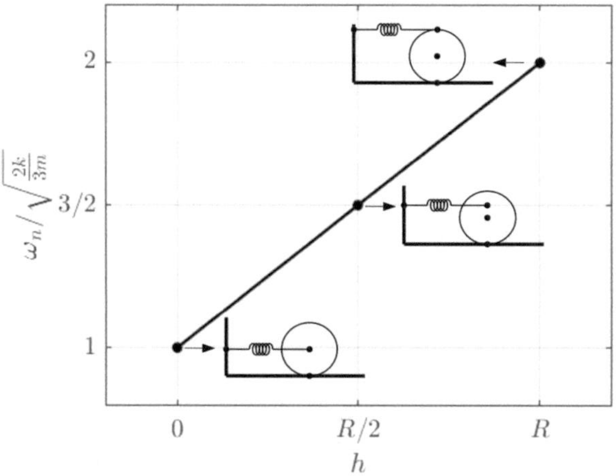

Fig. 2.4 Natural frequency of the rolling cylinder for different placements of the spring

2.2.3.3 Stable and Unstable Motions of a Metronome

In Sect. 1.5.2, it was shown that the free vibration of the undamped metronome is governed by the following ODE:

$$\ddot{\gamma} + \left[\frac{k}{m}\left(\frac{a}{b}\right)^2 - \frac{g}{b}\right]\gamma = 0. \tag{2.31}$$

For sufficiently stiff spring, the coefficient

$$\frac{k}{m}\left(\frac{a}{b}\right)^2 - \frac{g}{b} > 0 \tag{2.32}$$

and the metronome oscillates with natural frequency

$$\omega_n = \sqrt{\frac{k}{m}\left(\frac{a}{b}\right)^2 - \frac{g}{b}}. \tag{2.33}$$

Using the same data employed to generate Fig. 1.38a, it can be verified that

$$T = \frac{2\pi}{\omega_n} = 1.14\,\text{s}. \tag{2.34}$$

This result matches the period obtained through integration of the nonlinear ODE (1.96).

When the spring is not sufficiently stiff

$$\frac{k}{m}\left(\frac{a}{b}\right)^2 < \frac{g}{b}. \tag{2.35}$$

This results in what is known as "*negative stiffness*". Negative stiffness is a phenomenon where the system can respond in an unstable manner. This situation occurred when $k = 10$ N/m (see Fig. 1.38b). With this parameter value,

$$\frac{k}{m}\left(\frac{a}{b}\right)^2 - \frac{g}{b} = -17.12. \tag{2.36}$$

The characteristic polynomial

$$\lambda^2 - 17.12 = 0 \tag{2.37}$$

has two roots real roots equal to

$$\lambda_{1,2} = \pm 4.1376. \tag{2.38}$$

The general solution becomes

$$\gamma(t) = Ae^{4.1376t} + Be^{-4.1376t}. \tag{2.39}$$

Oscillatory behavior is not observed in this case. In fact, the positive root implies that generally $|\gamma| \to \infty$ as $t \to \infty$. This results in the unstable motion depicted in Fig. 2.5.

2.2.3.4 A Rocking Semicylinder

Consider the wood semicylinder shown in Fig. 2.6. It rests in a state of static equilibrium, with its flat surface positioned horizontally. Upon displacing the semicylinder from this position, it initiates the oscillatory motion illustrated in Fig. 2.7. It can be observed that the angular amplitude of oscillation remains nearly constant for a few cycles, as observed by the graduated scale in the background, after which its gradual decay is noticeable. Furthermore, the period of oscillation remains the same under different small release angles. This system exhibits behavior akin to that of a harmonic oscillator. Hence, the focus of this example revolves around establishing a model and applying the aforementioned theory to analyze its free vibration response.

Kinematic Analysis

According to Fig. 2.8, the semicylinder is modeled as a rigid body that is rolling without slipping on a horizontal surface. This problem has a close relationship to the rolling cylinder example from Sect. 1.5.1.

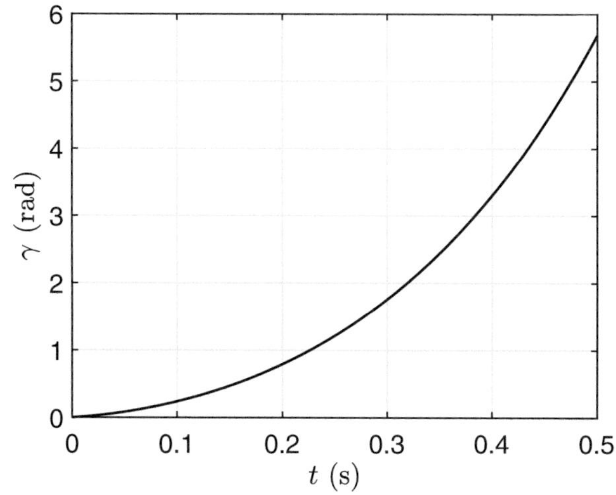

Fig. 2.5 Unstable motion of a metronome with $k = 10$ N/m and other parameters the same used in Sect. 1.5.2. The constants from Eq. (2.39) are determined with the ICs $\gamma(0) = 0$ and $\dot{\gamma}(0) = 1$ rad/s

Fig. 2.6 Semicylinder in equilibrium

The position vector of point A can be parameterized by the Cartesian coordinate x according with

$$\mathbf{x}_A = x\mathbf{i}. \tag{2.40}$$

In turn, the position vector of the mass center C may be represented by

$$\mathbf{x}_C = \mathbf{x}_A - h\mathbf{e}_2, \tag{2.41}$$

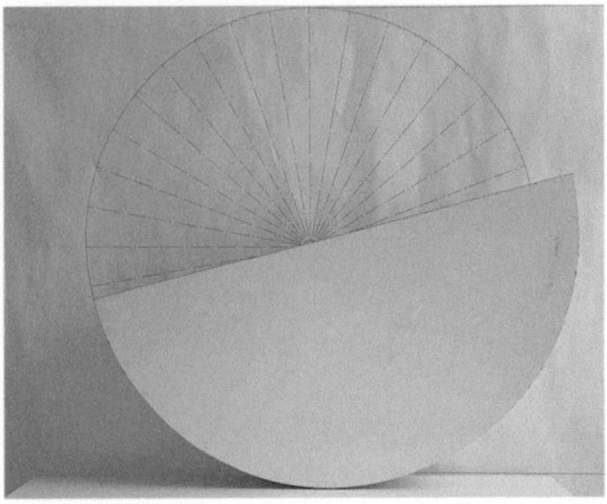

Fig. 2.7 A snapshot of the semicylinder oscillating when displaced from equilibrium

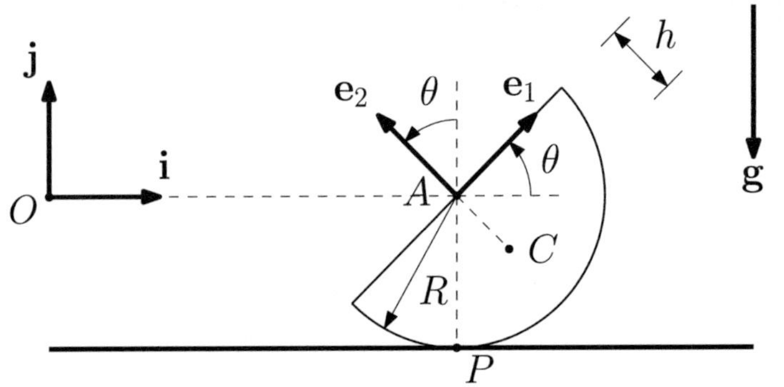

Fig. 2.8 A rigid semicylinder with radius R, mass m, and moment of inertia I_{zz} about the axis $\mathbf{e}_3 = \mathbf{k}$ passing through its center of mass C. The gravitational force acts along the $-\mathbf{j}$ direction

where

$$h = \frac{4R}{3\pi}. \tag{2.42}$$

The motion is characterized by two holonomic constraints. The first stems from the rectilinear motion of A:

$$\psi_1 = \mathbf{x}_A \cdot \mathbf{j} = 0, \tag{2.43}$$

and the second from the rolling condition involving the instantaneous point of contact P:

$$\dot{x} + R\dot{\theta} = 0, \quad \ddot{x} + R\ddot{\theta} = 0. \tag{2.44}$$

The semicylinder has one DOF, and its motion may be described by either x or θ.

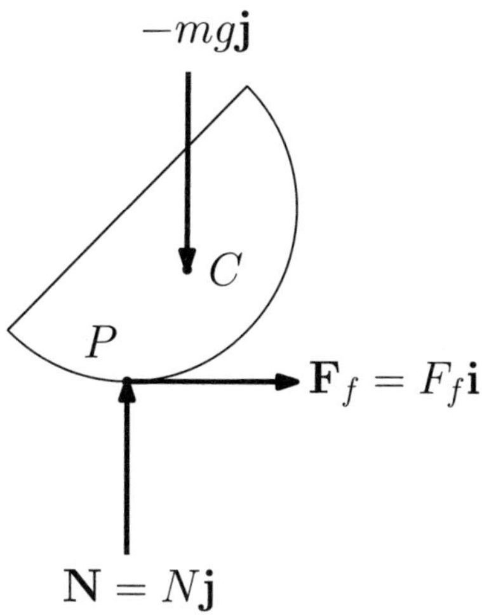

Fig. 2.9 FBD of the semicylinder

Obtaining the Equation of Motion

The FBD of the semicylinder is shown in Fig. 2.9. The normal force $\mathbf{N}$ and static friction force $\mathbf{F}_f$ act on the instantaneous point of contact P. The weight $-mg\mathbf{j}$ acts on the center of mass C. Application of Euler's first law together with the rolling constraint (2.44) yields

$$\mathbf{F}_f + \mathbf{N} = m\left[\ddot{\theta}(h\mathbf{e}_1 - R\mathbf{i}) + h\dot{\theta}^2\mathbf{e}_2 + g\mathbf{j}\right]. \tag{2.45}$$

Application of Euler's second law with respect to the mass center gives

$$I_{zz}\ddot{\theta} = (\mathbf{x}_P - \mathbf{x}_C) \times (\mathbf{F}_f + \mathbf{N}) \cdot \mathbf{k}. \tag{2.46}$$

Using $\mathbf{x}_P - \mathbf{x}_C = h\mathbf{e}_2 - R\mathbf{j}$ together with the shifting property of vector triple products, this equation becomes

$$I_{zz}\ddot{\theta} = (R\mathbf{i} - h\mathbf{e}_1) \cdot (\mathbf{F}_f + \mathbf{N}). \tag{2.47}$$

Substitute Eq. (2.45) into (2.47) to arrive at

$$[I_{zz} + m(R^2 + h^2) - 2mRh\cos(\theta)]\ddot{\theta} + mRh\dot{\theta}^2\sin(\theta) + mgh\sin(\theta) = 0. \tag{2.48}$$

The equation of motion is a second-order nonlinear ODE, and it can only be solved numerically. The MATLAB code `Chapt2_code1` provided in the *Electronic Supplementary Material* integrates Eq. (2.48) and produces the plot of $\theta(t)$ illustrated in Fig. 2.10 for the ICs $\theta(t = 0) = 10°$ and $\dot{\theta}(t = 0) = 0$. The plot shows that the semicylinder oscillates periodically with constant amplitude.

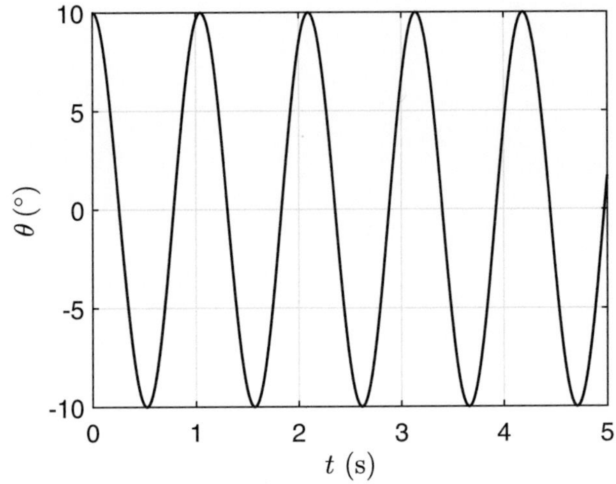

Fig. 2.10 Solution of Eq. (2.48) obtained numerically using MATLAB

Equilibrium Position

The equilibrium equation for this system is

$$\sin\theta_e = 0. \tag{2.49}$$

The only physically meaningful solution is $\theta_e = 0$. The MATLAB code `Chapt2_code1` from the previous section can be used to generate the phase portrait depicted in Fig. 2.11. It indicates that all solutions are periodic oscillations around the stable equilibrium.

Linearization

The objective is to linearize

$$\ddot{\theta} + \frac{mRh\dot{\theta}^2\sin(\theta) + mgh\sin(\theta)}{[I_{zz} + m(R^2 + h^2) - 2mRh\cos(\theta)]} = 0. \tag{2.50}$$

around the equilibrium $(\theta, \dot{\theta}) = (0, 0)$. Use the MATLAB code `Chapt2_code2` from the *Electronic Supplementary Material* to produce the linearized equation of motion for small oscillations:

$$\ddot{\theta} + \omega_n^2\theta = 0, \tag{2.51}$$

where the natural frequency is given by

$$\omega_n = \sqrt{\frac{mgh}{I_{zz} + m(R - h)^2}}. \tag{2.52}$$

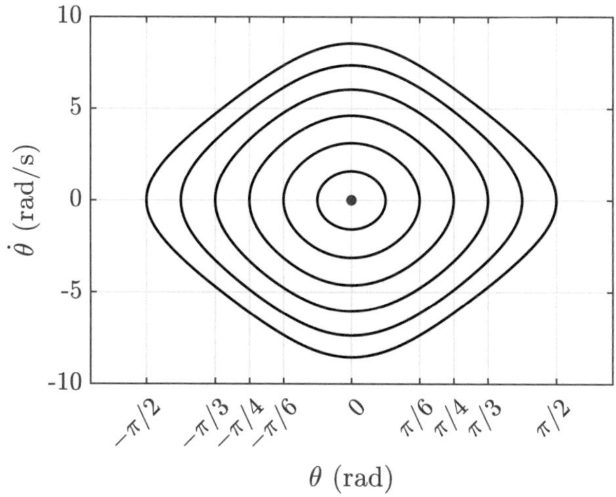

Fig. 2.11 Phase portrait of the semycilinder with stable equilibrium (•) shown

The error due to linearization may be computed according with

$$e(\theta, \dot{\theta}) = \left| \frac{\omega_n^2 \theta - \dfrac{mRh\dot{\theta}^2 \sin(\theta) + mgh \sin(\theta)}{[I_{zz} + m(R^2 + h^2) - 2mRh \cos(\theta)]}}{\dfrac{mRh\dot{\theta}^2 \sin(\theta) + mgh \sin(\theta)}{[I_{zz} + m(R^2 + h^2) - 2mRh \cos(\theta)]}} \right|. \tag{2.53}$$

Figure 2.12 shows a plot of $e(\theta, \dot{\theta})$ for various θ and $\dot{\theta}$ values, together with the trajectories from the phase portrait from Fig. 2.11. The validity of the approximation presented in Eq. (2.51) becomes apparent as the error remains small for trajectories close to the origin.

Energy Approximation

The total mechanical energy of the semicylinder is

$$E = \frac{1}{2}\left[I_{zz} + m(R^2 + h^2) - 2mRh \cos(\theta)\right]\dot{\theta}^2 - mgh \cos(\theta). \tag{2.54}$$

The kinetic energy has the form

$$\frac{1}{2}m(\theta)\dot{\theta}^2, \tag{2.55}$$

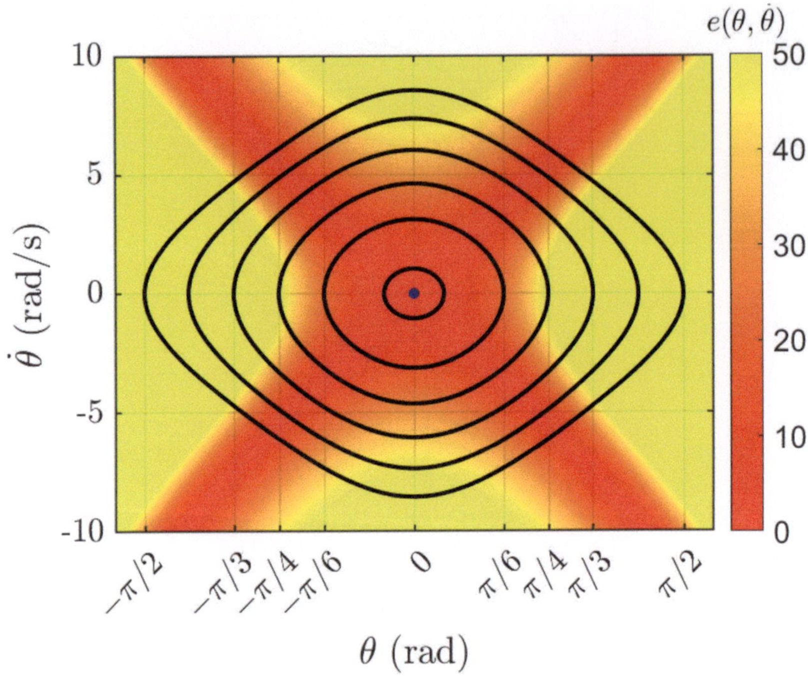

Fig. 2.12 Error $e(\theta, \dot{\theta})$ due to linearization, expressed in %, for various θ and $\dot{\theta}$ values. The trajectories in state space are shown, and the stable equilibrium is represented by (•)

where $m(\theta)$ is a nonlinear function of θ. An approximation for E under small motion assumption may be obtained by using a multivariate Taylor expansion of Eq. (2.54) around $(\theta, \dot{\theta}) = (0, 0)$ up to second-order terms, resulting in

$$
\begin{aligned}
E &\approx \frac{1}{2}m(0)\dot{\theta}^2 + U(0) + \frac{1}{2}\frac{d^2U}{d\theta^2}(0)\theta^2 \\
&= \frac{1}{2}\left[I_{zz} + m(R^2 - h^2)\right]\dot{\theta}^2 - mgh + \frac{1}{2}mgh\theta^2.
\end{aligned}
\tag{2.56}
$$

It can be checked that this approximation is a first integral of motion associated with Eq. (2.50).

Experimental Verification

If the semicylinder is given a small angular displacement $\theta(0) = \theta_0$ and released from rest, it oscillates harmonically according with

$$
\theta(t) = \theta_0 \cos(\omega_n t). \tag{2.57}
$$

To obtain an explicit expression for the period of oscillation, recall that the moment of inertia for a homogeneous semicylinder is

$$I_{zz} = \left(\frac{1}{2} - \frac{16}{9\pi^2}\right) mR^2. \tag{2.58}$$

The natural period becomes

$$T = 2\pi\sqrt{\frac{R}{4g}\left(\frac{9\pi}{2} - 8\right)}. \tag{2.59}$$

The period of oscillation is only affected by the radius of the semicylinder, not its mass. As the radius increases, so does the period of oscillation.

The result from Eq. (2.59) can be validated experimentally. A wood semicylinder was constructed with radius $R = 0.175$ m. Equation (2.59) predicts it executes "small" oscillatory motions with period $T = 1.0395$ s. This prediction can now be compared with experimental measurements. The semicylinder is released from rest at various initial angles θ_0, and a stopwatch is used to measure the time for three oscillations (before damping effects are noticeable), providing the experimental period T^*. Table 2.1 compares the analytical prediction with the measurements for different release angles. The theoretical and measured results align well for small angles, with a greater deviation observed for larger amplitudes, as expected. Even greater angles can be tested if there is enough friction to prevent the cylinder from slipping on the surface.

Alternative Derivation of the Equation of Motion

Assuming that the states $(\theta, \dot{\theta})$ are contained in a small neighborhood around the equilibrium $(0, 0)$, it is possible to introduce the approximation

$$\mathbf{e}_1 \approx \mathbf{i} + \theta\mathbf{j}, \quad \mathbf{e}_2 \approx -\theta\mathbf{i} + \mathbf{j}. \tag{2.60}$$

The position of the mass center can be linearized according with

$$\mathbf{x}_C = \mathbf{x}_A - h\mathbf{e}_2 \approx (x + h\theta)\mathbf{i} - h\mathbf{j}, \tag{2.61}$$

Table 2.1 Period of oscillation of semicylinder. Comparison between measured period of oscillation T^* and analytical period T predicted by Eq. (2.59)

θ_0	T^* (s)	$100 \times (T - T^*)/T$
10°	1.0518	−1.1833
20°	1.0679	−2.7321
30°	1.1113	−6.9072
40°	1.1530	−10.9187

i.e., C moves horizontally. Differentiate this expression two times and use the rolling constraint (2.44) to obtain

$$\mathbf{a}_C \approx -(R-h)\ddot{\theta}\mathbf{i}. \tag{2.62}$$

Application of Euler's laws yields

$$\mathbf{F}_f + \mathbf{N} = mg\mathbf{j} - m(R-h)\ddot{\theta}\mathbf{i}, \tag{2.63}$$

$$[-h\theta\mathbf{i} - (R-h)\mathbf{j}] \times (\mathbf{F}_f + \mathbf{N}) \cdot \mathbf{k} = I_{zz}\ddot{\theta}. \tag{2.64}$$

Solve this system of equations for θ to obtain the linearized equation of motion

$$\left[I_{zz} + m(R-h)^2\right]\ddot{\theta} + mgh\theta = 0. \tag{2.65}$$

2.2.3.5 A Bobbing Buoy

Figure 2.13 illustrates a buoy floating on water. The buoy is a cylindrical plastic bottle filled with coins. In equilibrium, a portion of the bottle remains submerged due to buoyancy. When disturbed slightly, the buoy oscillates vertically with nearly constant amplitude for a some time. It is observed that this period increases with the addition of more coins. This setup leads to the primary goal of the example: formulating an expression for the frequency governing this oscillatory motion.

Mathematical Model

Model the buoy as a rigid cylinder with a cross-sectional area A and mass m. This cylindrical buoy is partially submerged in an upright orientation within a fluid of density ρ. Referring to Fig. 2.14a, assume that the fluid surface remains unaffected

Fig. 2.13 A cylindrical buoy floating in bucket filled with water

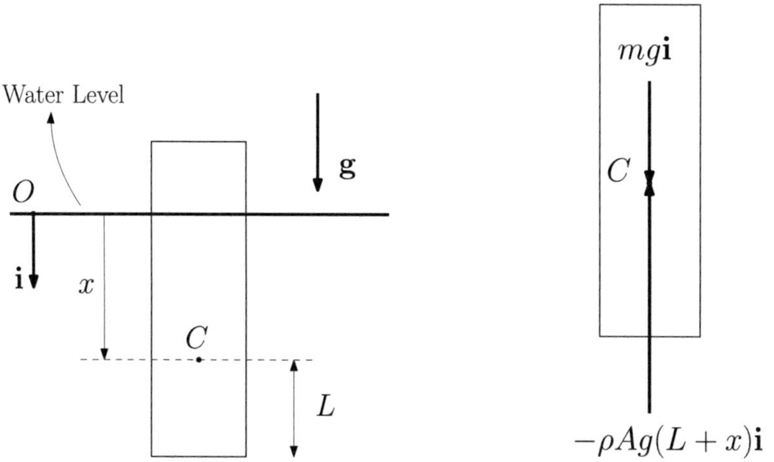

(a) Reference system.
(b) FBD of bobbing buoy.

Fig. 2.14 Model for the bobbing buoy

by the buoy's motion, and let x represent the position of the center of mass C relative to the fluid surface. The total submerged length of the buoy is $x + L$.

Neglecting viscous effects, the forces acting upon the buoy encompass the gravitational force and the buoyancy force, as shown in Fig. 2.14b. The buoyancy force has magnitude equal to the weight of fluid displaced by the submerged portion of the buoy.

Equation of Motion

Apply Euler's first law to obtain the equation of motion

$$m\ddot{x} = mg - \rho Ag(x + L). \tag{2.66}$$

The associated equilibrium equation is

$$mg = \rho Ag(x_e + L), \tag{2.67}$$

where $(x_e + L)$ is the submerged length of the buoy when in static equilibrium. Use the change of coordinates $x = x_e + q$ to obtain an equation of motion describing oscillations about equilibrium:

$$\ddot{q} + \frac{\rho Ag}{m} q = 0. \tag{2.68}$$

This implies that, once perturbed from equilibrium, the buoy oscillates harmonically with period

$$T = 2\pi \sqrt{\frac{m}{\rho A g}}. \tag{2.69}$$

Testing the Theory

To validate this theoretical result, it would be necessary to measure ρ, A and m. A simpler way can be obtained by recasting the equilibrium equation in the form

$$\frac{m}{\rho A g} = \frac{x_e + L}{g}. \tag{2.70}$$

In this manner, the period of oscillation can be expressed as

$$T = 2\pi \sqrt{\frac{x_e + L}{g}}. \tag{2.71}$$

Therefore, it depends on the submerged length of the buoy when in equilibrium. This is trivial to measure using a ruler, for example.

To verify Eq. (2.71) experimentally, a few coins are put inside the bottle, and its equilibrium length above water is measured. With this measurement, the theoretical period T may be computed. Then, the buoy is set into motion, and the period T^* is measured with a stopwatch. In fact, it is easier to measure the period using the time the buoy takes to complete the first four oscillations (before dissipative effects are perceptible). Table 2.2 indicates that the values are in good agreement.

2.2.3.6 The Bifiliar Pendulum

The moment of inertia influences torques and forces necessary for angular acceleration, making it an important parameter for consideration during engineering design. The determination of this parameter can be challenging for complex structures such as aerial vehicles. The presence of onboard systems with their specific mass distributions adds layers of complexity to this task, often leading engineers to rely

Table 2.2 Period of oscillation of the cylindrical buoy. Comparison between measured period of oscillation T^* and analytical period T predicted by Eq. (2.71)

$x_e + L$ (cm)	T (s)	T^* (s)	$100 \times (T - T^*)/T$
8.4	0.5814	0.5967	−2.5564
7.7	0.5567	0.5733	−2.9080
5.7	0.4789	0.4983	−3.8913

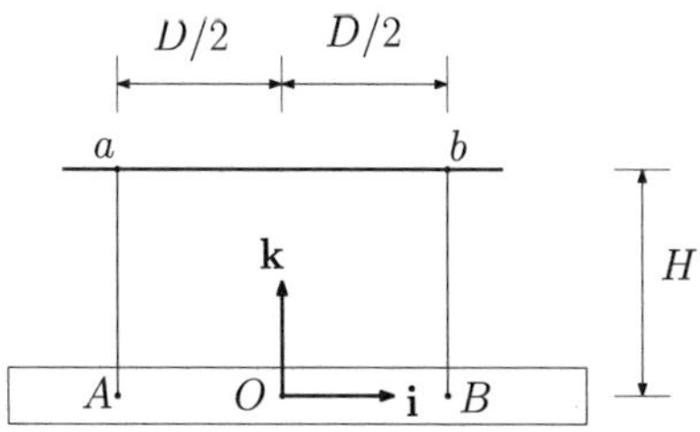

(a) Lateral view of the set up.

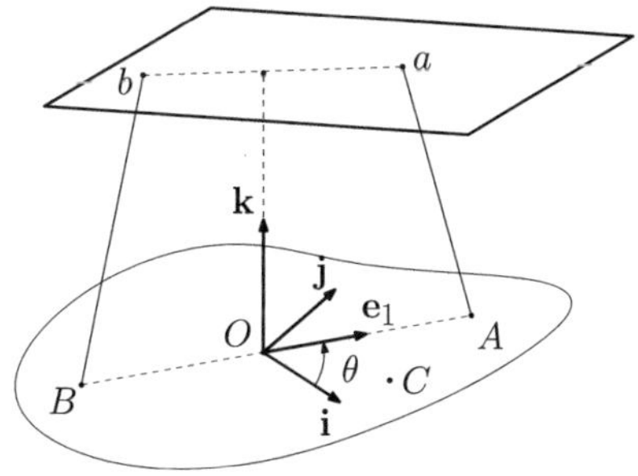

(b) Test body oscillating.

Fig. 2.15 The bifiliar pendulum

on experimental techniques to obtain precise and reliable values for the moment of inertia.

Common experimental methods for determining moments of inertia involve oscillating the structure like a pendulum. One such technique is the *bifilar pendulum*, depicted in Fig. 2.15. In this setup, a test body is suspended by two identical and equidistant cables, each of length H. The cables are positioned at a distance D apart, connecting points A and B of the body to points a and b of a fixed support, respectively. The test body undergoes oscillatory motion after it is rotated around the axis passing through the midpoint O between the cables and subsequently released from rest. The moment of inertia can then be estimated using the measured period of oscillation.

The objective of this example is to apply the theory of undamped free vibration to establish a relationship between easily measurable parameters (such as the period of oscillation T, H, and D) and the moment of inertia of the test body.

Mathematical Model

The dynamics of the bifilar pendulum may initially appear complex due to its multidimensional motion, but the problem can be significantly streamlined by making the following simplifying assumptions. The cables will be treated as inextensible and their mass neglected. In addition, the angle of rotation is assumed to be sufficiently small so that the test body practically executes planar rotations around the fixed axis passing through the midpoint O. Lastly, the body is modeled as a rigid body with mass m and moment of inertia I_{zz}^O about the rotation axis.

Given these assumptions, the fixed origin is placed at O together with the set of Cartesian axes shown in Fig. 2.15. The midpoint point is not necessarily coincident with the body's center of mass C. The corotational basis $\{\mathbf{e}_1,\ \mathbf{e}_2,\ \mathbf{e}_3\}$ is attached to the test body in such a way that the rotation axis aligns with $\mathbf{e}_3 = \mathbf{k}$. The position vectors of the points of interest are given by

$$\mathbf{r}_a = H\mathbf{k} + \frac{D}{2}\mathbf{i}, \tag{2.72}$$

$$\mathbf{r}_b = H\mathbf{k} - \frac{D}{2}\mathbf{i}, \tag{2.73}$$

$$\mathbf{x}_A = \frac{D}{2}\mathbf{e}_1 \approx \frac{D}{2}\mathbf{i} + \frac{D}{2}\theta\mathbf{j}, \tag{2.74}$$

$$\mathbf{x}_B = -\frac{D}{2}\mathbf{e}_1 \approx -\frac{D}{2}\mathbf{i} - \frac{D}{2}\theta\mathbf{j}, \tag{2.75}$$

$$\mathbf{x}_C = X_1\mathbf{e}_1 + X_2\mathbf{e}_2 + X_3\mathbf{k}, \tag{2.76}$$

where the coordinates (X_1, X_2, X_3) of the center of mass are constant.

Equation of Motion

The FBD of the test body is shown in Fig. 2.16. The weight $-mg\mathbf{k}$ acts on the mass center C and the cables exert traction forces given by

$$\mathbf{S}_A = S_A\frac{\mathbf{r}_a - \mathbf{x}_A}{\|\mathbf{r}_a - \mathbf{x}_A\|} \approx \frac{S_A}{\sqrt{H^2 + \frac{D^2}{4}}}\left(H\mathbf{i} - \frac{D}{2}\theta\mathbf{j}\right), \tag{2.77}$$

$$\mathbf{S}_B = S_B\frac{\mathbf{r}_b - \mathbf{x}_B}{\|\mathbf{r}_b - \mathbf{x}_B\|} \approx \frac{S_B}{\sqrt{H^2 + \frac{D^2}{4}}}\left(H\mathbf{i} + \frac{D}{2}\theta\mathbf{j}\right). \tag{2.78}$$

Applying Euler's second law with respect to O reveals that

$$I_{zz}^{O}\ddot{\theta} = [\mathbf{x}_A \times \mathbf{S}_A + \mathbf{x}_B \times \mathbf{S}_B + \mathbf{x}_C \times (-mg\mathbf{k})] \cdot \mathbf{k}. \tag{2.79}$$

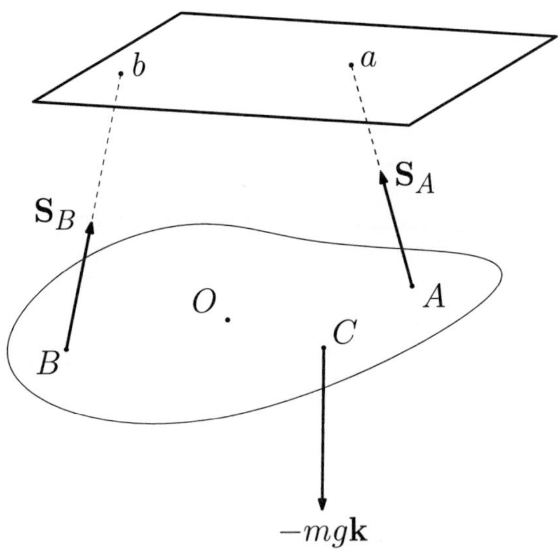

Fig. 2.16 FBD of the test body

Equations (2.74)–(2.76) can be used to rewrite this expression as

$$I_{zz}^{O}\ddot{\theta} + \frac{D^2/4}{\sqrt{H^2 + \dfrac{D^2}{4}}}\,(S_A + S_B)\,\theta = 0. \tag{2.80}$$

The unknown $S_A + S_B$ can be eliminated using Euler's first law. Taking the projection of $\mathbf{F} = m\mathbf{a}_C$ along the $\mathbf{k}$ direction yields

$$\frac{H}{\sqrt{H^2 + \dfrac{D^2}{4}}}\,(S_A + S_B) = mg. \tag{2.81}$$

Equation (2.81) can then be substituted into Eq. (2.80) to produce the differential equation governing the oscillatory motion of the test body:

$$\ddot{\theta} + \frac{mgD^2}{4HI_{zz}^{O}}\theta = 0. \tag{2.82}$$

Estimating the Moment of Inertia

The objective of the bifiliar pendulum experiment is to estimate I_{zz}^{O}. The equation of motion (2.82) dictates that, once perturbed from equilibrium, the test body oscillates harmonically with natural frequency given by

$$\omega_n = \sqrt{\frac{mgD^2}{4HI_{zz}^{O}}}. \tag{2.83}$$

If one measures the oscillation period T, the distance D, and the cable length H, the moment of inertia I_{zz}^{O} can be computed according with

$$I_{zz}^{O} = \frac{mgD^2T^2}{16H\pi^2}. \tag{2.84}$$

This relation provides an effective way to determine the moment of inertia based on the period of oscillation of the body.

Experimental Verification

The result predicted by Eq. (2.84) can be validated through a simple experiment. Select a test body with a simple geometry, whose moments of inertia can be easily calculated. An example is the rectangular cuboid shown in Fig. 2.17.

The selected test body is made of wood, has a mass of $m = 284$ g, and dimensions

$$a = 1.5\,\text{cm}, \quad b = 10\,\text{cm}, \quad c = 22\,\text{cm}. \tag{2.85}$$

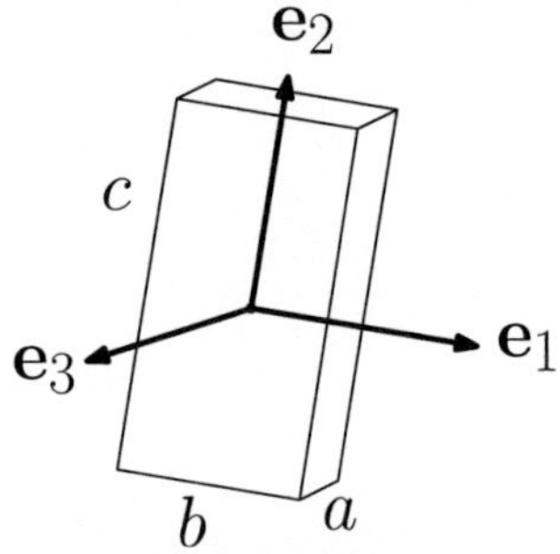

Fig. 2.17 Principal axes of a test body in the shape of a rectangular cuboid. The sides have lengths a, b, and c

Fig. 2.18 Setup of a bifilar pendulum for determining the principal moments of inertia of a rectangular cuboid

Using the axes shown in Fig. 2.17, the principal moments of inertia are

$$I_{11} = \frac{1}{12}m(a^2 + c^2) = 1.1508 \times 10^{-3}\,\text{kg.m}^2, \tag{2.86}$$

$$I_{22} = \frac{1}{12}m(a^2 + b^2) = 0.2420 \times 10^{-3}\,\text{kg.m}^2, \tag{2.87}$$

$$I_{33} = \frac{1}{12}m(b^2 + c^2) = 1.3821 \times 10^{-3}\,\text{kg.m}^2. \tag{2.88}$$

These values can now be compared with those obtained from Eq. (2.84) using the experimental setup shown in Fig. 2.18. Metal brackets were symmetrically screwed into the wooden body so that it could be easily suspended by strings. The brackets should be centered and aligned to ensure that the strings remain as parallel as possible.

Once the experimental apparatus was assembled, the body was set to oscillate from a small initial angle. The period T^* was determined from the time it took for the body to complete five full oscillations. The result was then substituted into Eq. (2.84), along with the dimensions D and H, to obtain an experimental value I^* for the moment of inertia. The results shown in Table 2.3 demonstrate that the theoretical and experimental results are in agreement.

Table 2.3 Principal moments of inertia of a rectangular cuboid. Comparison between moment of inertia I^* computed with experimental period T^* using Eq. (2.84) and moment of inertia I computed analytically using Eqs. (2.86)–(2.88)

Axis	H and D (cm)	T^* (s)	I^* (kg.m^2 × 10^3)	$100 \times (I - I^*)/I$
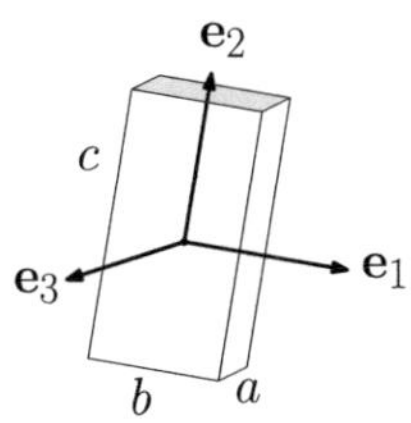	$H = 47, D = 11$	0.71	0.2290	-5.3825
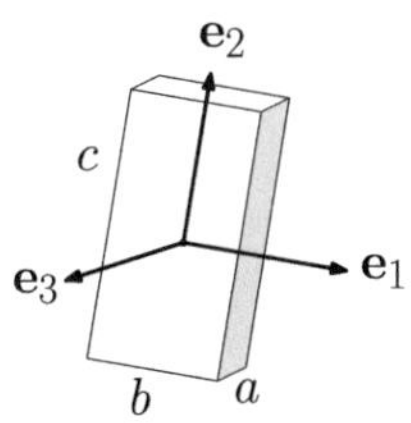	$H = 47, D = 22$	0.83	1.2516	8.7614
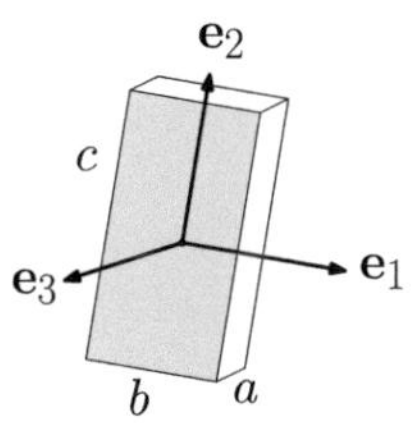	$H = 47, D = 11$	1.81	1.4880	7.6620

2.3 Free Vibration in the Presence of Viscous Damping

In this section, the effect of viscous damping on the free response is investigated to gain a more comprehensive understanding of the system's behavior.

2.3.1 The Mass-Spring-Damper System

The vibration of a mass-spring-damper system is described by the equation of motion:

$$m\ddot{q} + c\dot{q} + kq = 0, \tag{2.89}$$

where c is the viscous damping coefficient. This equation can be rearranged into the standard form

$$\ddot{q} + 2\zeta\omega_n\dot{q} + \omega_n^2 q = 0, \tag{2.90}$$

where $\zeta = c/(2m\omega_n)$ is a nondimensional parameter referred to as the *damping ratio*. The objective is to assess the system behavior across multiple values of ζ.

The general solution of Eq. (2.90) also assumes the form $q(t) = Ae^{\lambda t}$, where $A, \lambda \in \mathbb{C}$ are constants to be determined. Consequently, the characteristic polynomial associated with Eq. (2.90) is

$$\lambda^2 + 2\zeta\omega_n\lambda + \omega_n^2 = 0. \tag{2.91}$$

The nature of the roots depends on the value of ζ, leading to four distinct cases.

2.3.1.1 Case 1: Undamped ($\zeta = 0$)

This is the harmonic oscillator covered in Sect. 2.2.1. The roots of the characteristic polynomial form a purely imaginary complex conjugate pair

$$\lambda_1 = \bar{\lambda}_2 = i\omega_n. \tag{2.92}$$

The system then partakes in harmonic oscillations with frequency of ω_n:

$$q(t) = Q\cos(\omega_n t - \phi), \tag{2.93}$$

for which a typical response is depicted in Fig. 2.2.

2.3.1.2 Case 2: Underdamped ($0 < \zeta < 1$)

Since $\zeta < 1$, the roots of Eq. (2.91) materialize as a complex conjugate pair

$$\lambda_1 = \bar{\lambda}_2 = -\zeta\omega_n + i\omega_d, \tag{2.94}$$

where the imaginary part $\left|Im(\lambda_{1,2})\right| = \omega_d = \omega_n\sqrt{1-\zeta^2}$ has units of rad/s and is called *damped frequency*. Notably, the roots have negative real part: $Re(\lambda_{1,2}) = -\zeta\omega_n < 0$. The general solution becomes

$$q(t) = A_1 e^{\lambda_1 t} + A_2 e^{\lambda_2 t}, \tag{2.95}$$

or, alternatively,

$$q(t) = (a\cos\omega_d t + b\sin\omega_d t)\, e^{-\zeta\omega_n t}. \tag{2.96}$$

Impose the ICs $q(0) = q_0$ and $\dot{q}(0) = \dot{q}_0$ to show that

$$q(t) = \left[q_0 \cos \omega_d t + \left(\frac{\dot{q}_0 + \zeta \omega_n q_0}{\omega_d}\right) \sin \omega_d t\right] e^{-\zeta \omega_n t}. \tag{2.97}$$

Analogous to the harmonic oscillator, the response may be reformulated as

$$q(t) = Qe^{-\zeta \omega_n t} \cos(\omega_d t - \phi) \tag{2.98}$$

wherein

$$Q = \sqrt{q_0^2 + \left(\frac{\dot{q}_0 + \zeta \omega_n q_0}{\omega_d}\right)^2}, \quad \phi = \arctan \frac{\dot{q}_0 + \zeta \omega_n q_0}{\omega_d q_0}. \tag{2.99}$$

Figure 2.19 shows that the free response of an underdamped system comprises harmonic oscillations with damped frequency ω_d within an exponentially decaying envelope $\pm Qe^{-\zeta \omega_n t}$ that modulates the amplitude and brings the system back to equilibrium. Since the damped frequency satisfies $\omega_d < \omega_n$, the damped period of oscillation $T_d = 2\pi/\omega_d$ is greater than the natural period $T = 2\pi/\omega_n$. Many real-life systems are underdamped with $\zeta < 20\%$, including buildings, bridges, dams, nuclear power plants, and offshore structures.

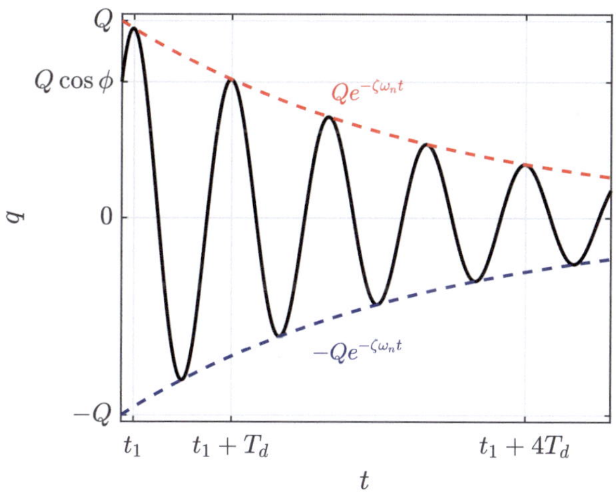

Fig. 2.19 Typical free vibration response of underdamped SDOF systems

2.3.1.3 Case 3: Overdamped ($\zeta > 1$)

The two roots are real, distinct, and negative:

$$\lambda_1, \lambda_2 = -\zeta\omega_n \pm \omega_n\sqrt{\zeta^2 - 1}, \tag{2.100}$$

with $\lambda_2 < \lambda_1 < 0$. The response takes the form

$$q(t) = A_1 e^{\lambda_1 t} + A_2 e^{\lambda_2 t}, \tag{2.101}$$

where the constants A_1 and A_2 are determined from the ICs. A typical response is shown in Fig. 2.20: it is non-oscillatory and exponentially decaying. Instruments used to measure steady-state values, like a scale measuring stationary weight, are usually overdamped by design.

2.3.1.4 Case 4: Critically Damped ($\zeta = 1$)

One can think of a critically damped system as representing the smallest value of damping that produces non-oscillatory motion. In this case, the roots are real, equal, and negative:

$$\lambda_1 = \lambda_2 = -\omega_n < 0. \tag{2.102}$$

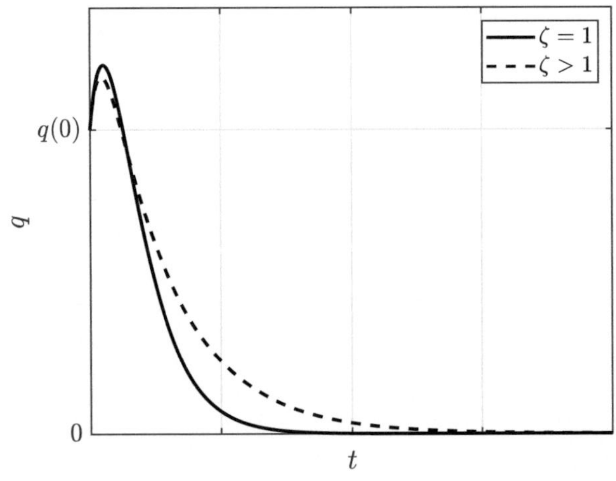

Fig. 2.20 Typical free vibration response of critically damped and overdamped SDOF systems

The general free response for the special case of equal roots is given by

$$q(t) = (A_1 + tA_2)e^{-\omega_n t}, \tag{2.103}$$

where A_1 and A_2 are also determined from the ICs. Even though there is a linear term t in the solution, the exponential $e^{-\omega_n t}$ dominates the response as t increases. As a result, the motion is non-oscillatory and the system exponentially decays back to equilibrium. Figure 2.20 demonstrates that a critically damped system generally reaches equilibrium faster than an overdamped one.

2.3.1.5 Relation Between Solution and Roots

The preceding discussion illustrated that the type of motion experienced is contingent upon the nature of the characteristics roots λ_1 and λ_2. An effective means of comprehending this is by examining how these roots behave as the damping ratio ζ varies. This understanding can be facilitated by employing a root locus diagram, which involves mapping the positions of λ_1 and λ_2 in the complex plane for different values of damping, as depicted in Fig. 2.21.

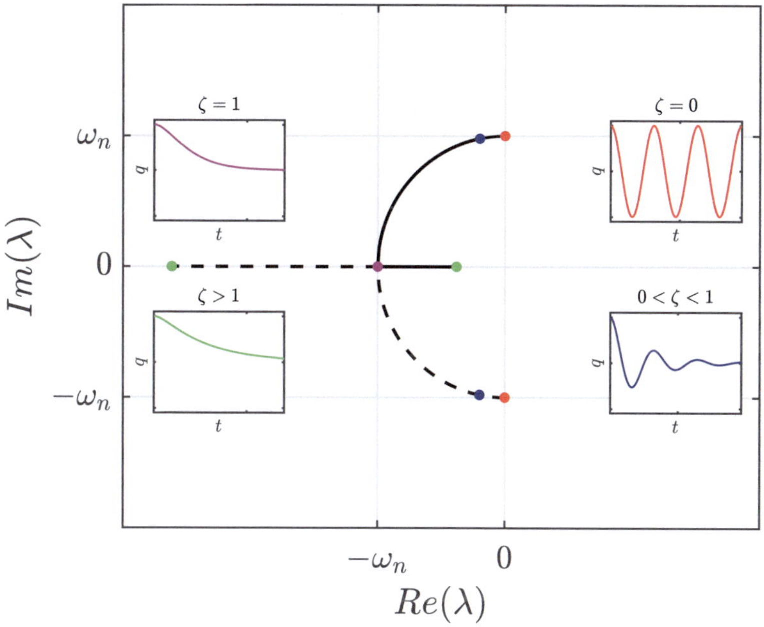

Fig. 2.21 Root locus diagram. The solid black curve (—) represents the locus of points corresponding to λ_1, while the dashed curve (- -) represents the locus of points for λ_2. The accompanying subplots display typical responses associated with the different locations of λ_1 and λ_2 indicated by corresponding colored locations on the main diagram

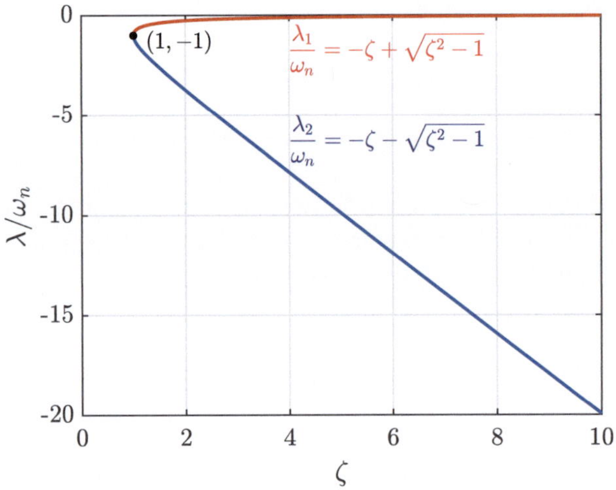

Fig. 2.22 Real roots of overdamped systems

When $\zeta = 0$, the the characteristic roots are $\lambda_1, \lambda_2 = \pm i\omega_n$. They lie on the imaginary axis and the typical response is purely oscillatory with frequency ω_n. As ζ increases from 0 to 1, these roots evolve into a complex conjugate pair $\lambda_1, \lambda_2 = -\zeta\omega_n \pm i\omega_d$ and each traces a quadrant of a circular path of radius ω_n. The resulting motion is oscillatory with frequency ω_d while exhibiting exponentially decaying amplitude. When $\zeta = 1$, these roots collapse on the real axis, becoming equal and negative: $\lambda_1 = \lambda_2 = -\omega_n$. At this point, the system transitions from oscillatory to non-oscillatory behavior, when the response decays exponentially. As ζ continues to increase beyond the unit value, the roots remain negative with $\lambda_2 < \lambda_1 < 0$. If $\zeta \to \infty$, the larger root $\lambda_1 \to 0$ and the smaller one $\lambda_2 \to -\infty$. This progression becomes even clearer when examining Fig. 2.22, which shows the roots of an overdamped systems as a function of ζ.

These observations are summarized in Fig. 2.23. An undamped system oscillates periodically, and a trajectory in the phase plane forms a closed path. Since an underdamped system oscillates with decaying amplitude, a typical trajectory spirals toward the origin. A critically damped or overdamped system approaches the origin without oscillations.

2.3.2 *Illustrative Examples*

Three examples are presented to demonstrate how the theory of damped free motion can be applied.

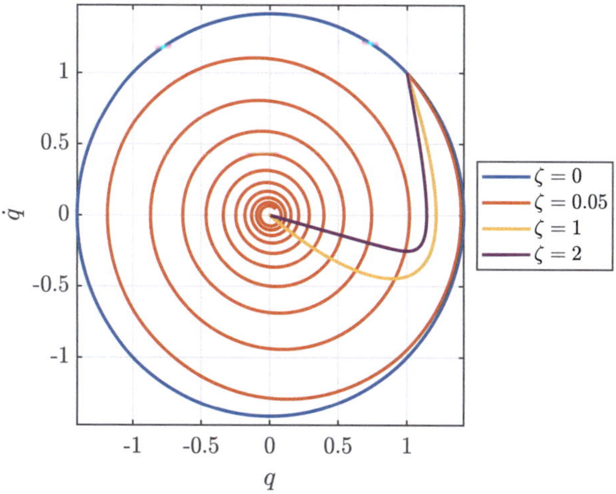

Fig. 2.23 Typical trajectories in the phase plane for an SDOF system in free vibration

2.3.2.1 The Damped Metronome

Recall that the equation of motion of the damped metronome, discussed in Sect. 1.5.2, is given by

$$\ddot{\gamma} + \frac{c}{m}\dot{\gamma} + \left[\frac{k}{m}\left(\frac{a}{b}\right)^2 - \frac{g}{b}\right]\gamma = 0, \tag{2.104}$$

where the natural frequency is given by Eq. (2.33). The data from Eq. (1.97) together with $k = 200$ N/m and $c = 0.1$ Ns/m imply that the damping ratio is given by

$$\zeta = \frac{c}{2m\omega_n} = \frac{cb}{2\sqrt{m(a^2k - mgb)}} = 0.0091. \tag{2.105}$$

The system is underdamped and it oscillates with damped frequency

$$\omega_d = \omega_n\sqrt{1-\zeta^2} = \frac{\sqrt{4m(a^2k - mgb) - c^2b^2}}{2mb} = 5.5116\,\text{rad/s}. \tag{2.106}$$

In turn, the damped period is

$$T_d = \frac{2\pi}{\omega_d} \approx 1.14\,\text{s}. \tag{2.107}$$

This offers a good approximation to the period $T_d \approx 1.147$ s obtained from Fig. 1.39b, which is constructed by numerically integrating the nonlinear ODE (1.96).

2.3.2.2 The Damped Pendulum

Consider the damped pendulum analyzed in Sect. 1.3.4. Recall that the equation of motion for small oscillations is

$$\ddot{\theta} + \frac{c}{m}\dot{\theta} + \frac{g}{l}\theta = 0. \tag{2.108}$$

With the parameter values from Sect. 1.3.4, this system is underdamped with $\zeta = 0.075$, and it oscillates with frequency $\omega_d = 0.9972$ rad/s. Because of the exponential decaying amplitude, the trajectories in the neighborhood of $(\theta, \dot{\theta}) = (0, 0)$ shown in the phase portrait from Fig. 1.23 spiral toward the equilibrium.

2.3.2.3 A Rolling Cylinder Connected to a Damper

A uniform rigid cylinder with mass m, moment of inertia I_{zz} with respect to axis $\mathbf{e}_3 = \mathbf{k}$ passing through its mass center C, and radius R, rolls without slipping on a horizontal surface, as illustrated in Fig. 2.24. The cylinder is connected at point A to a spring with stiffness k and unstretched length l_0. At point B, it is connected to a viscous damper with coefficient c. The objective of this example is to derive the equation of motion for small oscillations of the cylinder.

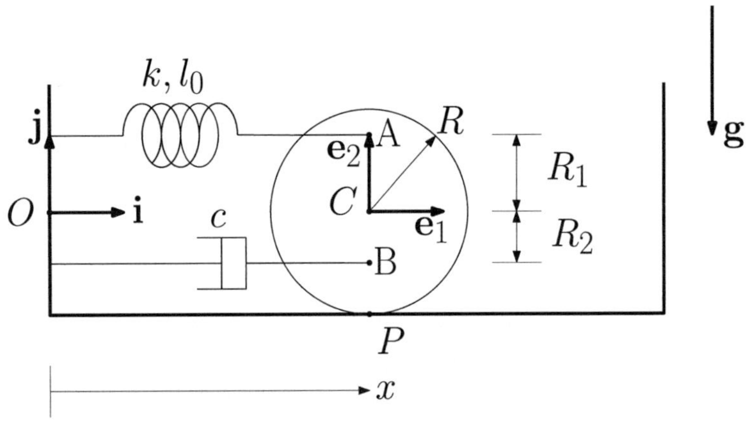

Fig. 2.24 Cylinder rolling without slipping. In the configuration shown, $\theta = 0$ and $x = l_0$. The gravitational force acts along $-\mathbf{j}$

Kinematic Analysis
The cylinder has one DOF because it moves under the constraints

$$\psi_1 = \mathbf{x}_C \cdot \mathbf{j} = 0, \tag{2.109}$$

$$\psi_2 = x - l_0 + R\theta = 0. \tag{2.110}$$

The rolling constraint ψ_2 implies that $\dot{x} = -R\dot{\theta}$ and $\ddot{x} = -R\ddot{\theta}$.

Assuming that the states $(\theta, \dot{\theta})$ are close to the equilibrium (0, 0), one can introduce the approximations

$$\mathbf{x}_A = x\mathbf{i} + R_1\mathbf{e}_2 \approx (x - R_1\theta)\mathbf{i} + R_1\mathbf{j} = [l_0 - (R + R_1)\theta]\mathbf{i} + R_1\mathbf{j}, \tag{2.111}$$

$$\mathbf{x}_B = x\mathbf{i} - R_2\mathbf{e}_2 \approx (x + R_2\theta)\mathbf{i} - R_2\mathbf{j} = [l_0 - (R - R_2)\theta]\mathbf{i} - R_2\mathbf{j}. \tag{2.112}$$

This means

$$\mathbf{v}_B \approx -(R - R_2)\dot{\theta}\mathbf{i}. \tag{2.113}$$

Equation of Motion
The FBD of the rolling cylinder is shown in Fig. 2.25. Because of the small motion assumption, points A and B move horizontally. Therefore, the elastic and damping forces acting on the cylinder are given by

$$\mathbf{F}_e = -k\epsilon\mathbf{i} = k(R + R_1)\theta\mathbf{i}, \quad \mathbf{F}_d = -c\mathbf{v}_B = c(R - R_2)\dot{\theta}\mathbf{i}, \tag{2.114}$$

respectively. Application of Euler's laws of motion results in the system of equations

$$\mathbf{N} + \mathbf{F}_f = -mR\ddot{\theta}\mathbf{i} + mg\mathbf{j} - k(R + R_1)\theta\mathbf{i} - c(R - R_2)\dot{\theta}\mathbf{i}, \tag{2.115}$$

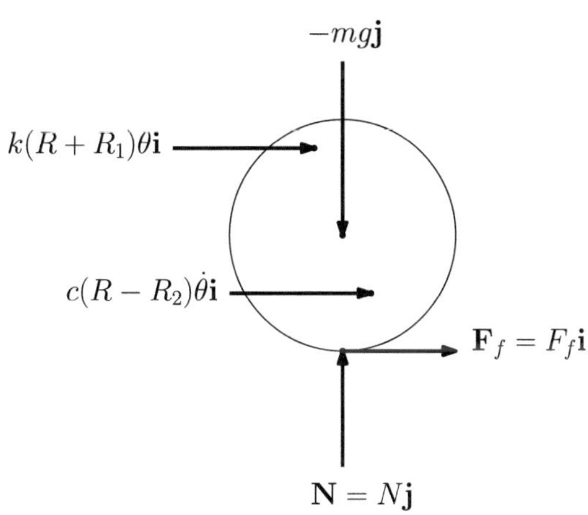

Fig. 2.25 FBD of the cylinder

$$I_{zz}\ddot{\theta} = [(-R\mathbf{j}) \times (\mathbf{N} + \mathbf{F}_f) \\ +R_1(-\theta\mathbf{i} + \mathbf{j}) \times \mathbf{F}_e + R_2(\theta\mathbf{i} - \mathbf{j}) \times \mathbf{F}_d] \cdot \mathbf{k}. \tag{2.116}$$

Use these expressions to obtain the equation of motion

$$\ddot{\theta} + \frac{c(R - R_2)^2}{I_{zz} + mR^2}\dot{\theta} + \frac{k(R + R_1)^2}{I_{zz} + mR^2}\theta = 0. \tag{2.117}$$

The parameters of interest can be readily inspected from this equation. The natural frequency is

$$\omega_n = (R + R_1)\sqrt{\frac{k}{I_{zz} + mR^2}} \tag{2.118}$$

and the damping ratio is

$$\zeta = \frac{c}{2\sqrt{k(I_{zz} + mR^2)}} \frac{(R - R_2)^2}{(R - R_1)}. \tag{2.119}$$

Estimating the Damping Ratio of the Rolling Cylinder

Engineers often encounter situations where they have access to the response of a system to specific stimuli but lack knowledge of modal parameters like the natural frequency and damping ratio. The situation where one seeks to deduce the system model based on observed response under controlled conditions is known as *system identification*.

One established technique is called the *method of logarithmic decrement*. This method aims to determine the damping ratio of an underdamped system by analyzing its free vibration response. To understand how this method works, consider the case when the cylinder is underdamped and exhibits the free response recorded in Fig. 2.26. Let $\theta_1 = \theta(t_1)$ and $\theta_2 = \theta(t_2) = \theta(t_1 + T_d)$ be two successive peaks occurring a full damped cycle apart, as indicated in the plot. The objective is to use the quantity

$$\delta = \ln \frac{\theta_1}{\theta_2}, \tag{2.120}$$

called logarithmic decrement, to estimate ζ. To achieve that, note that the cylinder vibrates according with

$$\theta(t) = \Theta \cos(\omega_d t - \phi)e^{-\zeta\omega_n t}. \tag{2.121}$$

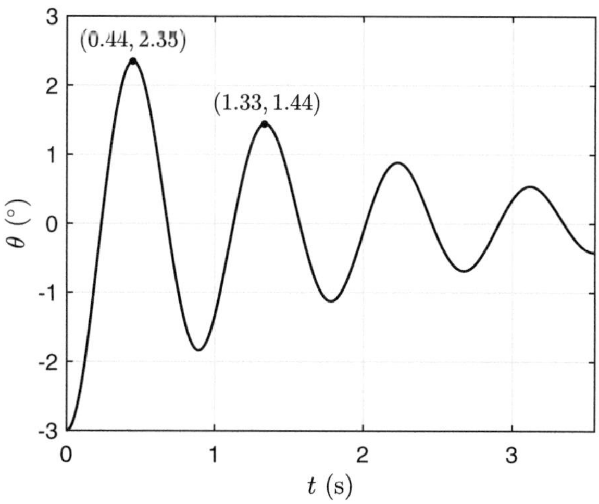

Fig. 2.26 Response of the cylinder in free vibration. The instant and value of two consecutive peaks are indicated in the plot

The ratio between the successive amplitudes becomes

$$\frac{\theta_1}{\theta_2} = \frac{\Theta \cos(\omega_d t_1 - \phi) e^{-\zeta \omega_n t_1}}{\Theta \cos(\omega_d t_2 - \phi) e^{-\zeta \omega_n t_2}} = e^{\frac{2\pi\zeta}{\sqrt{1-\zeta^2}}}. \tag{2.122}$$

Substitute this expression into Eq. (2.120) to conclude that

$$\delta = \frac{2\pi\zeta}{\sqrt{1-\zeta^2}}. \tag{2.123}$$

This equation allows one to relate the measurable quantity δ to the unknown ζ via

$$\zeta = \frac{\delta}{\sqrt{\delta^2 + (2\pi)^2}}. \tag{2.124}$$

Based on Fig. 2.26, the peak values $\theta_1 = 2.35°$ and $\theta_2 = 1.44°$ occur at $t_1 = 0.44$ s and $t_2 = 1.33$ s, respectively. Use Eq. (2.120) to compute $\delta = 0.4898$ and substitute this result into Eq. (2.124) to get $\zeta = 0.0777$.

2.4 Summary

The equation of motion describing the free vibration of SDOF systems is given by

$$\ddot{q} + 2\zeta\omega_n\dot{q} + \omega_n^2 q = 0,$$

where ω_n is called natural frequency, and ζ is the damping ratio. This equation can be solved with knowledge of the ICs

$$q(0) = q_0, \quad \dot{q}(0) = \dot{q}_0.$$

The type of solution depends on the value of ζ. If the system is undamped ($\zeta = 0$), it oscillates harmonically with frequency ω_n, and the solution is

$$q(t) = \sqrt{q_0^2 + \frac{\dot{q}_0^2}{\omega_n^2}} \cos\left(\omega_n t - \arctan\frac{\dot{q}_0}{q_0\omega_n}\right).$$

The period of oscillation $T = 2\pi/\omega_n$ is determined solely by the system's physical properties, and the amplitude of vibration remains constant over time. When the system is underdamped ($0 < \zeta < 1$), the solution is given by

$$q(t) = \sqrt{q_0^2 + \left(\frac{\dot{q}_0 + \zeta\omega_n q_0}{\omega_d}\right)^2} e^{-\zeta\omega_n t} \cos\left(\omega_d t - \arctan\frac{\dot{q}_0 + \zeta\omega_n q_0}{\omega_d q_0}\right).$$

In this case, the response is composed of harmonic oscillations with damped frequency $\omega_d = \omega_n\sqrt{1-\zeta^2}$ within an exponentially decaying envelope. If $\zeta \geq 1$, the response is non-oscillatory and exponentially decaying as shown by either Eq. (2.103) when $\zeta = 1$ or Eq. (2.101) when $\zeta > 1$.

2.5 Problems

2.1 Determine the constants of integration in Eq. (2.101) using the ICs $q(0) = q_0$ and $\dot{q}(0) = \dot{q}_0$.

2.2 Determine the constants of integration in Eq. (2.103) using the ICs $q(0) = q_0$ and $\dot{q}(0) = \dot{q}_0$.

2.3 What ICs should be chosen for an overdamped system to reach equilibrium more rapidly than a critically damped one?

2.4 Consider the function

$$x(t) = 7\sin 5t + 4\cos 5t. \tag{2.125}$$

(a) Write this function in the form

$$x(t) = A\cos(5t - \phi), \tag{2.126}$$

with $0 < \phi < 2\pi$.

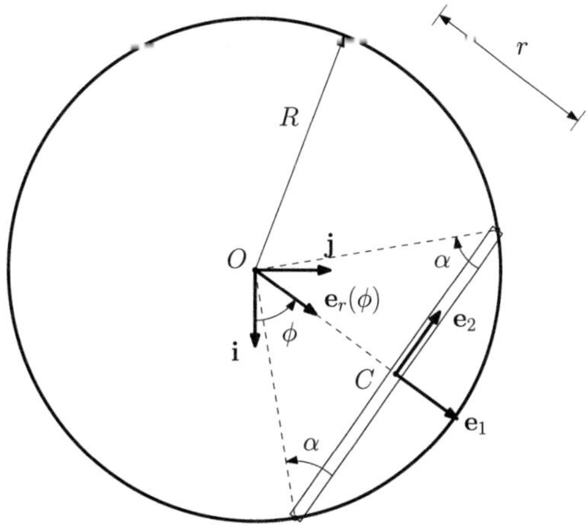

Fig. 2.27 Rigid bar sliding inside a smooth circular surface. The gravitational force acts along the **i** direction

(b) Write this function in the form

$$x(t) = B\sin(5t - \phi), \tag{2.127}$$

with $0 < \phi < 2\pi$.

2.5 Find the natural frequency of the system from Problem 1.10.

2.6 As shown in Fig. 2.27, a uniform thin rod with mass m and length $L = \sqrt{2}R$ slides inside a smooth circular surface of radius R. Determine the frequency of oscillation of the bar.

2.7 Figure 2.28 shows a mechanical system consisting of a thin rod of negligible mass connected to three linear springs and a particle with mass m. Using the indicated reference system, the position vectors of the points shown in the figure are represented by

$$\mathbf{x}_A = a_1\mathbf{e}_1, \tag{2.128}$$

$$\mathbf{x}_B = a_2\mathbf{e}_1, \tag{2.129}$$

$$\mathbf{x}_C = a_3\mathbf{e}_1, \tag{2.130}$$

$$\mathbf{r}_a = a_1\mathbf{i} + l_1\mathbf{j}, \tag{2.131}$$

$$\mathbf{r}_b = a_2\mathbf{i} - l_2\mathbf{j}. \tag{2.132}$$

Find the natural frequency of this system.

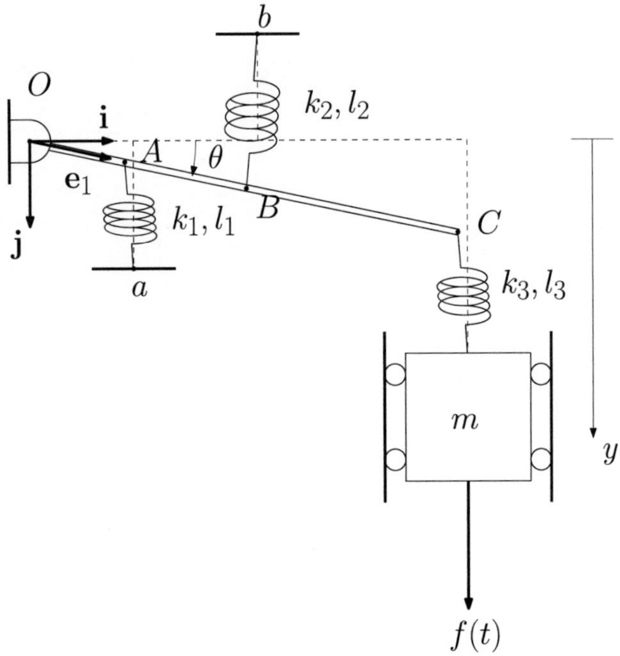

Fig. 2.28 Thin rod of negligible mass connected to three springs and a particle with mass m. The gravitational force acts along the $\mathbf{j}$ direction

Fig. 2.29 Thin rod that is free to rotate around the axis $\mathbf{k}$ passing through the fixed point O. The gravitational force acts along $-\mathbf{k}$

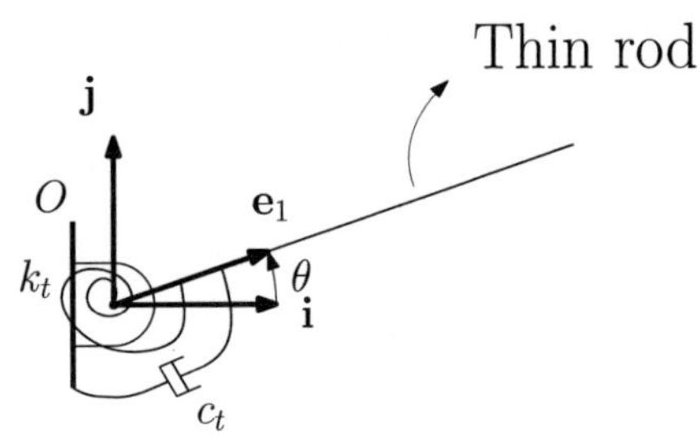

2.8 A thing rod, shown in Fig. 2.29, is free to rotate around the axis $\mathbf{k}$ passing through the fixed point O. The rod is connected to a torsional spring with stiffness k_t and a torsional damper with coefficient c_t. Assuming that the rod has moment of inertia I_{zz}^O with respect to the axis $\mathbf{e}_3 = \mathbf{k}$ passing through O, determine the equation of motion of the rod and provide expressions for the natural frequency and damping ratio.

2.9 A uniform rigid cylinder with mass m and radius R rolls without slipping on a horizontal surface, as illustrated in Fig. 2.30. The position vector of the center of mass C can be represented by

$$\mathbf{x}_C = x\mathbf{i}. \tag{2.133}$$

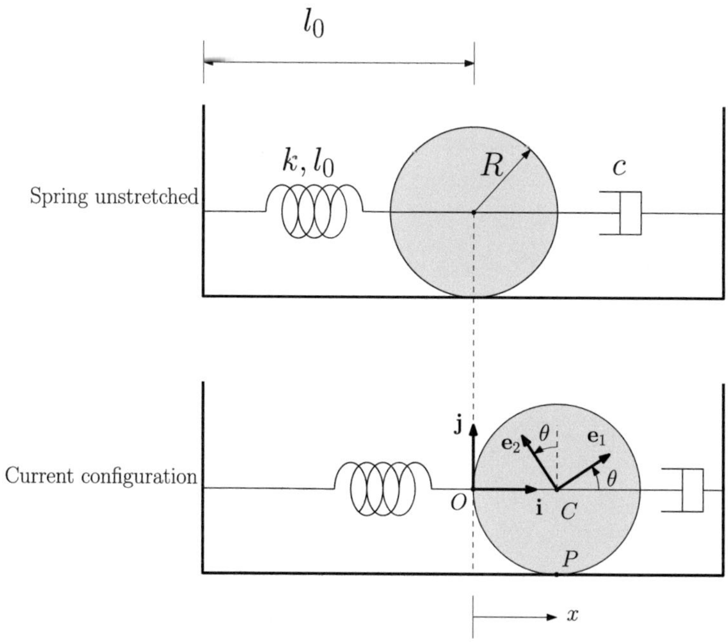

Fig. 2.30 Cylinder rolling without slipping. At position $x = 0$ (or $\theta = 0$), the spring is relaxed. The gravitational force acts in the $-\mathbf{j}$ direction

Suppose

$$m = 3\,\text{kg}, \quad k = 200\,\text{N/m}, \quad c = 1\,\text{Ns/m}, \quad R = 0.5\,\text{m}, \tag{2.134}$$

and that the cylinder is released from rest with $\theta(0) = 10°$. Determine the system's response $\theta(t)$ and estimate the number of periods it takes for the amplitude of oscillation to reduce to 5% of the first peak.

2.10 A particle with mass m is dropped from rest from a height h onto another particle with mass M, as shown in Fig. 2.31. The table behaves like an SDOF system with mass M, stiffness k, and damping c. Determine an expression for the force transmitted to the the foundation after the particles collide. Assume that the particles stick together after the collision and that the system is underdamped.

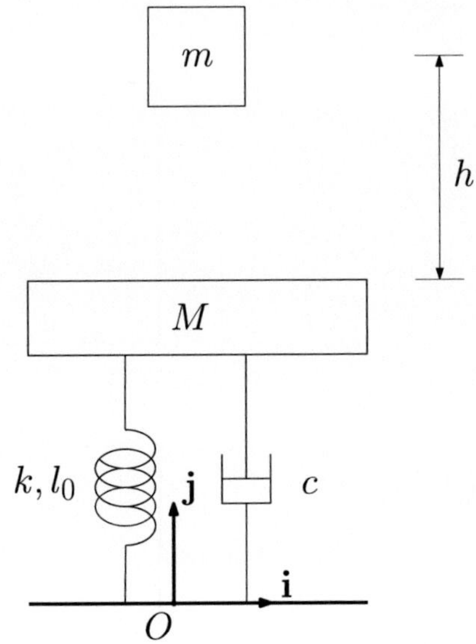

Fig. 2.31 Particle free-falling on top of a table

Chapter 3
Forced Vibration of Systems with One Degree of Freedom

This chapter is concerned with the response of single-degree-of-freedom systems subject to harmonic excitation. It introduces the complex exponential representation of harmonic forces to discuss phenomena such as resonance and beating in undamped systems. The focus then shifts to damped systems, highlighting key differences from the undamped case. The frequency response function is presented, with an emphasis on interpreting frequency response curves. MATLAB is integrated throughout the chapter to enhance understanding and visualization. Five illustrative examples are provided to reinforce the concepts discussed, including two simple experiments that readers can easily replicate to qualitatively test the theory.

3.1 The Forced Vibration Problem

The prototype SDOF system is illustrated in Fig. 3.1. Its vibratory behavior, represented by the generalized coordinate q, is governed by the second-order ODE

$$m\ddot{q} + c\dot{q} + kq = f(t), \tag{3.1}$$

that is subjected to the ICs $q(0) = q_0$ and $\dot{q}(0) = \dot{q}_0$. Free vibration problems were covered in Chapt. 2, and they involve solving this equation when the external force is $f = 0$, meaning the solution is determined solely by the ICs.

The aim of this chapter is to cover problems that fall into the category known as forced vibrations, i.e., when $f \neq 0$. The analysis of forced vibrations clarifies how external forces influence the vibration of a system. For example, engineers want to understand how road conditions excite a vehicle when designing a suspension

Supplementary Information The online version contains supplementary material available at (https://doi.org/10.1007/978-3-031-83995-5_3).

R. Gonçalves Salsa Junior, *Fundamentals of the Theory of Mechanical Vibrations*,
https://doi.org/10.1007/978-3-031-83995-5_3

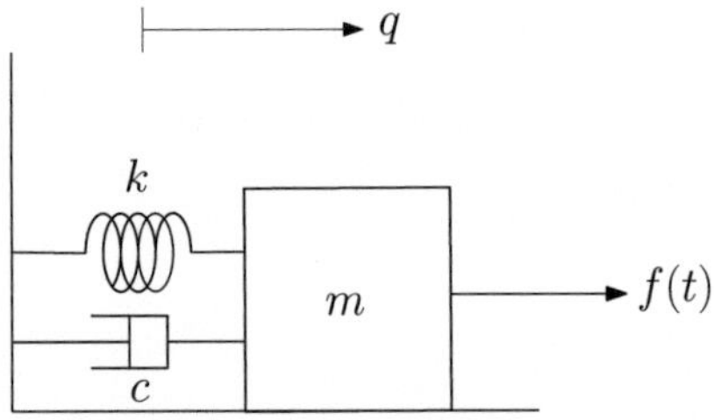

Fig. 3.1 Prototype SDOF system in forced vibration. The system consists of a mass m, damping c, stiffness k, and is acted upon by the external force $f(t)$. The motion is parameterized by the generalized coordinate q, measured from the equilibrium position

system that guarantees passenger comfort. Similarly, engineering components such as compressors, fans, and turbines experience forced vibrations due to unbalanced rotating parts. Proper knowledge of this phenomenon is necessary to implement effective balancing and isolation techniques to minimize vibrations. Additionally, engineers must comprehend how earthquakes, wind, or sea currents excite buildings, bridges, or dams, enabling them to build structures that can withstand these forces.

The task of computing the response in the presence of applied forces can be more evolved than that of the free vibration case because it is necessary to take into account not only the force but also the ICs. Nevertheless, linearity offers effective means to solve the equation. The process of computing the solution can be split into two parts by applying the principle of superposition. Firstly, the general homogeneous solution $q_h(t)$ that satisfies the equation

$$m\ddot{q}_h + c\dot{q}_h + kq_h = 0 \tag{3.2}$$

can be computed. Recall from Chap. 2 that this part of the solution depends on the level of damping according with

$$q_h(t) = \begin{cases} a\cos\omega_n t + b\sin\omega_n t, & \text{if } \zeta = 0, \\ e^{-\zeta\omega_n t}(a\cos\omega_d t + b\sin\omega_d t), & \text{if } 0 < \zeta < 1, \\ (a + bt)e^{-\omega_n t}, & \text{if } \zeta = 1, \\ ae^{\lambda_1 t} + be^{\lambda_2 t}, & \text{if } \zeta > 1, \end{cases} \tag{3.3}$$

where a and b are constants of integration. Secondly, a particular solution $q_p(t)$ that is independent of ICs and is maintained by the driving force $f(t)$ is sought. This part of the solution satisfies the differential equation

$$m\ddot{q}_p + c\dot{q}_p + kq_p = f(t). \tag{3.4}$$

There are many techniques to find q_p, the choice of which depends on the functional form of $f(t)$: it can be periodic, harmonic, or even arbitrary, as illustrated in Fig. 3.2.

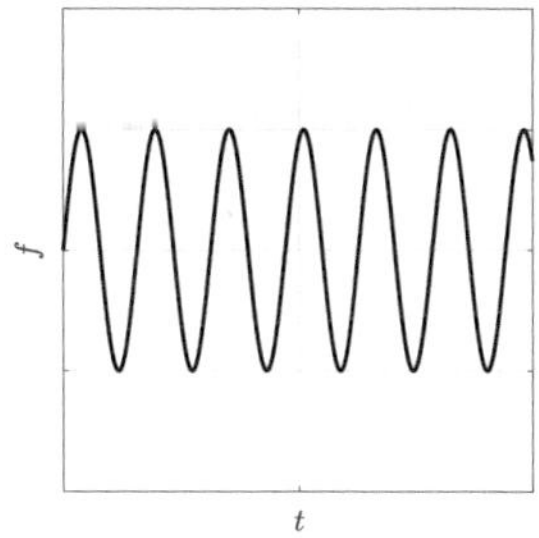

(a) A harmonic excitation, i.e., either a cosine or sine.

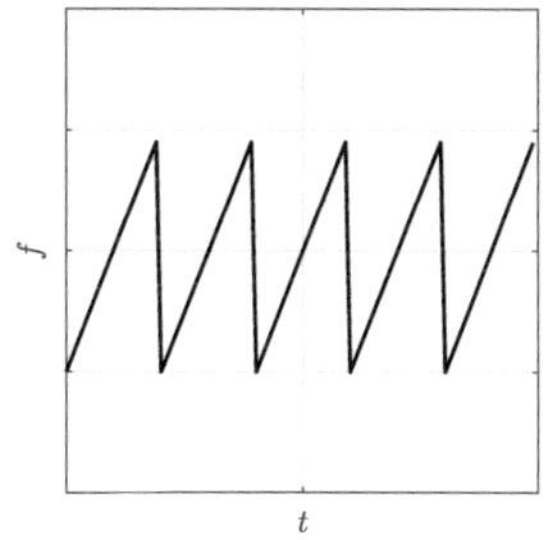

(b) A periodic excitation in the form of a sawtooth function.

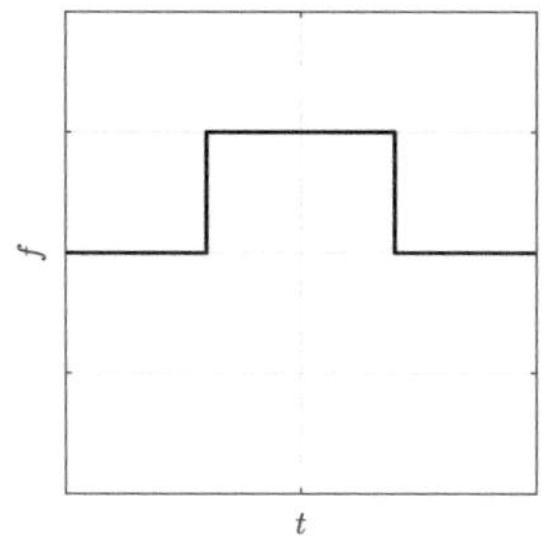

(c) An arbitrary excitation in the form of a rectangular pulse.

Fig. 3.2 Some types of excitation

Table 3.1 Guess for particular solution of Eq. (3.1) for some common forcing functions $f(t)$

$f(t)$	$q_p(t)$
$e^{\alpha t}$	$Ae^{\alpha t}$
$\sin \alpha t,\ \cos \alpha t$	$A \sin \alpha t + B \cos \alpha t$
$a_n t^n + a_{n-1} t^{n-1} + \ldots + a_1 t + a_0$	$A_n t^n + A_{n-1} t^{n-1} + \ldots + A_1 t + A_0$

The method of undetermined coefficients proves to be adequately effective and simple to use in many cases. In this method, a general form of q_p is guessed based on $f(t)$, and the unknown coefficients shown in Table 3.1 are determined once this guess is substituted into Eq. (3.4).

The complete solution of the linear non-homogeneous ODE is then given by

$$q(t) = q_h(t) + q_p(t). \tag{3.5}$$

It must satisfy the ICs $q(0) = q_0$ and $\dot{q}(0) = \dot{q}_0$ in order to determine the two arbitrary constants of integration contained in the homogeneous part. The organization of this chapter is as follows. The exponential representation of harmonic functions is introduced in Sect. 3.2. The application of the theory to undamped systems is covered in Sect. 3.3, introducing the concepts of beating and resonance. The effect of damping on the forced response is discussed in Sect. 3.4. Many illustrative examples are supplied in Sect. 3.5. A summary of the main results is presented in Sect. 3.6.

3.2 Representation of Harmonic Excitation

Harmonic loads hold particular significance due to their common occurrence in engineering, like forces resulting from unbalance in rotating machinery or those generated by reciprocating engines. For this reason, the primary focus of these notes is on harmonic forces. This emphasis is not restrictive because any function, such as $f(t)$, can be uniquely characterized in terms of its frequency content, achieved through either a Fourier series for periodic forces or a Fourier transform for arbitrary ones. Thus, understanding the response to harmonic excitation provides the foundation for addressing these other cases.

In this context, the objective of this chapter is to study the response of a system governed by

$$m\ddot{q} + c\dot{q} + kq = F_0 \cos(\omega t) \quad \text{or} \quad F_0 \sin(\omega t), \tag{3.6}$$

where ω is the frequency of the harmonic driving force, and F_0 is its amplitude. It is common to represent harmonic functions using the complex exponential form. The basic idea behind this representation is to introduce q_c as the particular solution to Eq. (3.6) when the excitation is of the form $f(t) = F_0 \cos \omega t$, and q_s when $f(t) = F_0 \sin \omega t$. Using the principle of superposition, these solutions satisfy

$$m(\ddot{q}_c + i\ddot{q}_s) + c(\dot{q}_c + i\dot{q}_s) + k(q_c + iq_s) = F_0 \cos \omega t + iF_0 \cos \omega t = F_0 e^{i\omega t}, \tag{3.7}$$

where $i = \sqrt{-1}$. Rather than solving Eq. (3.6), one seeks to solve

$$m\ddot{q} + c\dot{q} + kq = F_0 e^{i\omega t}. \tag{3.8}$$

The solution of Eq. (3.8) is complex because the excitation is also complex. However, the observed physical response must always be a real function. It corresponds to the real part of the complex solution when $f(t) = F_0 \cos \omega t$ or the imaginary part when $f(t) = F_0 \sin \omega t$.

At first glance, it might appear that the problem has been needlessly complicated. Yet the complex exponential representation offers an efficient method for handling harmonic functions, primarily due to the simplicity of differentiating them. This representation is also well-suited to the typical interest in steady-state behavior, when the focus lies on long-term system operation and the emphasis shifts to characteristics such as response amplitude and phase lag. In these cases, whether the excitation is a cosine or a sine is not important.

3.3 Response of Undamped Systems Subjected to Harmonic Force

Now that the necessary background material has been covered, one can proceed to calculate the forced response. The first case to consider is the undamped system, where Eq. (3.8) simplifies to:

$$m\ddot{q} + kq = F_0 e^{i\omega t}. \tag{3.9}$$

This section details the steps for computing the solution.

3.3.1 *Computing the Response*

Assuming that $\omega \neq \omega_n$, the method of undetermined coefficients postulates that the particular solution takes the form

$$q_p = Qe^{i\omega t}, \tag{3.10}$$

where Q is an unknown constant amplitude. Substitute the derivatives of the particular solution

$$\dot{q}_p = i\omega Q e^{i\omega t}, \quad \ddot{q}_p = -\omega^2 Q e^{i\omega t}, \tag{3.11}$$

into Eq. (3.9) to obtain

$$\left(-m\omega^2 + k\right) Qe^{i\omega t} = F_0 e^{i\omega t}. \tag{3.12}$$

Solving this equation for Q reveals that

$$Q = \frac{F_0}{k - m\omega^2}. \tag{3.13}$$

Without loss of generality, let the excitation be $f(t) = F_0 \cos \omega t$. This implies that the physical response is the real part of the complex solution:

$$q_p(t) = Re\left[Qe^{i\omega t}\right] = \frac{F_0}{k - m\omega^2} \cos \omega t. \tag{3.14}$$

The complete solution is the superposition of particular and homogeneous parts:

$$q(t) = a \cos \omega_n + b \sin \omega_n t + \frac{F_0/m}{\omega_n^2 - \omega^2} \cos \omega t. \tag{3.15}$$

The constants of integration are determined from the ICs. For illustration purposes, assume $q(0) = 0$ and $\dot{q}(0) = 0$. It can be verified that

$$a = -\frac{F_0/m}{\omega_n^2 - \omega^2}, \qquad b = 0. \tag{3.16}$$

In conclusion, the vibration of the system is given by

$$q(t) = \frac{F_0/m}{\omega_n^2 - \omega^2} \left(\cos \omega t - \cos \omega_n t \right). \tag{3.17}$$

The response predicted by this equation is shown in Fig. 3.3. Both the homogeneous and particular parts of the solution are harmonic and, consequently, periodic: q_h oscillates at the natural frequency ω_n and q_p at the excitation frequency ω. The complete solution $q = q_h + q_p$ fulfills the specified ICs and is in general neither harmonic nor periodic.

Figure 3.3c shows the associated trajectory in state space. The trajectory changes direction when it crosses itself at various points. This occurs because the time t appears explicitly in the equation of motion (3.9). Therefore, the state state $(q, \dot{q})$ alone does not fully determine the time evolution of the system. Such a dynamical system is called *non-autonomous*.

3.3.2 *The Frequency Response Function*

An alternative perspective on the particular solution holds significant importance in vibration analysis. Define the frequency response function (FRF), denoted by $H(\omega)$, to be the ratio between vibration output (q_p) and force input (f) in the form of complex exponential:

$$H(\omega) = \frac{Q e^{i\omega t}}{F_0 e^{i\omega t}} = \frac{Q}{F_0}. \tag{3.18}$$

The FRF can be expressed equivalently in the following forms:

$$H(\omega) = \frac{1}{k - m\omega^2} = \frac{1/m}{\omega_n^2 - \omega^2} = \frac{1/k}{1 - \left(\dfrac{\omega}{\omega_n}\right)^2}. \tag{3.19}$$

Fig. 3.3 Solution of the undamped system. The response was computed with $F_0 = 1$ N, $m = 1$ kg, $\omega_n = 1$ rad/s, and $\omega = \sqrt{2}$ rad/s

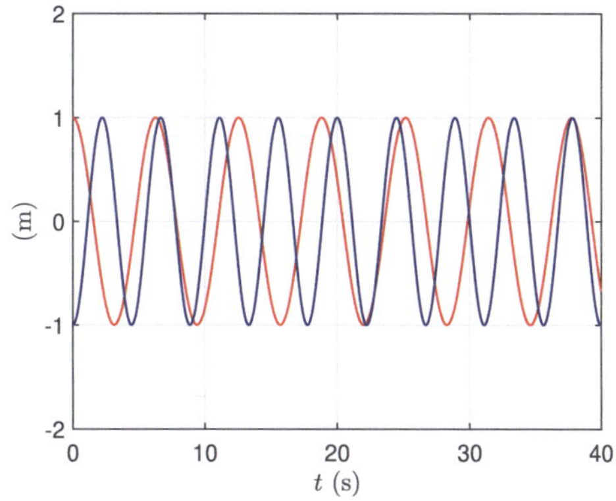

(a) Particular q_p (—) and homogeneous q_h (—) parts of the solution are harmonic.

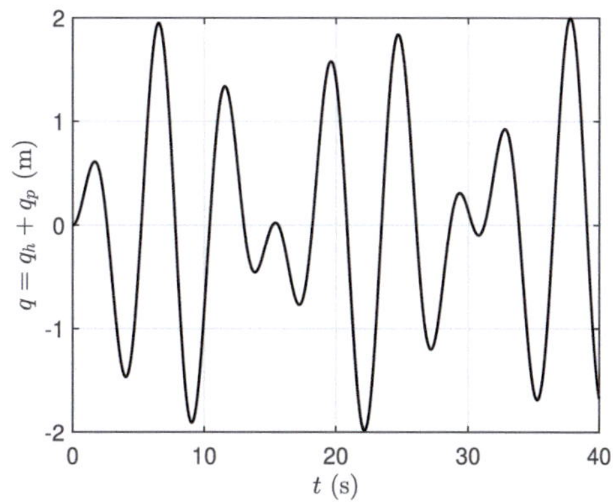

(b) The complete solution $q = q_h + q_p$ is, in general, neither harmonic nor periodic.

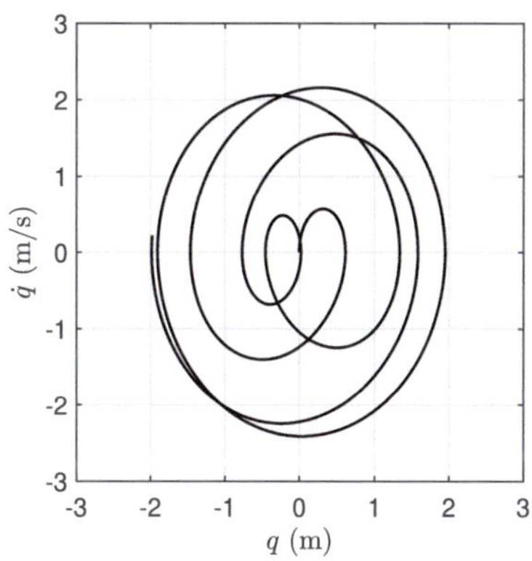

(c) Phase-plane trajectory.

The influence of this parameter on the system response can be better understood if it is expressed in polar form:[1]

$$H(\omega) = |H(\omega)| e^{-i\phi(\omega)}, \tag{3.20}$$

where the the absolute value is

$$|H(\omega)| = \frac{1/k}{\left|1 - \left(\frac{\omega}{\omega_n}\right)^2\right|} \tag{3.21}$$

and the phase angle is given by

$$\phi(\omega) = \begin{cases} 0, & \text{if } \omega < \omega_n, \\ \pi, & \text{if } \omega > \omega_n. \end{cases} \tag{3.22}$$

The phase angle ϕ exclusively takes the values 0 or π because the real-valued function $H(\omega)$ is either positive or negative. The polar form representation may seem unnecessarily elaborate for an undamped system, but this notation is emphasized even in this case to establish the framework applicable to damped systems, where the FRF has non-zero imaginary part.

The polar form (3.20) can be used to express the particular solution as

$$q_p(t) = |H(\omega)| F_0 e^{i(\omega t - \phi(\omega))}. \tag{3.23}$$

The physical becomes

$$\begin{aligned} q_p(t) &= |H(\omega)| F_0 \cos(\omega t - \phi(\omega)) \\ &= \frac{F_0/k}{\left|1 - \left(\frac{\omega}{\omega_n}\right)^2\right|} \cos(\omega t - 0 \, \text{or} \, -\pi) \end{aligned} \tag{3.24}$$

because $f(t) = F_0 \cos \omega t$. The FRF serves as a fundamental characterization of the system, as illustrated in Fig. 3.4. It determines whether the system amplifies or attenuates the force input and by how much it delays the response, all contingent upon the value of the driving frequency ω.

The influence of $H(\omega)$ is inferred from the plot depicted in Fig. 3.5, called *frequency response curve*. It displays the absolute value and phase of $H(\omega)$ against the excitation frequency. Several aspects may be observed from this plot. At low

[1] Appendix A reviews different forms of expressing complex numbers.

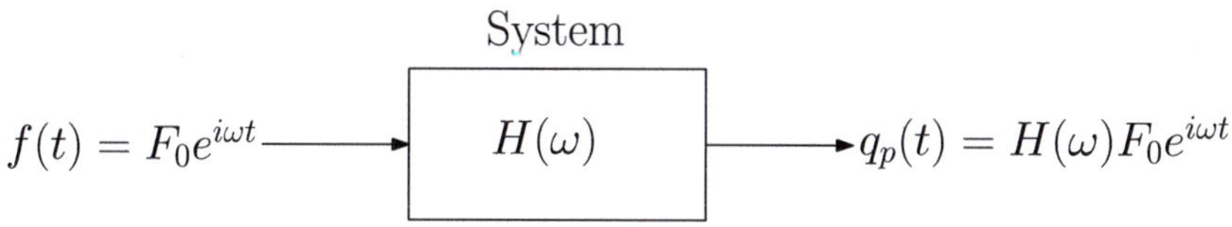

Fig. 3.4 The input and output are interconnected by means of the FRF. The absolute value $|H(\omega)|$ is responsible for magnifying or attenuating the amplitude, while the time delay in the response is determined by the phase angle $\phi(\omega)$

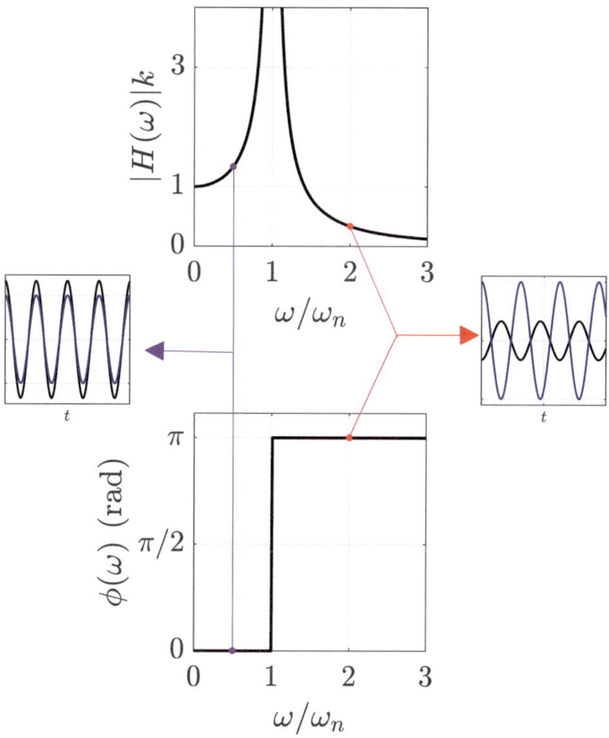

Fig. 3.5 Frequency response curve of $H(\omega)$ from Eq. (3.19). Two representative points are highlighted to showcase the dynamic behavior—one at a lower frequency and the other at a higher frequency—and the corresponding subfigures visually contrast the response $q_p(t)$ (—) with the excitation $f(t)$ (—) in the time domain

frequencies ($\omega/\omega_n << 1$), the absolute value and phase angle are

$$|H| \approx 1/k, \quad \phi = 0. \tag{3.25}$$

The particular solution becomes

$$q_p(t) \approx \frac{F_0}{k} \cos \omega t. \tag{3.26}$$

This solution has amplitude equal to the static deflection of the spring caused by F_0, and it consists of harmonic oscillations in phase with the excitation. "In phase" means that $q_p(t)$ and $f(t)$ reach maximum and minimum values simultaneously, i.e., the peaks and valleys in their oscillatory pattern align perfectly over time. This happens because inertial effects are negligible due to the slow variation of the excitation, providing ample time for the system to respond. On the other hand, the phase angle is $\phi = \pi$ when $\omega > \omega_n$. For any frequency in this interval,

$$q_p(t) = |H(\omega)| \cos(\omega t - \pi) = -|H(\omega)| \cos \omega t. \tag{3.27}$$

The negative sign in this expression implies that the peaks of $q_p(t)$ align with the valleys of $f(t)$, meaning that $q_p(t)$ is "out of phase" with respect to the excitation $f(t)$.

Figure 3.5 shows that the amplitude of vibration grows as ω increases from 0 to ω_n. It reaches maximum value at a *resonant frequency* ω_r, which equals ω_n in this case. The plot indicates that q_p has "infinite" amplitude. This is investigated in detail in Sect. 3.3.3. The amplitude of vibration decreases as the driving frequency is increased beyond ω_n, but it is still magnified ($|H|k > 1$) as long as $\omega < \sqrt{2}\omega_n$. The response at high frequencies ($\omega/\omega_n >> 1$) is

$$q_p(t) \approx -\frac{F_0/m}{\omega^2} \cos \omega t. \tag{3.28}$$

The amplitude of $q_p(t)$ is quite small when $\omega/\omega_n >> 1$: the excitation is varying rapidly, but the inertia of the system limits its ability to keep up with these rapid changes and causes a reduction in amplitude.

3.3.3 Beating and Resonance

An interesting behavior becomes apparent in Fig. 3.5: as ω approaches the resonant frequency, the absolute value $|H|$ tends to infinity. This phenomenon, known as *resonance*, denotes a condition where the system vibrates violently with substantial amplitude. However, the infinite amplitude predicted by the FRF is not physically possible. One reason is that springs cannot deflect infinitely without breaking.

Additionally, the material properties of the spring would change at large amplitudes, and the assumption of small vibrations would be violated. A linear model would no longer be able to predict the behavior of the system.

To accurately understand how the system behaves when $\omega = \omega_n$, use the trigonometric identity

$$\cos \omega t - \cos \omega_n t = 2 \sin \left(\frac{\omega_n + \omega}{2} t \right) \sin \left(\frac{\omega_n - \omega}{2} t \right) \tag{3.29}$$

to reframe Eq. (3.17) more conveniently as

$$q(t) = \frac{F_0/m}{\omega_n + \omega} \sin \left(\frac{\omega + \omega_n}{2} t \right) \frac{\sin(\Delta t)}{\Delta}, \tag{3.30}$$

where $\Delta = (\omega_n - \omega)/2$. There are two cases to consider. The first one is when the frequencies are close ($\omega \approx \omega_n$) but **not** equal. Under this condition, they satisfy $\omega_n + \omega \approx 2\omega_n$ and Eq. (3.30) can be approximated by

$$q(t) \approx \frac{F_0/m}{2\omega_n \Delta} \sin(\omega_n t) \sin(\Delta t). \tag{3.31}$$

Figure 3.6 illustrates the typical response predicted by Eq. (3.31), showing a fluctuating amplitude pattern known as *beating*. Beating results from the constructive and destructive interference of two harmonics with slightly different frequencies. The amplitude fluctuates at the *beat frequency* $\omega_b = |\omega_n - \omega|$.

Beating is present in the tuning of musical instruments, radio communication, and signal processing. Acoustic applications like these, which are closely related to the field of mechanical vibrations, provide a clear way to visualize (or hear) the beating phenomenon. For example, the MATLAB code `Chapt3_code1` provided in the *Electronic Supplementary Material* plays the musical note C, which has a frequency of approximately 261.6 Hz. The other code `Chapt3_code2` plays the superposition of this wave with another one of frequency close to 261.6 Hz. Note how beating gives the sound a pulsating or throbbing quality, overlaid on the regular tone of the original musical note.

It is now appropriate to investigate the behavior of the system at resonance. As ω approaches the resonant frequency,

$$\frac{\omega + \omega_n}{2} \to \omega_n, \qquad \Delta \to 0. \tag{3.32}$$

But it is known that

$$\lim_{\Delta \to 0} \frac{\sin \Delta t}{\Delta} = t. \tag{3.33}$$

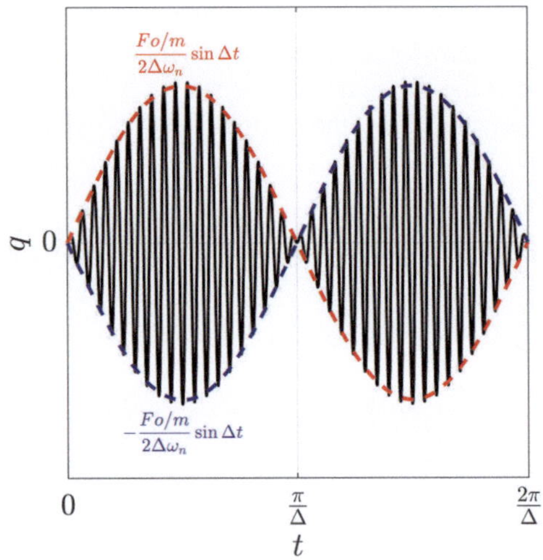

(a) A plot of function (3.31). Beating is observed when two harmonics of slightly different frequencies overlap.

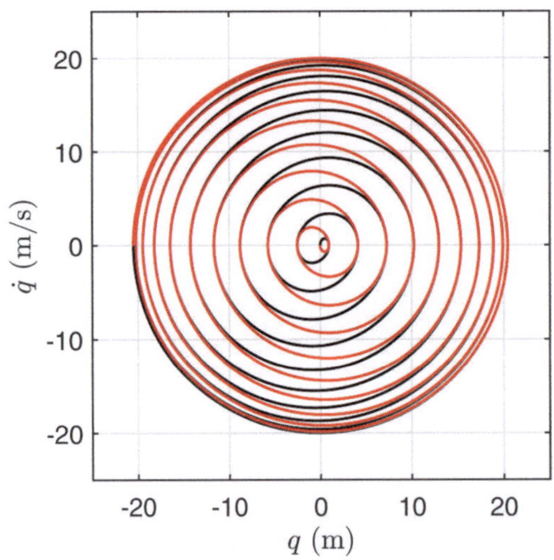

(b) Phase-plane trajectory. The curve (—) represents the part of the trajectory for $t \in (0, \pi/\Delta)$ and (—) corresponds to $t \in (\pi/\Delta, 2\pi/\Delta)$. Plot constructed with $m = 1$ kg, $k = 1$ N/m, $F_0 = 1$ N, and $\omega = 0.95 \times \omega_n$.

Fig. 3.6 Response when the system is excited with $\omega \approx \omega_n$

Therefore, Eq. (3.30) converges to

$$q(t) = \frac{F_0/m}{2\omega_n} t \sin \omega_n t, \tag{3.34}$$

when $\omega \to \omega_n$. This equation reveals that the amplitude grows linearly with the envelope $\pm F_o t/(2m\omega_n)$, but $|q| \to \infty$ only when $t \to \infty$ as shown in Fig. 3.7. The synchronization between the external force and the system natural frequency continuously increases the amplitude. The peaks and valleys of q during resonance coincide with the instants when $f = 0$ because the force was assumed to be $f(t) = F_0 \cos \omega t$. This results in a phase lag between q and f of $\pi/2$ rad.

3.4 Response of Damped Systems Subjected to Harmonic Force

This section focuses on analyzing the forced response of damped systems, which offers a more realistic representation of physical systems. A detailed solution process is provided, highlighting the differences from the undamped case.

3.4.1 Particular Solution and the Frequency Response Function

The particular solution of Eq. (3.8), with $c \neq 0$, has the form

$$q_p = Qe^{i\omega t}. \tag{3.35}$$

Differentiate this expression two times and substitute into the equation of motion to obtain

$$\left[-m\omega^2 + c(i\omega) + k\right] Qe^{i\omega t} = F_0 e^{i\omega t}. \tag{3.36}$$

The amplitude of the particular solution is now complex and given by

$$Q = \frac{F_0}{k - m\omega^2 + ic\omega}. \tag{3.37}$$

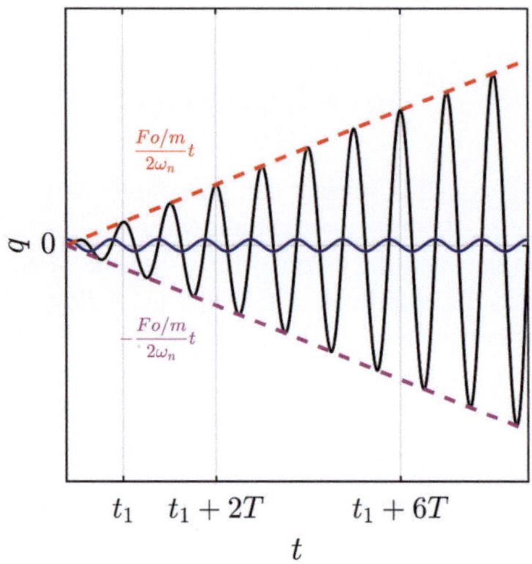

(a) A comparison between q (—) and f (—). The amplitude of the response grows linearly.

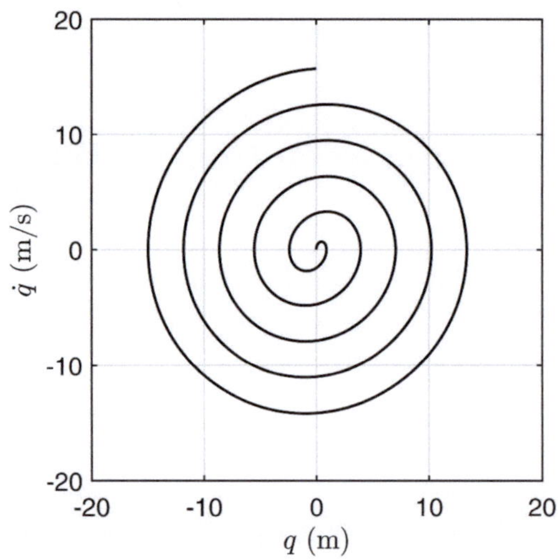

(b) Phase-plane trajectory. Plot constructed with m = 1 kg, k = 1 N/m, F_0 = 1 N, and $\omega = \omega_n$.

Fig. 3.7 Response when the system is exited with $\omega = \omega_n$

The FRF $H(\omega) = Q/F_0$ can be expressed in the following ways:

$$H(\omega) = \frac{1}{k - m\omega^2 + i\omega c} = \frac{1/m}{\omega_n^2 - \omega^2 + i2\zeta\omega_n\omega} = \frac{1/k}{1 - \left(\frac{\omega}{\omega_n}\right)^2 + i2\zeta\left(\frac{\omega}{\omega_n}\right)}. \tag{3.38}$$

The main difference to the undamped case is that the FRF has nonzero imaginary part. Use the polar form $H(\omega) = |H(\omega)|e^{-i\phi(\omega)}$ to express the particular solution as

$$q_p(t) = H(\omega)F_0e^{i\omega t} = |H(\omega)|F_0e^{i(\omega t - \phi(\omega))}, \tag{3.39}$$

where

$$|H(\omega)| = \frac{1/k}{\sqrt{\left[1 - \left(\frac{\omega}{\omega_n}\right)^2\right]^2 + \left[2\zeta\left(\frac{\omega}{\omega_n}\right)\right]^2}} \tag{3.40}$$

and

$$\phi(\omega) = \arctan\frac{2\zeta\left(\frac{\omega}{\omega_n}\right)}{1 - \left(\frac{\omega}{\omega_n}\right)^2}. \tag{3.41}$$

If $f(t) = F_0\cos\omega t$, the physical response is the real part of the complex solution:

$$q_p(t) = \frac{F_0/k}{\sqrt{\left[1 - \left(\frac{\omega}{\omega_n}\right)^2\right]^2 + \left[2\zeta\left(\frac{\omega}{\omega_n}\right)\right]^2}} \times \cos\left(\omega t - \arctan\frac{2\zeta\left(\frac{\omega}{\omega_n}\right)}{1 - \left(\frac{\omega}{\omega_n}\right)^2}\right). \tag{3.42}$$

3.4.2 Steady-State Response

The complete solution consists of

$$q(t) = q_h(t) + q_p(t), \tag{3.43}$$

where $q_h(t)$ adopts one of the three forms outlined in Eq. (3.3) for $\zeta > 0$. It also contains the two constants of integration used to satisfy the ICs. Unlike the undamped case, the homogeneous solution decays exponentially, as shown in Fig. 3.8. These plots provide a comparison between the the complete solution and the particular part, illustrating that q_h represents only an initial transient in q that eventually subsides, leaving the steady-state solution $q \approx q_p$ that is sustained while the force is applied.

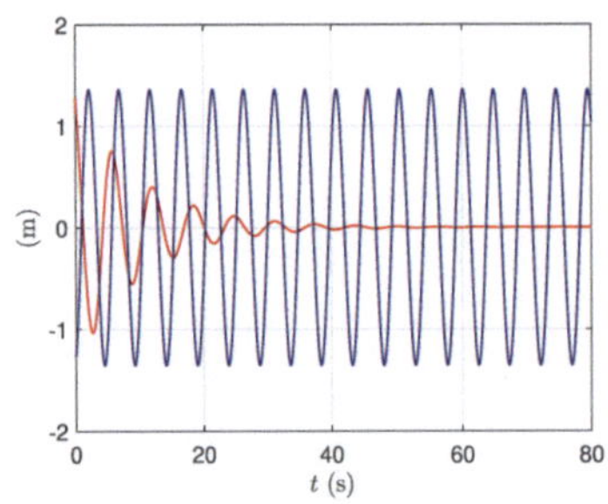

(a) Particular q_p (—) and homogeneous q_h (—) parts of the solution. The homogeneous part decays exponentially and eventually vanishes, leaving only the particular solution q_p.

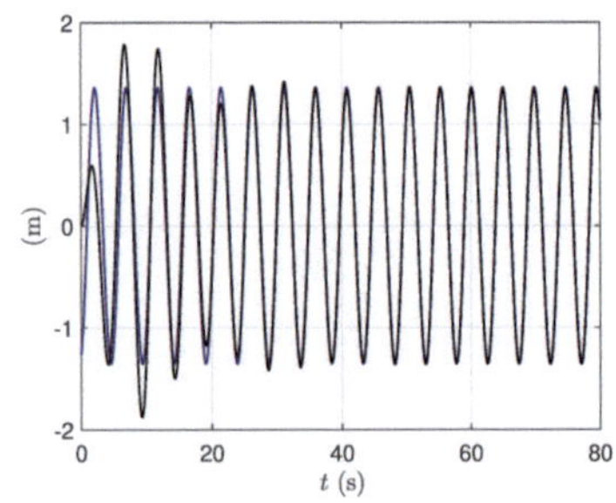

(b) Complete solution $q = q_h + q_p$ (—) and particular solution q_p (—). The homogeneous solutions represents an initial transient in the response that vanishes and leaves the steady-state response $q = q_p$.

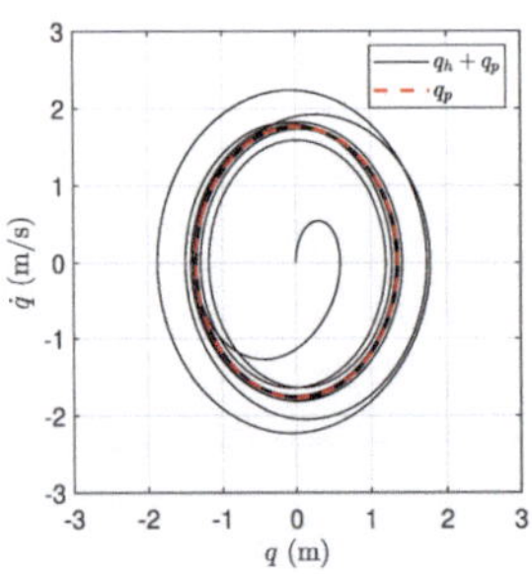

(c) Phase-plane trajectories. The decay of the transient part of the solution makes the trajectories spiral towards a steady-state closed trajectory (- -).

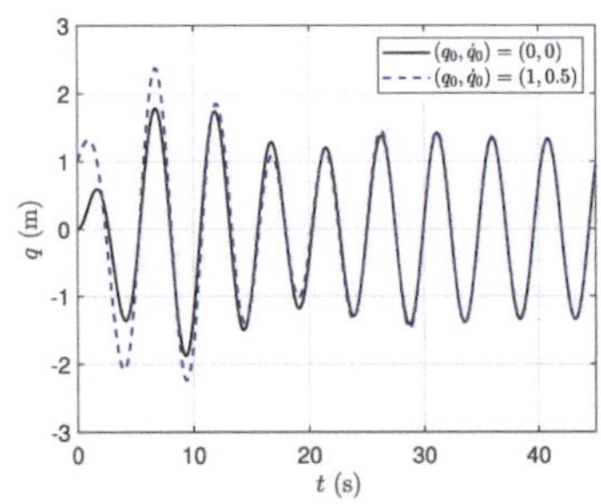

(d) Solution for two sets of ICs. Different ICs affect only the transient part of the response because only the integration constants would change.

Fig. 3.8 Solution of the damped system. The response was computed with $F_0 = 1$ N, $m = 1$ kg, $\omega_n = 1$ rad/s, $\zeta = 0.1$, and $\omega = 1.3$ rad/s. The ICs $q(0) = 0$ and $\dot{q}(0) = 0$ were used to produce subfigures (a)–(c)

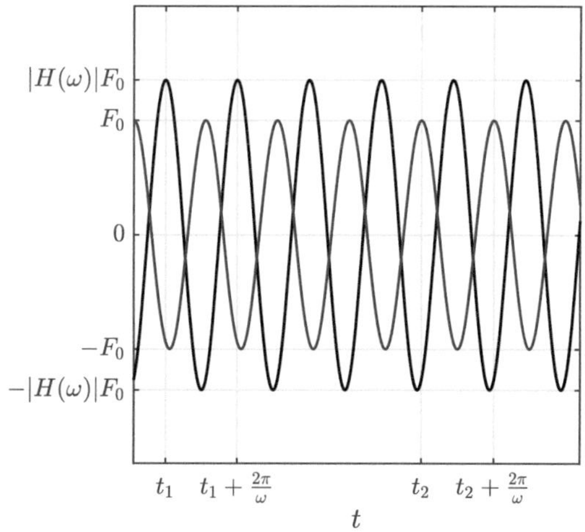

Fig. 3.9 Steady-state response $q = q_p$ (—) and applied force $f(t)$ (—)

The transient q_h is discarded if the emphasis lies on long-term system operation. In this type of analysis, attention is given solely to frequency-dependent parameters illustrated in Fig. 3.9. This plot compares the steady-state response and the applied force. Both q and f oscillate at the same frequency ω. The amplitude of the response is determined by the absolute value of the FRF, and the temporal delay is given by the phase angle.

3.4.3 Response Dependence on the Driving Frequency

The frequency response curve shown in Fig. 3.10 summarizes how the excitation frequency ω affects the steady-state behavior for different values of ζ. This plot was constructed with the MATLAB code `Chapt3_code3` supplied in the *Electronic Supplementary Material*. The plot reveals that every curve of $|H|$ with $\zeta > 0$ is located below that of the undamped case. In addition, all the curves are very close to each other at low and high frequencies. This indicates that the system behaves similarly to an undamped one, allowing the analysis from Sect. 3.3.2 to remain applicable within these frequency ranges. A striking distinction from the undamped case is at $\omega = \omega_n$, where

$$|H(\omega_n)| = \frac{1/k}{2\zeta}, \quad \phi(\omega_n) = \frac{\pi}{2}. \tag{3.44}$$

The amplitude is now finite and decreases as ζ grows. Highly damped systems do not have a large amplitude of vibration at this frequency. The phase angle is $\pi/2$ irrespective of the value of ζ when $\omega = \omega_n$, but can take any value from 0 to π rad at other frequencies.

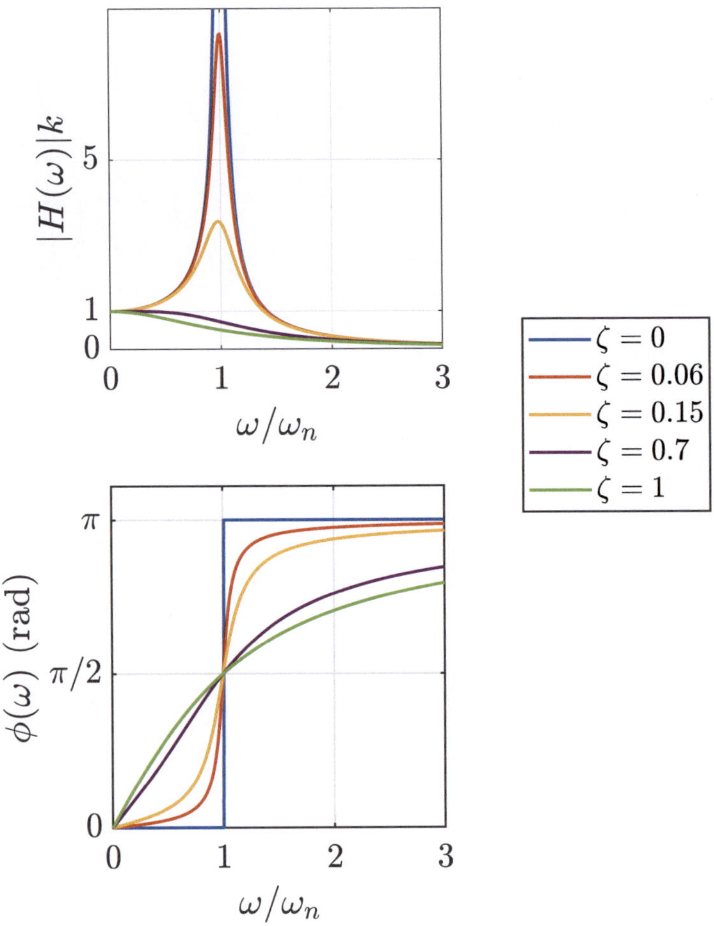

Fig. 3.10 Frequency response curve of $H(\omega)$ from Eq. (3.38) for different values of ζ

This plot shows that the resonant frequency ω_r, where $|H|$ is maximum, is not coincident with the natural frequency ω_n when the system is damped. This frequency can be found by solving the equation

$$\left.\frac{d|H|}{d\omega}\right|_{\omega=\omega_r} = 0. \tag{3.45}$$

Proceeding with the differentiation reveals that the resonant frequency is the solution of

$$\left(\frac{\omega_r}{\omega_n}\right)^2 = 1 - 2\zeta^2. \tag{3.46}$$

In order for the left side of this equation to be positive, a resonant frequency $\omega_r > 0$ can be computed for values of damping satisfying

$$\zeta < \frac{\sqrt{2}}{2} \approx 0.707. \tag{3.47}$$

With this in mind, solve Eq. (3.46) to obtain

$$\omega_r = \omega_n \sqrt{1 - 2\zeta^2}. \tag{3.48}$$

The maximum value of $|H|$ is

$$|H(\omega_r)| = \frac{1/k}{2\zeta\sqrt{1 - \zeta^2}}. \tag{3.49}$$

Figure 3.11 depicts the variation of $|H(\omega_r)|$ and ω_r for different values of ζ. The resonant frequency is coincident with the natural frequency for undamped systems,

Fig. 3.11 Dependence of response amplification and resonant frequency on damping ratio

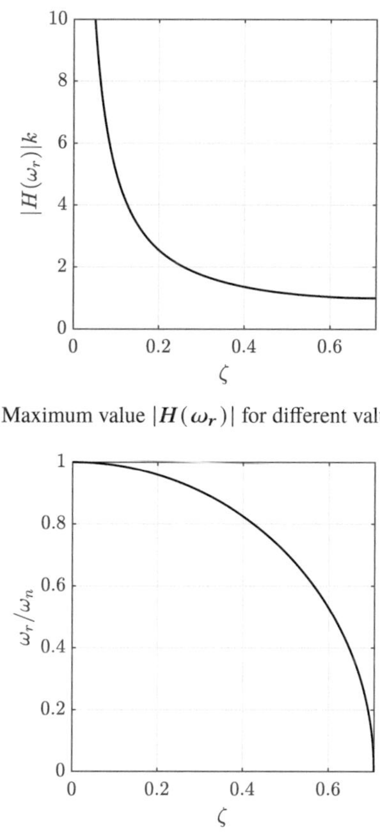

(a) Maximum value $|H(\omega_r)|$ for different values of ζ.

(b) Resonant frequency ω_r for different values of ζ.

but shifts to $\omega_r < \omega_n$ when $\zeta > 0$. In practice, the difference between ω_r and ω_n is small because many engineering applications have $\zeta < 20\%$. For instance, $0.9592 < \omega_r/\omega_n < 0.9999$ when $0.01 < \zeta < 0.2$. The maximum magnification factor $|H(\omega_r)|$ decreases with increasing ζ and is very sensitive to variations in this parameter when damping is small: $2.5516 < |H|_{\max} \times k < 50.0025$ in the range $0.01 < \zeta < 0.2$. If damping is too high, the response shows a peak, but it is more moderate compared to the peak observed when $\zeta << 1$.

A summary of how the steady-state response is affected by the driving frequency is depicted in Fig. 3.12. At low frequencies, the system closely follows the excitation, resulting in a small phase lag and an amplitude approximately equals to the static deflection of the spring caused by the force application. If $\zeta < \sqrt{2}/2$, the amplitude of vibration increases as the frequency rises. It reaches maximum value at the resonant frequency $\omega_r < \omega_n$ and begins to decrease as ω grows beyond ω_r. When $\omega = \omega_n$, the system vibrates with phase lag of $\pi/2$ rad. At high frequencies, the amplitude is attenuated, and the system vibrates out of phase.

3.5 Illustrative Examples

This section provides five illustrative examples. Two of these contain easy-to-replicate demonstrations that help validate and clarify the theoretical concepts discussed in this chapter.

3.5.1 *An Oscillating Bar*

Figure 3.13 shows a uniform bar of length L and mass m that is free to rotate around the axis $\mathbf{k}$ passing through the pin joint at O. The end of the bar at point B is connected to a viscous damper with coefficient c, while the opposite end is acted upon by a harmonic force $f(t)$ with amplitude F_0 and frequency ω. At point A on the bar, which is at a distance l from O, a linear spring with stiffness k and unstretched length l_0 is attached. The objective of this example is to analyze how the distance l affects the amplitude of steady-state vibration of the bar using the following parameter values: $m = 20$ kg, $l_0 = 0.2$ m, $k = 4 \times 10^4$ N/m, $c = 100$ Ns/m, $L = 2$ m, and $\omega = 50$ rad/s.

3.5.1.1 Kinematic Analysis

The derivation of the equation of motion requires knowledge of the position vectors of all points indicated in Fig. 3.13. Using the fixed reference system, the position vector of point a on the fixed support is

$$\mathbf{r}_a = l\mathbf{i} + l_0\mathbf{j}. \tag{3.50}$$

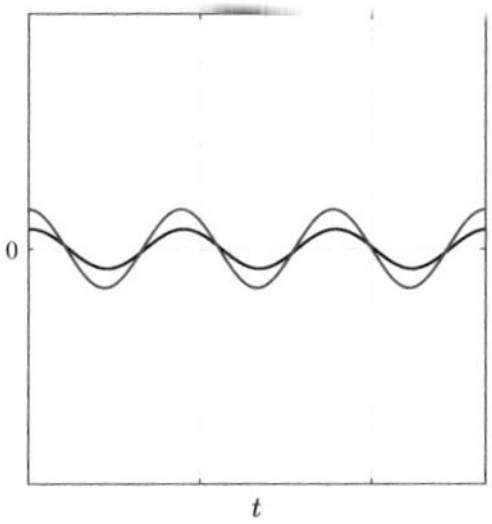

(a) Vibration when $\omega/\omega_n = 0.1$. Motion in phase with excitation and amplitude is equal to F_0/k.

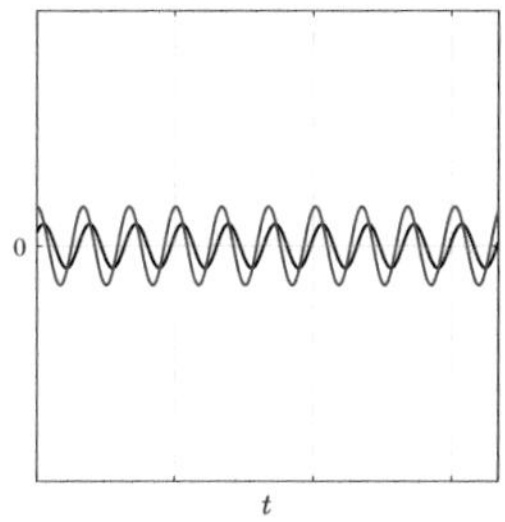

(b) Vibration at resonant frequency $\omega = \omega_r$. At this frequency, the maximum amplitude of vibration is observed.

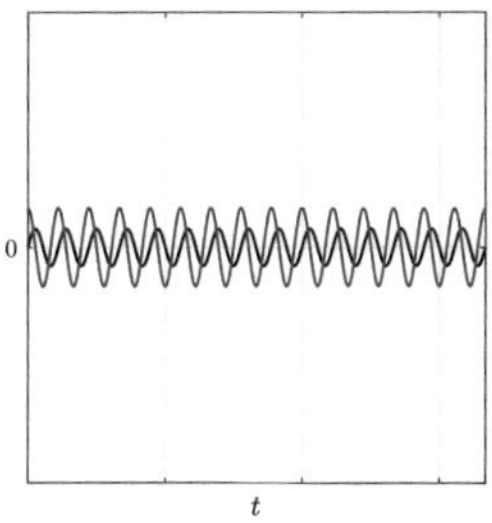

(c) Vibration when $\omega = \omega_n$. When excited at the natural frequency, the system vibrates with phase lag of $\phi = \pi/2$ rad.

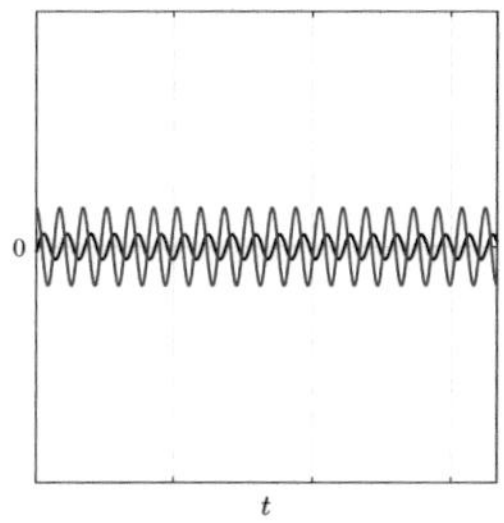

(d) Vibration when $\omega/\omega_n = 1.3$ rad/s. Amplitude is still amplified and response is delayed with $\pi/2 < \phi < \pi$.

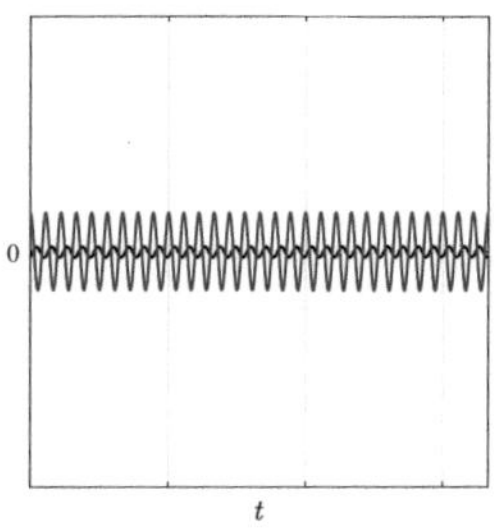

(e) Vibration when $\omega/\omega_n = 2$ rad/s. At high frequencies, amplitude is attenuated and response is out of phase: $\phi = \pi$ rad.

Fig. 3.12 Steady state response $q = q_p$ (—) and applied force $f(t)$ (—) for different values of ω. These plots were generated with $F_0 = 1$ N, $m = 1$ kg, $k = 2$ N/m, and $c = 1$ Ns/m. For these parameter values, it can be checked that $\zeta = 0.3536$, $\omega_n = 1.4142$ rad/s, and $\omega_r = 1.2247$ rad/s

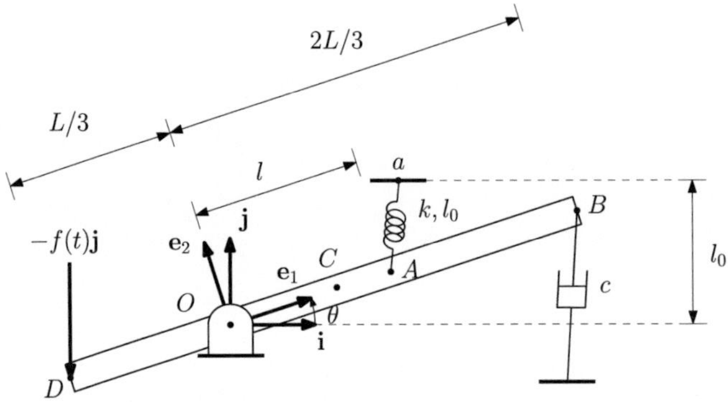

Fig. 3.13 A uniform bar of length L and mass m. Its center of mass is located at C, and the gravitational force acts along the $-\mathbf{j}$ direction

The positions of points A, B, and D of the bar are given by the vectors

$$\mathbf{x}_A = l\mathbf{e}_1, \quad \mathbf{x}_B = \frac{2L}{3}\mathbf{e}_1, \quad \mathbf{x}_D = -\frac{L}{3}\mathbf{e}_1, \tag{3.51}$$

respectively. The position vector of the mass center C is

$$\mathbf{x}_C = \frac{L}{6}\mathbf{e}_1. \tag{3.52}$$

One way to proceed with solving this problem is to first derive the equation of motion and then linearize it. However, to avoid this laborious process (and given that the focus is already on the linear differential equations of vibration), it is simpler to assume from the outset that the system undergoes small oscillations. The main challenge is that the static equilibrium position of the bar, which is essential for formulating the equations for small oscillations around it, is not yet determined. A possible pathway is to assume that the system is sufficiently rigid, such that the equilibrium position θ_e is close to the horizontal, i.e., $\theta = 0$. Assuming any state $(\theta, \dot{\theta})$ to be contained in a small neighborhood around $(0, 0)$, the position vectors can be approximated by

$$\mathbf{x}_A \approx l\mathbf{i} + l\theta\mathbf{j}, \tag{3.53}$$

$$\mathbf{x}_B \approx \frac{2L}{3}\mathbf{i} + \frac{2L}{3}\theta\mathbf{j}, \tag{3.54}$$

$$\mathbf{x}_D \approx -\frac{L}{3}\mathbf{i} - \frac{L}{3}\theta\mathbf{j}, \tag{3.55}$$

$$\mathbf{x}_C \approx \frac{L}{6}\mathbf{i} + \frac{L}{6}\theta\mathbf{j}. \tag{3.56}$$

Differentiate Eq. (3.54) to obtain a linearized expression for the velocity of point B:

$$\mathbf{v}_B \approx \frac{2L}{3}\dot{\theta}\mathbf{j}. \tag{3.57}$$

The angular momentum of the bar with respect to the fixed point O is necessary for applying Euler's second law. Use the parallel axis theorem[2] to compute it according with

$$\mathbf{H}^O = \frac{1}{9}mL^2\dot{\theta}\mathbf{k}. \tag{3.58}$$

3.5.1.2 Equation of Motion

The FBD of the bar is shown in Fig. 3.14. A reaction force $\mathbf{R}_O$ acts at O, and the weight acts at the mass center C. The small motion approximation from Eq. (3.53) implies that point A moves only vertically, so the elastic force is

$$\mathbf{F}_e = -kl\theta\mathbf{j}. \tag{3.59}$$

Similarly, the damping force is given by

$$\mathbf{F}_d = -c\mathbf{v}_B = -\frac{2L}{3}c\dot{\theta}\mathbf{j}. \tag{3.60}$$

The applied force is harmonic and has the representation $f(t) = F_0e^{i\omega t}$.

Fig. 3.14 FBD of the bar

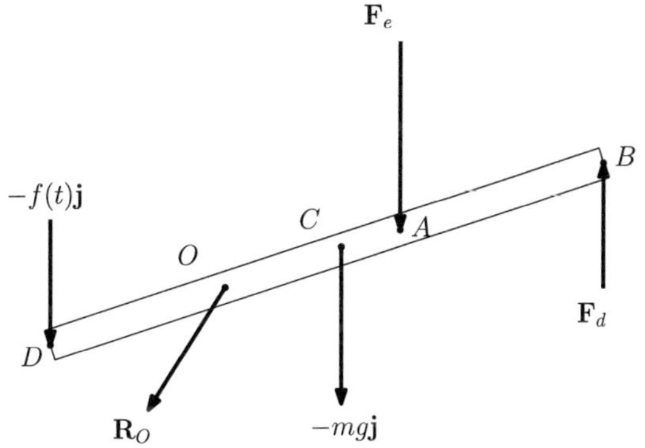

[2] See Appendix D.

Application of Euler's second law with respect to point O yields

$$\begin{aligned}\frac{1}{9}mL^2\ddot{\theta} = &\left(-\frac{L}{3}\mathbf{i} - \frac{L}{3}\theta\mathbf{j}\right) \times (-f\mathbf{j}) \cdot \mathbf{k} + (l\mathbf{i} + l\theta\mathbf{j}) \times (-kl\theta\mathbf{j}) \cdot \mathbf{k} \\ &+ \left(\frac{2L}{3}\mathbf{i} + \frac{2L}{3}\theta\mathbf{j}\right) \times \left(-\frac{2L}{3}c\dot{\theta}\mathbf{j}\right) \cdot \mathbf{k} \\ &+ \left(\frac{L}{6}\mathbf{i} + \frac{L}{6}\theta\mathbf{j}\right) \times (-mg\mathbf{j}) \cdot \mathbf{k}.\end{aligned} \tag{3.61}$$

Expand the vector triple products to obtain the differential equation

$$\frac{1}{9}mL^2\ddot{\theta} + \frac{4}{9}cL^2\dot{\theta} + kl^2\theta = \frac{1}{3}Lf(t) - \frac{1}{6}mgL. \tag{3.62}$$

When the system is not acted upon by $f(t)$, the static equilibrium position θ_e is given by

$$\theta_e = -\frac{mgL}{6kl^2} = -0.0147\,\text{rad}. \tag{3.63}$$

Define a new generalized coordinate γ by $\gamma = \theta - \theta_e$. This coordinate measures the angular position of the system with respect to the equilibrium position. It can be used to rewrite the equation of motion as follows:

$$\ddot{\gamma} + \frac{4c}{m}\dot{\gamma} + \frac{9kl^2}{mL^2}\gamma = \frac{3F_0}{mL}e^{i\omega t}. \tag{3.64}$$

The initial assumption that the bar performs small oscillations around the position $\theta = 0$ allowed the approximation $\theta_e = -mgL/(6kl^2)$ for the equilibrium position to be obtained. Although this equilibrium position does not exactly coincide with the linearization point of the equation of motion, this discrepancy does not significantly affect the final approximation for small vibrations, as the equilibrium and linearization points are sufficiently close.

3.5.1.3 Analysis

The steady-state solution has the form

$$\gamma(t) = \Gamma e^{i\omega t}. \tag{3.65}$$

Differentiating twice and substituting into Eq. (3.64) yields

$$\Gamma = \frac{LF_0/3}{kl^2 - \frac{1}{9}mL^2\omega^2 + i\frac{4}{9}cL^2\omega}. \tag{3.66}$$

Suppose the excitation is $f(t) = F_0 \cos \omega t$. If the complex amplitude is represented in polar form by $\Gamma = |\Gamma| e^{-i\phi}$, the physical solution is simply

$$\gamma(t) = |\Gamma| \cos(\omega t - \phi), \tag{3.67}$$

where the amplitude of vibration is

$$|\Gamma| = \frac{L F_0/3}{\sqrt{\left(kl^2 - \frac{1}{9} m L^2 \omega^2\right)^2 + \frac{16}{81} c^2 L^4 \omega^2}}. \tag{3.68}$$

Figure 3.15 shows how $|\Gamma|$ varies for different choices of l. This plot indicates there is a value l^* in $\in (0,\ 2L/3)$ where the amplitude is maximum. This value can be

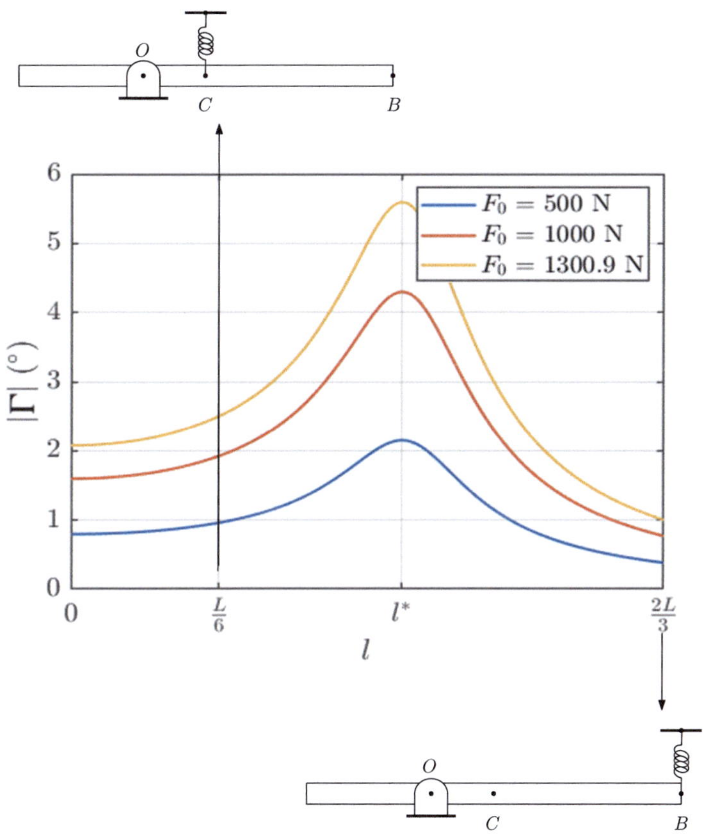

Fig. 3.15 Variation of steady-state amplitude $|\Gamma|$ for different spring placements. Configurations for two positions are illustrated

found by solving the equation

$$\left.\frac{d|\Gamma|}{dl}\right|_{l=l^*} = 0. \tag{3.69}$$

It can be verified that

$$l^* = \frac{\omega L}{3\sqrt{k/m}}. \tag{3.70}$$

In addition, the smallest amplitude of vibration occurs when $l = 2L/3$, i.e., when the spring is attached at point B. In this configuration, the range of forces that result in amplitude less than a chosen value of $|\Gamma|_{\max}$ can be computed with

$$F_0 \leq |\Gamma|_{\max} L \sqrt{\left(4k - m\omega^2\right)^2 + \frac{16}{9}c^2\omega^2}. \tag{3.71}$$

For example, the force needs to be $F_0 \leq 1300.9$ N if one desires $|\Gamma| \leq 1^\circ$. This restriction is valid only for the steady-state solution. Figure 3.16 shows that the system may attain amplitudes larger than 1° when the transient part of the solution is present.

3.5.2 *Exciting a Mass-Spring System*

Figure 3.17 presents a simple vibration demonstration involving a linear spring attached to a bottle filled with coins. Hold the free end of the spring in your

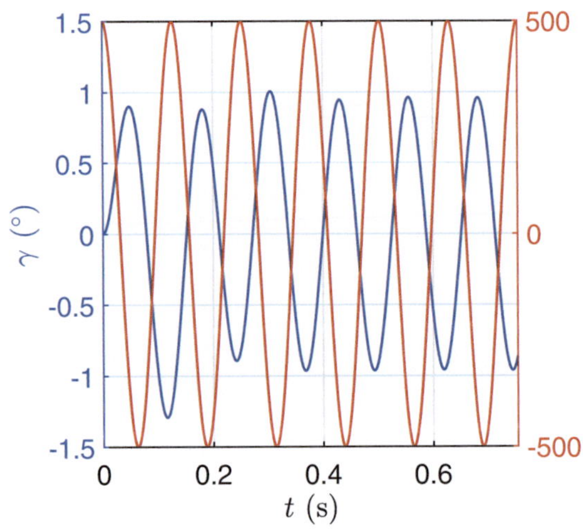

Fig. 3.16 Complete solution of Eq. (3.64). Plot constructed using $F_0 = 500$ N with ICs $\gamma(0) = 0$ and $\dot{\gamma}(0) = 0$

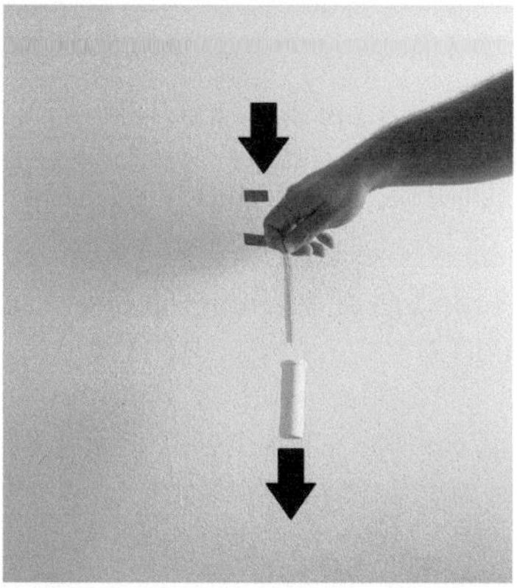

(a) Hand and bottle moving in phase at low frequency.

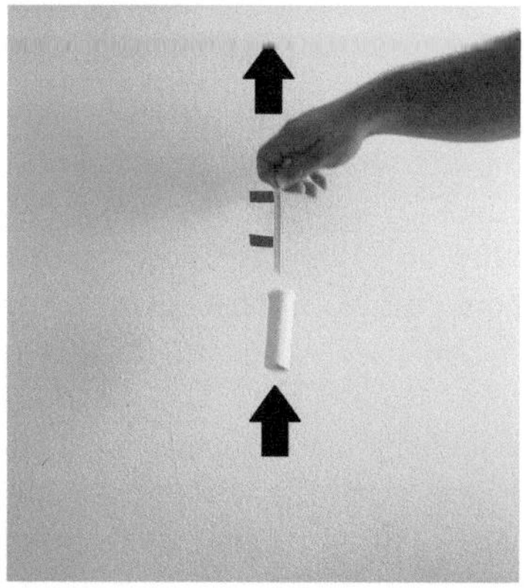

(b) Hand and bottle moving in phase at low frequency.

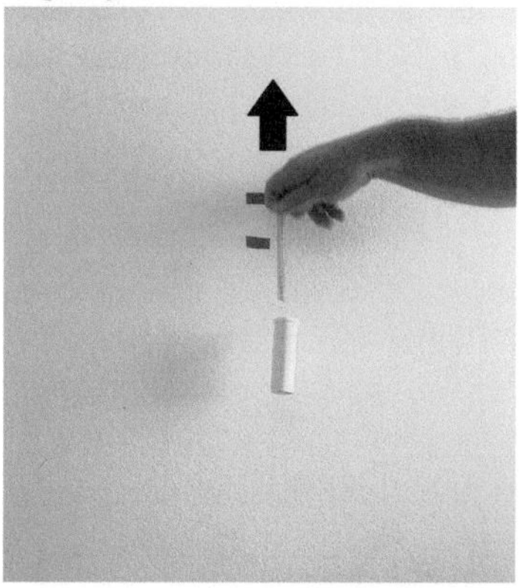

(c) Bottle barely moves at high frequencies.

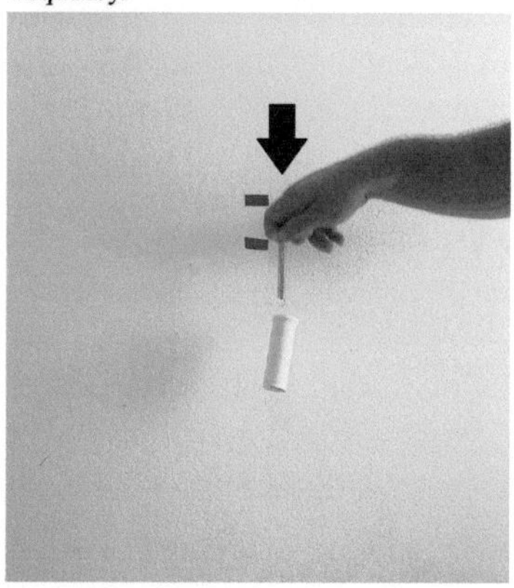

(d) Bottle barely moves at high frequencies.

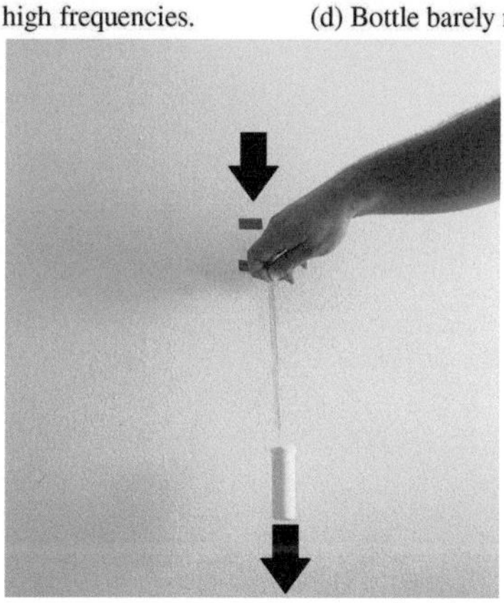

(e) Bottle with greatest amplitude (resonance).

Fig. 3.17 Snapshots of the motion. The colored tapes on the wall delineate the amplitude of the imposed motion, and the arrows show the direction of motion

hand and observe the motion of the bottle as you move your hand up and down rhythmically. This set up is often termed *base excitation*, where the excitation is not a force/moment directly applied to the object but involves the transmission of motion from the ground or a support to the system itself.

Examine how the frequency of the imposed motion affects the oscillatory behavior of the bottle. To begin, move your hand slowly. It can be observed that the bottle moves in phase with the imposed motion, maintaining an amplitude of oscillation nearly identical to that of the hand (refer to Figs. 3.17a and b). On the other hand, Figs. 3.17c and d shows that the bottle exhibits small amplitude and oscillates out of phase when the hand moves very rapidly. An intermediate frequency can be found where the amplitude of oscillation of the bottle is very large. A snapshot of the motion at this resonant frequency is shown in Fig. 3.17e.

The objective of this example is to apply the material from this chapter to develop a model capable of replicating these observations, providing a deeper understanding of the dynamics involved. Despite the fact that the base excitation is not a force directly applied to the bottle, it will be demonstrated how the mathematical machinery presented in Sects. 3.3 and 3.4 can still be applied.

3.5.2.1 Mathematical Model

Assuming damping to be small, a model for the bottle-spring system can be devised according to Fig. 3.18a. It consists of a particle with mass m connected to a linear spring with stiffness k and unstretched length l_0. The motion imposed by the hand is represented by the known coordinate $y_A(t)$ of the support A.

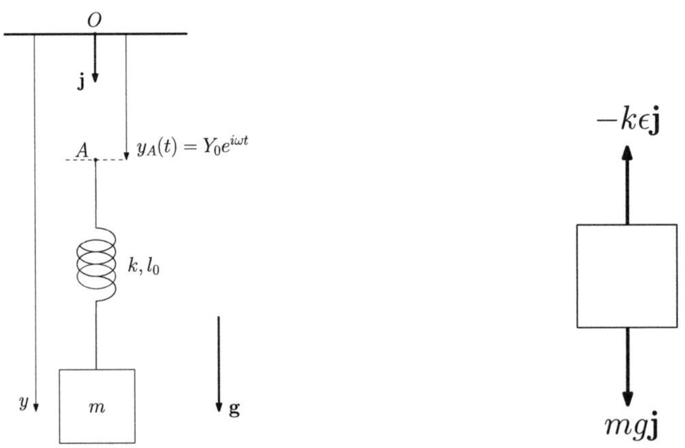

(a) Reference system. (b) FBD.

Fig. 3.18 Mathematical model of the demonstration from Fig. 3.17

Using the reference system indicated in Fig. 3.18a, the position, velocity, and acceleration vectors of the particle have the general representation

$$\mathbf{r} = y\mathbf{j}, \quad \mathbf{v} = \dot{y}\mathbf{j}, \quad \mathbf{a} = \ddot{y}\mathbf{j}, \tag{3.72}$$

respectively.

The FBD is shown in Fig. 3.18b. The only forces acting on the particle are the weight, $-mg\mathbf{j}$, and the elastic force, $\mathbf{F}_e = -k\epsilon\mathbf{j}$. The change of length of the spring is

$$\epsilon = y - y_A - l_0. \tag{3.73}$$

Use Newton's second law to show that coordinate y satisfies the differential equation

$$m\ddot{y} = mg - k\epsilon. \tag{3.74}$$

To rewrite Eq. (3.74) in terms of a generalized coordinate q measured from the static equilibrium position, substitute the change of coordinates depicted in Fig. 3.19, which is given by

$$y = q + l_0 + \frac{mg}{k}, \tag{3.75}$$

into Eq. (3.74) to obtain

$$\ddot{q} + \omega_n^2 q = \omega_n^2 y_A(t), \tag{3.76}$$

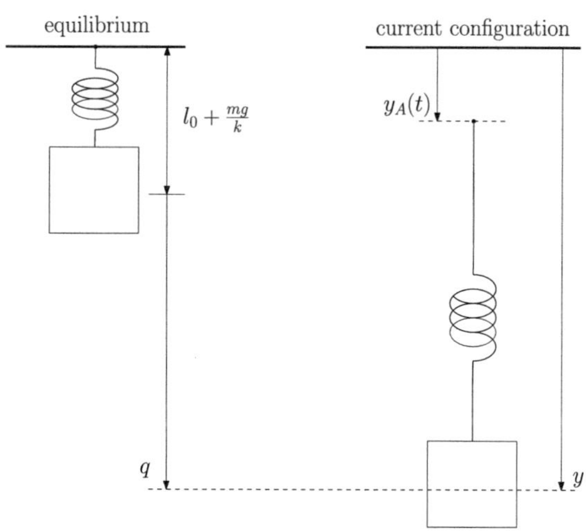

Fig. 3.19 Geometric representation of the change of coordinate used in the example

where $\omega_n = \sqrt{k/m}$ is the natural frequency. The imposed support motion is translated into a forcing term, allowing all the analytical tools developed in this chapter to be employed in the investigation of the system response.

3.5.2.2 Analysis of the Motion

The equation of motion (3.76) allows for the investigation of the effect of the driving frequency. Assume the imposed motion $y_A(t)$ to be harmonic with amplitude Y_0 and frequency ω: $y_A(t) = Y_0 e^{i\omega t}$. Since the real system is lightly damped, the transient part of the solution can be ignored, and the complete solution can be taken as $q(t) = q_p(t) = Qe^{i\omega t}$. Upon substitution into Eq. (3.76), the *transmissibility* of motion $G(\omega) = Q/Y_0$ is computed:

$$G(\omega) = \frac{\omega_n^2}{\omega_n^2 - \omega^2}. \tag{3.77}$$

The transmissibility $G(\omega)$ is a nondimensional parameter that connects the imposed motion Y_0 (input) to resulting vibration Q (output). The frequency response curve of $G(\omega)$ is depicted in Fig. 3.20. Since the transmissibility is a real-valued function, it was plotted directly with the signal taken into account instead of separating it in distinct plots for magnitude and phase angle.

This plot shows that, at low frequencies, $G(\omega)$ assumes the value $G \approx 1$. The bottle, consequently, oscillates with amplitude $Q \approx Y_0$ and remains in phase with the support motion ($G > 0$). Essentially, the spring behaves akin to a rigid rod, merely transmitting motion from the hand to the bottle. At high frequencies, the spring almost does not transmit motion, and the bottle has very low amplitude ($|G| \approx 0$) while moving out of phase ($G < 0$). When $\omega = \omega_n$, the transmissibility satisfies $|G| \to \infty$. This behavior arises due to the omission of damping in the model. In reality, this system is lightly damped, so the bottle oscillates with a finite yet large amplitude. These deductions closely align with the qualitative observations made during the demonstration, reaffirming the theoretical predictions derived from the analysis.

The frequency response curve can provide other interesting information on the system response. If ω is not significantly smaller than ω_n, the bottle oscillates in phase with the support motion, but with amplitude greater than that of the support. Conversely, if ω is not much greater than ω_n, the amplitude of the bottle can be greater than or equal to the amplitude of the support motion, while it oscillates out of phase with respect to the support.

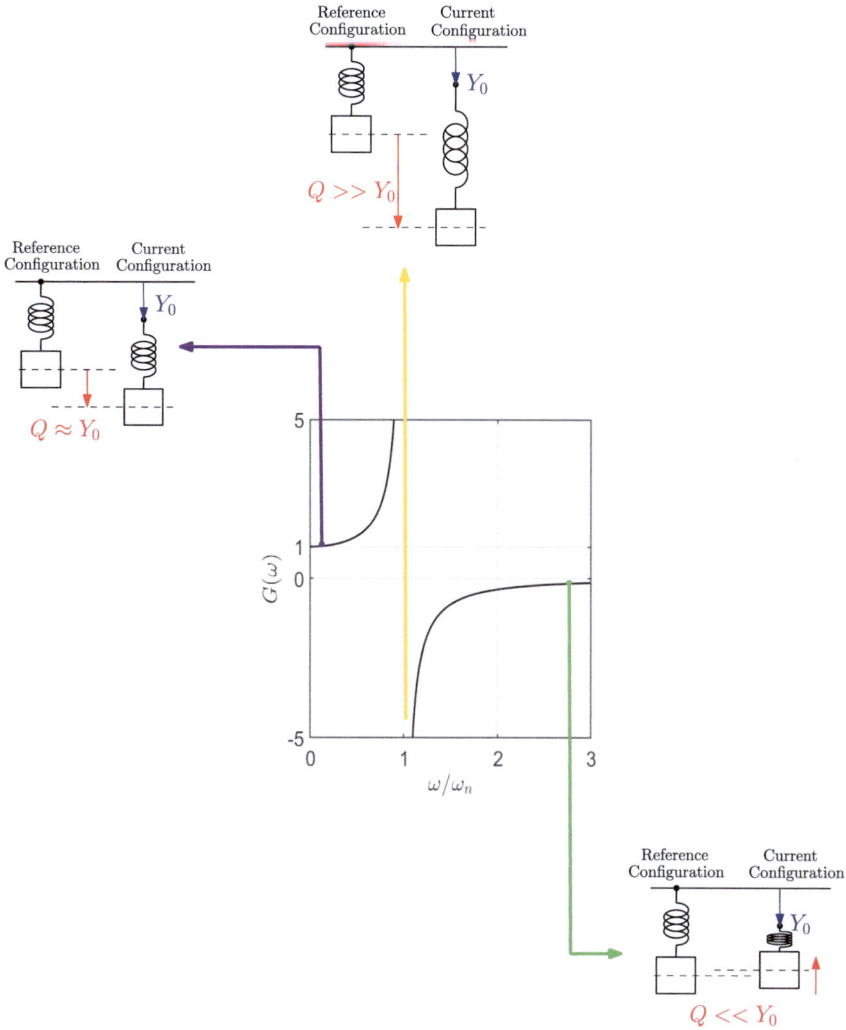

Fig. 3.20 Frequency response curve of the transmissibility from Eq. (3.77). This plot juxtaposes the current system configuration against a reference configuration across three values of driving frequency. The direction of the support (—) and bottle (—) motions are shown

3.5.3 *Exciting a Pendulum*

In this next example, consider a pendulum consisting of a particle with mass m attached to an inextensible cable of negligible mass and length l. The cable is free to rotate around the pin joint A of a moving support, as illustrated in Fig. 3.21a. The primary goal of this example is to evaluate how the motion of the support $y_A(t)$ influences the oscillatory behavior of the pendulum.

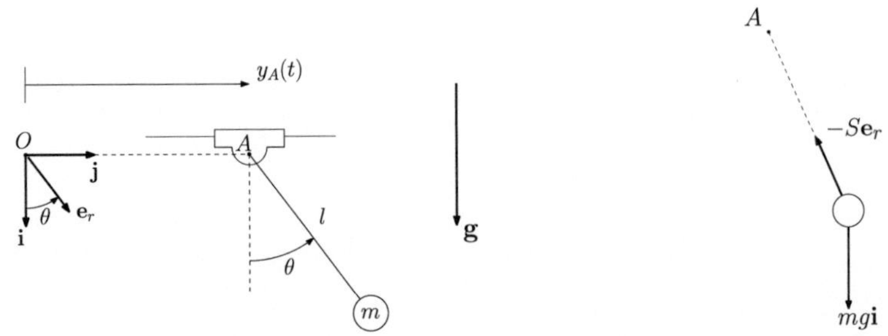

(a) Reference system. (b) FBD of the pendlum.

Fig. 3.21 Mathematical model of a pendulum excited by support motion

3.5.3.1 Mathematical Model

With reference to the coordinate system from Fig. 3.21a, the position vector of the pendulum bob can be represented by[3]

$$\mathbf{r} = y_A(t)\mathbf{j} + l\mathbf{e}_r. \tag{3.78}$$

The velocity and acceleration vectors of this particle are

$$\mathbf{v} = \dot{y}_A(t)\mathbf{j} + l\dot{\theta}\mathbf{e}_\theta, \quad \mathbf{a} = \ddot{y}_A(t)\mathbf{j} + l\ddot{\theta}\mathbf{e}_\theta - l\dot{\theta}^2\mathbf{e}_r, \tag{3.79}$$

respectively.

The FBD is shown in Fig. 3.21b, where S is the traction force that acts along the direction of the cable. Application of Newton's second law yields

$$m\ddot{y}_A(t)\mathbf{j} + ml(\ddot{\theta}\mathbf{e}_\theta - \dot{\theta}^2\mathbf{e}_r) = mg\mathbf{i} - S\mathbf{e}_r. \tag{3.80}$$

This is a vector equation that is composed of two scalar equations for the unknowns S and θ. By taking the projection of this equation along the $\mathbf{e}_\theta$ direction, a nonlinear differential equation for θ is obtained:

$$\ddot{\theta} + \frac{g}{l}\sin\theta = -\frac{\ddot{y}_A(t)}{l}\cos\theta. \tag{3.81}$$

[3] Appendix D contains a brief review of the cylindrical-polar system.

If the pendulum executes small oscillations around the equilibrium $(\theta, \dot{\theta}) = (0, 0)$, this equation can be linearized according with

$$\ddot{\theta} + \omega_n^2 \theta = -\frac{\ddot{y}_A(t)}{l}, \tag{3.82}$$

where $\omega_n = \sqrt{g/l}$ is the natural frequency of a simple pendulum.

3.5.3.2 Influence of Support Motion

Assume that the support motion is harmonic with amplitude Y_0 and frequency ω: $y_A(t) = Y_0 e^{i\omega t}$. The second derivative of y_A with respect to time is

$$\ddot{y}_A(t) = -\omega^2 Y_0 e^{i\omega t}. \tag{3.83}$$

Substituting this into the linearized equation of motion (3.82) yields

$$\ddot{\theta} + \omega_n^2 \theta = \frac{\omega^2 Y_0}{l} e^{i\omega t} = \frac{\omega^2}{l} y_A(t). \tag{3.84}$$

The harmonic support motion is translated into an applied load with frequency-dependent amplitude.

The steady-state response is given by

$$\theta(t) = \Theta e^{i\omega t}. \tag{3.85}$$

Substitute this into Eq. (3.84) and solve for the steady-state amplitude:

$$\Theta = \frac{\left(\dfrac{\omega}{\omega_n}\right)^2}{1 - \left(\dfrac{\omega}{\omega_n}\right)^2} \frac{Y_0}{l}. \tag{3.86}$$

A frequency response curve for Θ is shown in Fig. 3.22. At lower frequencies, the pendulum oscillates in phase with the support ($\Theta > 0$), but with small amplitude ($\Theta \approx 0$). When $\omega = \omega_n$, the system is in resonance, and the pendulum attains the largest amplitude of oscillation. The small amount of damping present in the system prevents the oscillations to grow unbounded. The pendulum and support motion are out of phase ($\Theta < 0$) at frequencies $\omega > \omega_n$. However, the pendulum oscillates with amplitude $|\Theta| \approx Y_0/l$ when $\omega >> \omega_n$.

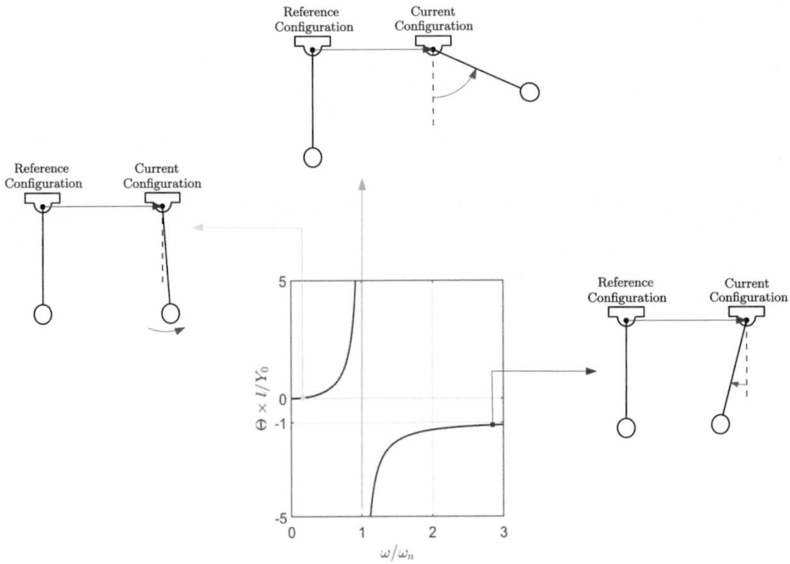

Fig. 3.22 Frequency response curve for the amplitude Θ from Eq. (3.86). This plot juxtaposes the current configuration against a reference configuration across three values of driving frequency. The direction of the support (—) and pendulum (—) motions are shown

3.5.3.3 Experimental Verification

The frequency-dependent behavior depicted in Fig. 3.22 can be observed and validated qualitatively through a straightforward demonstration. Tie a shoelace to a ball and grab the free end. Swing your hand horizontally and take note of the behavior of the pendulum. If the imposed motion is very slow, the pendulum barely oscillates, and it accompanies the hand as a rigid assembly (see Figs. 3.23a and b). As the speed of the hand increases, the pendulum traces a larger arc, remaining consistently in phase with the imposed motion. There is a resonant frequency, illustrated in Figs. 3.23c and d, where the pendulum has a very large amplitude of oscillation. If the speed of the hand is increased even further, the pendulum starts moving in the opposite direction. Figures 3.23e and e captures snapshots at a high frequency, where the pendulum swings out of phase and maintains amplitude $|\Theta| \approx Y_0/l$. This demonstration mirrors the frequency-dependent behavior predicted by the mathematical model.

3.5.4 Vehicle Suspension

The suspension of a vehicle is designed to absorb and dampen the shocks and vibrations caused by uneven road surfaces, thereby providing stability, better handling,

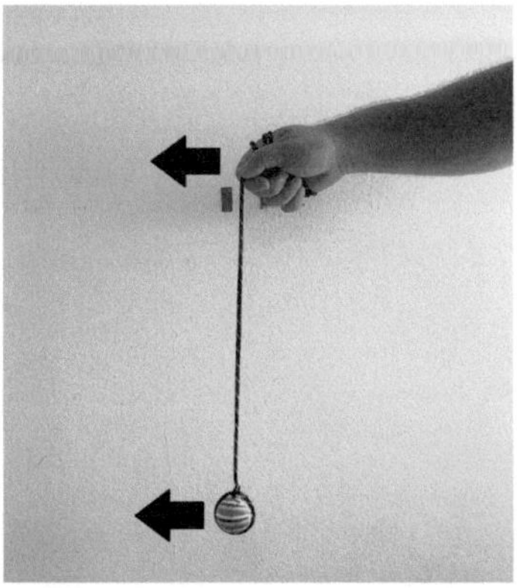

(a) Hand and pendulum moving in phase at low frequency.

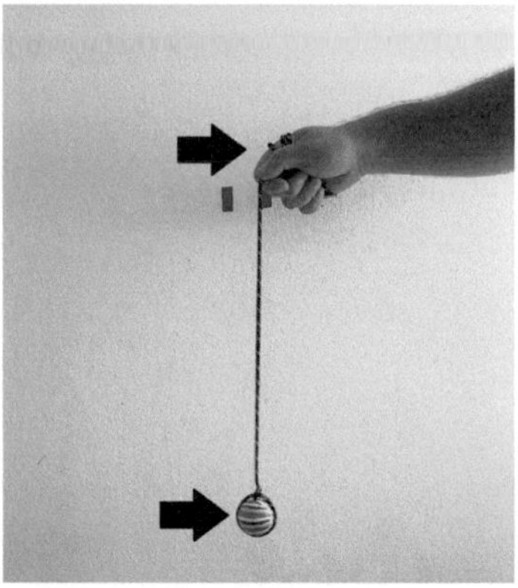

(b) Hand and pendulum moving in phase at low frequency.

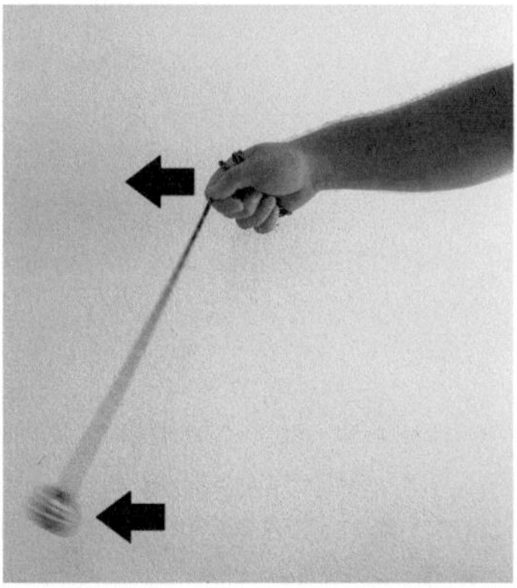

(c) Hand and pendulum moving near resonance.

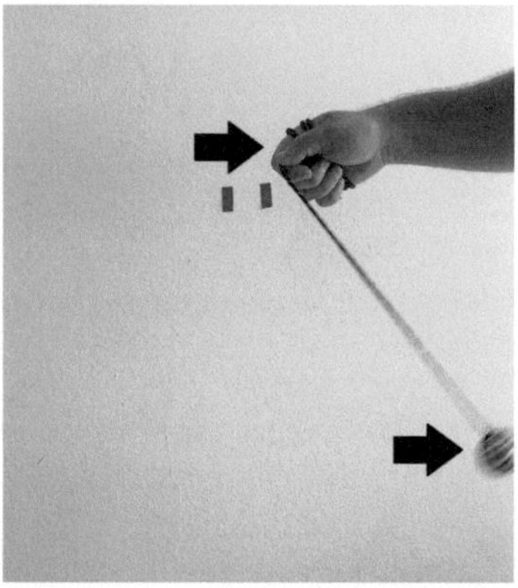

(d) Hand and pendulum moving near resonance.

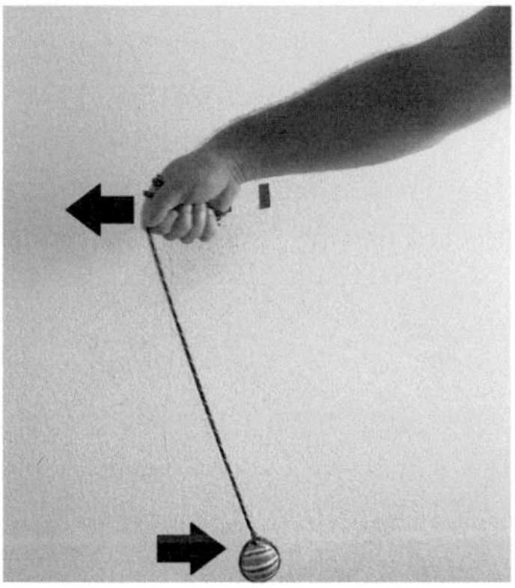

(e) Hand and pendulum moving out of phase at high frequency.

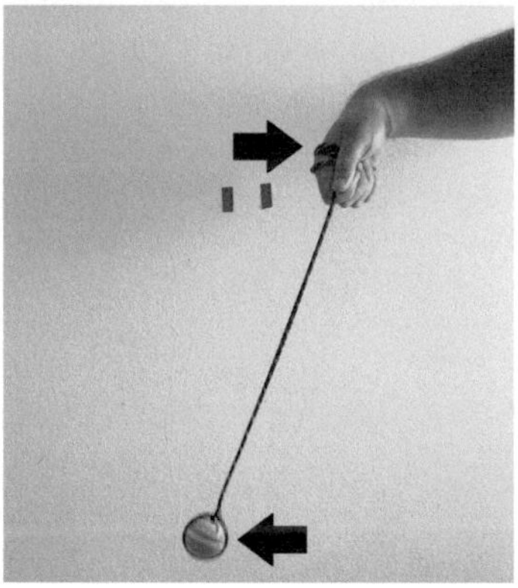

(f) Hand and pendulum moving out of phase at high frequency.

Fig. 3.23 Snapshots of the motion. The colored tapes on the wall delineate the amplitude of the imposed motion, and the arrows show the direction of motion

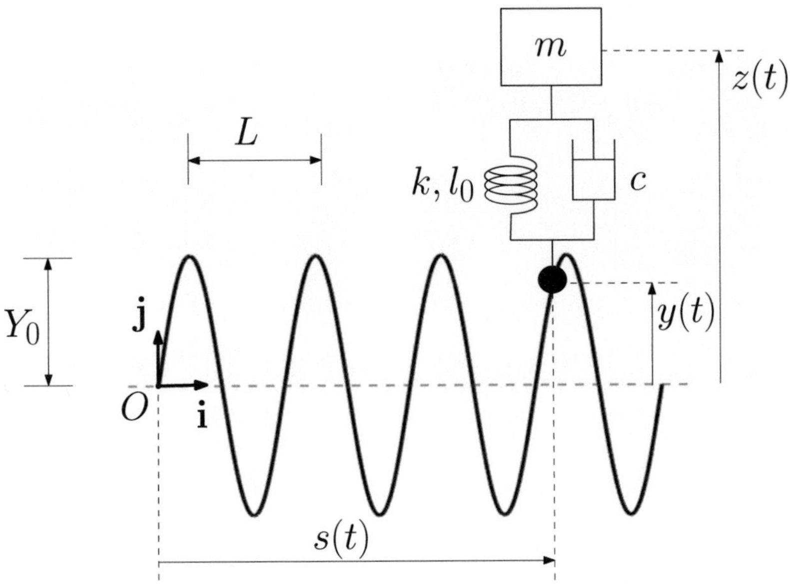

Fig. 3.24 Simplified model of a motorcycle traveling with constant speed on an uneven road. Gravitational force acts along the $-\mathbf{j}$ direction

and a smoother ride for passengers. Two crucial components in a suspension system are the springs and the shock absorbers (or dampers). The springs support the weight of the vehicle and reduce the impact felt by passengers due to road irregularities. The dampers control the oscillations, ensuring a smoother ride and minimizing vibrations. The goal of this example is to characterize the influence of the spring and damper on the vibrations transmitted to the passenger of a motorcycle.

A simplified model of a motorcycle is shown in Fig. 3.24. The motorcycle travels to the right with a constant speed v. The motorcycle and passenger are represented by a particle with mass m connected to a suspension comprising a viscous damper with coefficient c and a linear spring with stiffness k and unstretched length l_0. The uneven road has sinusoidal shape with roughness of height Y_0 and *wavelength* L.

3.5.4.1 Kinematic Analysis

The position, velocity, and acceleration vectors of the particle are represented by

$$\mathbf{r} = s\mathbf{i} + z\mathbf{j}, \quad \dot{\mathbf{r}} = v\mathbf{i} + \dot{z}\mathbf{j}, \quad \ddot{\mathbf{r}} = \ddot{z}\mathbf{j}, \tag{3.87}$$

respectively. The position vector of the wheel is

$$\mathbf{r}_w = s\mathbf{i} + y\mathbf{j} \tag{3.88}$$

and the associated velocity vector becomes

$$\dot{\mathbf{r}}_w = v\mathbf{i} + \dot{y}\mathbf{j}. \tag{3.89}$$

The component y of the position of the wheel can be expressed in terms of the geometry of the road and the motorcycle speed. When the motorcycle travels a distance s to the right, the vertical coordinate of the wheel is

$$y = Y_0 \sin\left(2\pi \frac{s}{L}\right). \tag{3.90}$$

For constant speed, the distance traveled is $s = vt$. As a consequence, coordinate y becomes

$$y(t) = Y_0 \sin \omega t, \tag{3.91}$$

where the driving frequency is given in terms of the speed and wavelength:

$$\omega = 2\pi \frac{v}{L}. \tag{3.92}$$

3.5.4.2 Equation of Motion

Figure 3.25 depicts the FBD of this system. The change of length of the spring is $\epsilon = z - y - l_0$ and the elastic force acting on the particle is

$$\mathbf{F}_e = -k(z - y - l_0)\mathbf{j}. \tag{3.93}$$

Fig. 3.25 FBD of the motorcycle

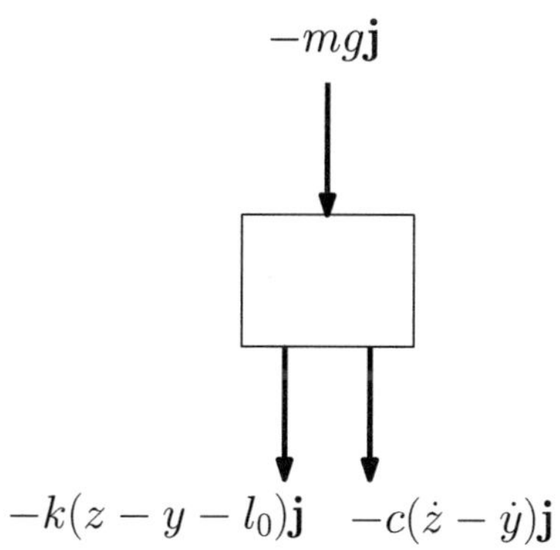

The velocity difference between the two ends of the damper is $\dot{\mathbf{r}} - \dot{\mathbf{r}}_w = (\dot{z} - \dot{y})\mathbf{j}$. Therefore, the damping force on the particle can be expressed as

$$\mathbf{F}_d = -c(\dot{z} - \dot{y})\mathbf{j}. \tag{3.94}$$

Use Newton's second law to obtain the differential equation of motion

$$m\ddot{z} + c\dot{z} + k(z - l_0) = c\dot{y} + ky - mg. \tag{3.95}$$

A new generalized coordinate q, measured from the static equilibrium position according with

$$q = z + \frac{mg}{k} - l_0, \tag{3.96}$$

can be used to rewrite the equation of motion as

$$m\ddot{q} + c\dot{q} + kq = ky + c\dot{y} = kY_0 \sin(\omega t) + c\omega Y_0 \cos(\omega t). \tag{3.97}$$

Use the complex exponential form $y(t) = Y_0 e^{i\omega t}$ and express the equation of motion in standard form:

$$\ddot{q} + 2\zeta\omega_n\dot{q} + \omega_n^2 q = (\omega_n^2 + i2\zeta\omega_n\omega)Y_0 e^{i\omega t}, \tag{3.98}$$

where the natural frequency and damping ratio are

$$\omega_n = \sqrt{\frac{k}{m}}, \quad \zeta = \frac{c}{2m\omega_n}, \tag{3.99}$$

respectively. The road roughness is represented in Eq. (3.98) by a harmonic excitation with complex, frequency-dependent amplitude. The imaginary part of this expression represents the physical excitation.

3.5.4.3 Steady-State Solution

The steady-state solution has the form $q(t) = Qe^{i\omega t}$. Substitute this solution into Eq. (3.98) to compute the transmissibility of motion:

$$G(\omega) = \frac{Q}{Y_0} = \frac{1 + i2\zeta\left(\dfrac{\omega}{\omega_n}\right)}{1 - \left(\dfrac{\omega}{\omega_n}\right)^2 + i2\zeta\left(\dfrac{\omega}{\omega_n}\right)}. \tag{3.100}$$

The polar form of the transmissibility is $G = |G|e^{-i\phi}$, where

$$|G| = \sqrt{\frac{1 + 4\zeta^2 \left(\frac{\omega}{\omega_n}\right)^2}{\left[1 - \left(\frac{\omega}{\omega_n}\right)^2\right]^2 + 4\zeta^2 \left(\frac{\omega}{\omega_n}\right)^2}} \tag{3.101}$$

and

$$\phi = \arctan \frac{2\zeta \left(\frac{\omega}{\omega_n}\right)^3}{1 - \left(\frac{\omega}{\omega_n}\right)^2 + 4\zeta^2 \left(\frac{\omega}{\omega_n}\right)^2}. \tag{3.102}$$

The solution becomes $q(t) = G(\omega)Y_0 e^{i\omega t} = |G|Y_0 e^{i(\omega t - \phi)}$. The dependence of these two parameters on ω is shown in the frequency response curve from Fig. 3.26. This plot indicates that $|G| \approx 1$ when $\omega << \omega_n$. Consequently, at low frequencies, the passenger experiences vibrations with the same amplitude as the roughness of the road. The amplitude of vibration increases as the frequency increases, and the system experiences violent vibration (resonance) at $\omega = \omega_n$ when damping is small.

It can be checked that $|G| = 1$ when $\omega = \sqrt{2}\omega_n$, a result that is independent of ζ. The suspension amplifies the vibration when $\omega < \sqrt{2}\omega_n$. If the objective is passenger comfort, then the suspension should be designed to operate withing the range $\omega > \sqrt{2}\omega_n$. The amount of reduction also depends on ζ. Systems with lower ζ are more effective at isolating the passenger from the excitation.

3.5.4.4 Design Considerations

The frequency of excitation depends on the motorcycle speed through Eq. (3.92). This expression can be used to determine the speed when the suspension system makes the trip most uncomfortable for the passenger. To do this, assume that the motorcycle has mass of 150 kg and the passenger, 80 kg. Use the following parameter values:

$$k = 6000\,\text{N/m}, \quad c = 1500\,\text{Ns/m}, \quad L = 30\,\text{m}. \tag{3.103}$$

The MATLAB code `Chapt3_code4` from the *Electronic Supplementary Material* computes the amplitude of vibration for $v \in [0, 200]$ km/h and produces the plot shown in Fig. 3.27. The trip is most uncomfortable when $v = 71$ km/h. In fact, this suspension is poorly designed because the amplitude of vibration is amplified when the motorcycle travels with $v \leq 124$ km/h. This happens because $\omega < \sqrt{2}\omega_n$ in this interval.

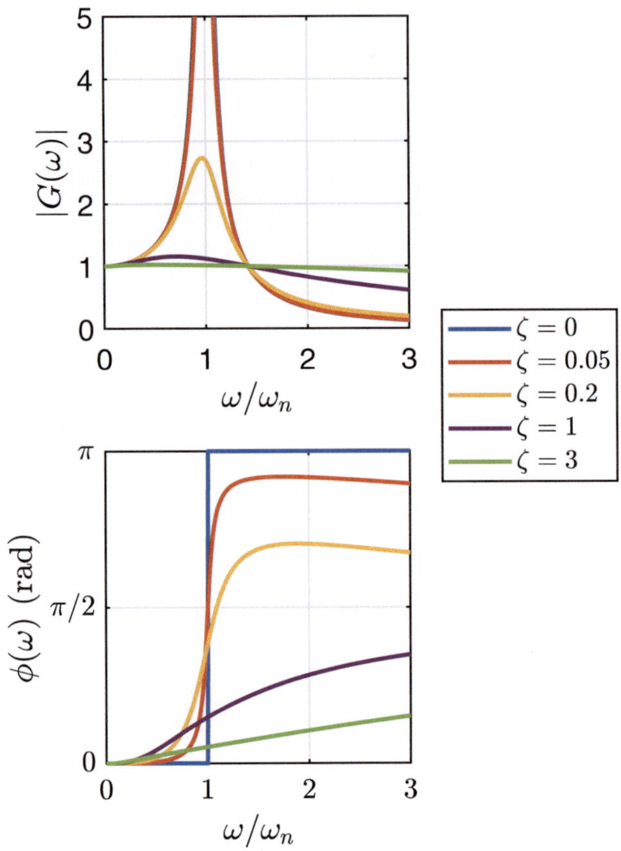

Fig. 3.26 Frequency response curve of the transmissibility from Eq. (3.100)

If one desires to drive comfortably at $v = 71$ km/h, this system needs to be redesigned. It would be necessary to reduce ω_n so that the excitation falls within the interval $\omega > \sqrt{2}\omega_n$. Usually, engineering efforts aim to avoid increasing the mass of a vehicle. Consequently, the remaining option is to use a more flexible spring. Once $\omega > \sqrt{2}\omega_n$, the amplitude of vibration can be reduced further by using a damper with lower ζ (see Fig. 3.26). The result of such a redesign is shown in Fig. 3.28. The amplitude of vibration is reduced when the motorcycle travels with $v > 62$ km/h, and it is approximately 80% of Y_0 when $v = 71$ km/h.

This example shows that there are conflicting factors engineers must consider when designing an effective suspension. One of the primary considerations is the stiffness of the spring. While a more flexible spring can improve ride comfort at greater speeds, it can also negatively impact the handling of the vehicle. A spring that is too flexible may result in excessive body roll, reducing the vehicle's stability and making it more difficult to control. Another critical factor is the damping ratio.

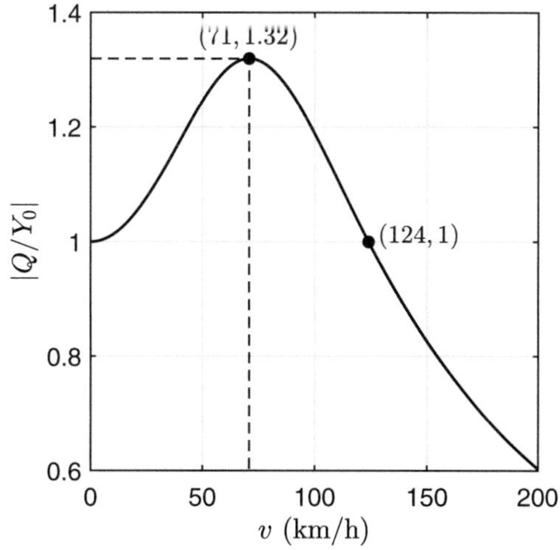

Fig. 3.27 Transmissibility of vibration for different speeds

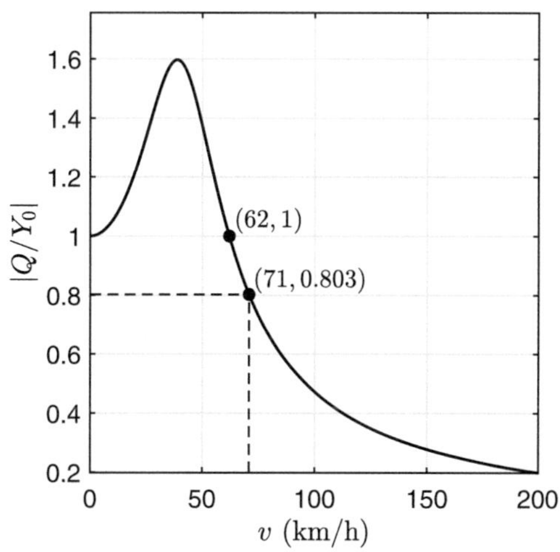

Fig. 3.28 Redesigned suspension system with $k = 1500$ N/m and $c = 500$ Ns/m

If ζ is too low, the suspension system will not effectively dampen oscillations, causing transients to take too long to decay. This prolonged oscillation can lead to an uncomfortable ride and can be particularly problematic when driving over a series of bumps or rough surfaces. Additionally, a low damping ratio can result in a resonant peak at the natural frequency of the system. This can be especially troublesome if the vehicle encounters roads where ω is lower than expected and ends up matching ω_n.

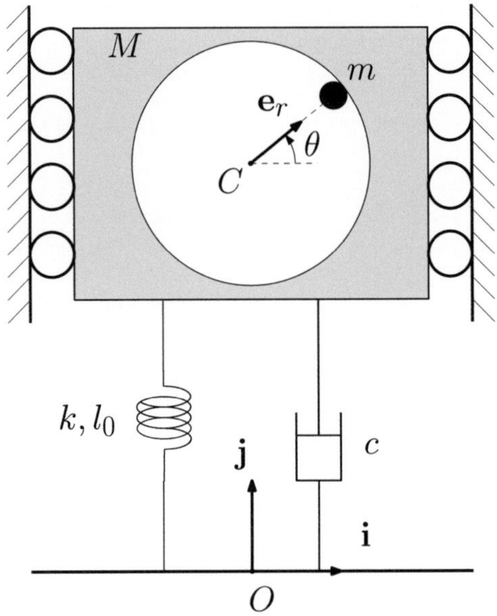

Fig. 3.29 Model for a washing machine. The machine has mass M and spins with constant angular velocity ω. The laundry has mass m and is clumped on the drum. The machine support has equivalent stiffness k and viscous damping coefficient c. The gravitational force acts along the $-\mathbf{j}$ direction

3.5.5 *Vibrations of a Washing Machine*

The objective of this example is to analyze the vibrations of a washing machine. As the drum of a washing machine approaches its peak spin speed, any imbalance caused by clumped heavy items can lead to harmonic forces that shake the machine. Systems of this type fall into the category known as *rotating unbalance* problems.

Consider the model of a washing machine shown in Fig. 3.29. The machine is represented as a rigid body with mass M and center of mass at C, constrained to vertical translation. The laundry, with mass m, is modeled as a particle clumped on the internal surface of the drum, which spins with constant angular speed ω. The machine suspension is represented by an equivalent spring with stiffness k and unstretched length l_0 and a viscous damper with coefficient c. The distance between the *unbalanced mass* of the laundry and the center C is e, also termed the *eccentricity*.

3.5.5.1 Kinematic Analysis

The position, velocity, and acceleration vectors of C can be represented as

$$\mathbf{x}_C = y\mathbf{j}, \quad \mathbf{v}_C = \dot{y}\mathbf{j}, \quad \mathbf{a}_C = \ddot{y}\mathbf{j}, \tag{3.104}$$

respectively. The position, velocity, and acceleration vectors of the unbalanced mass can be parameterized using the cylindrical-polar system according with

$$\mathbf{r} = \mathbf{x}_C + e\mathbf{e}_r, \quad \mathbf{v} = \mathbf{v}_C + e\omega\mathbf{e}_\theta, \quad \mathbf{a} = \mathbf{a}_C - e\omega^2\mathbf{e}_r, \tag{3.105}$$

where the eccentricity e and $\dot{\theta} = \omega$ are constant.

3.5.5.2 Equation of Motion

The FBDs displayed in Fig. 3.30 illustrate the forces acting on each component of the system "washing machine+laundry." A reaction force $N\mathbf{i}$ maintains the vertical motion constraint on the machine. The elastic and damping forces from the suspension are $\mathbf{F}_e = -k(y - l_0)\mathbf{j}$ and $\mathbf{F}_d = -c\dot{y}\mathbf{j}$, respectively. Weight forces are also present, acting on each component. Lastly, the drum exerts a normal force $-N_r\mathbf{e}_r$ and static friction force $F_\theta\mathbf{e}_\theta$ on the clumped laundry. Their reaction is shown on the washing machine.

Apply Euler's first law for the washing machine to obtain

$$N\mathbf{i} - Mg\mathbf{j} - k(y - l_0)\mathbf{j} - c\dot{y}\mathbf{j} + N_r\mathbf{e}_r - F_\theta\mathbf{e}_\theta = M\ddot{y}\mathbf{j}. \tag{3.106}$$

Newton's second law reveals that the particle satisfies

$$-mg\mathbf{j} - N_r\mathbf{e}_r + F_\theta\mathbf{e}_\theta = m\ddot{y}\mathbf{j} - me\omega^2\mathbf{e}_r. \tag{3.107}$$

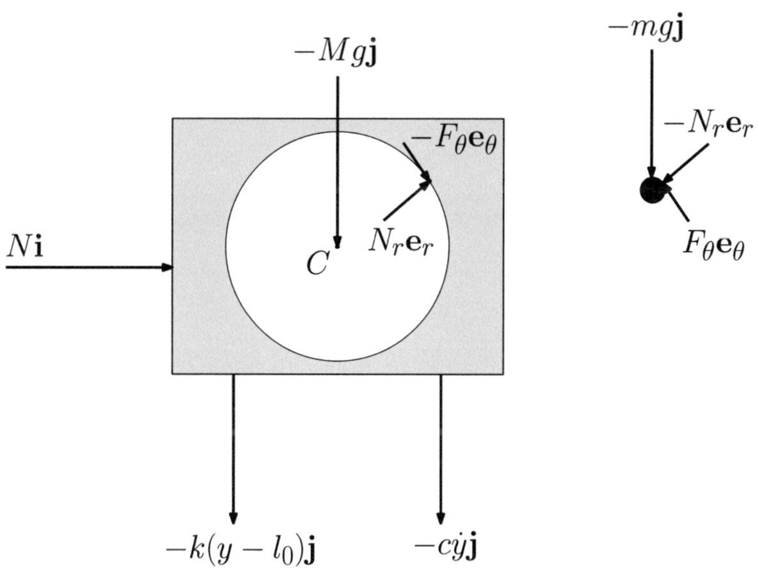

Fig. 3.30 FBDs of the washing machine and clumped laundry

Add Eqs. (3.106) and (3.107) to eliminate the internal forces:

$$N\mathbf{i} - (M+m)g\mathbf{j} - k(y-l_0)\mathbf{j} - c\dot{y}\mathbf{j} = (M+m)\ddot{y}\mathbf{j} - me\omega^2\mathbf{e}_r. \tag{3.108}$$

This vector equation contains two scalar equations that can be solved for the unknowns N and θ. Taking the projection in the $\mathbf{j}$ direction yields

$$(M+m)\ddot{y} + c\dot{y} + k(y-l_0) = -(M+m)g + me\omega^2 \sin\omega t. \tag{3.109}$$

Use a new generalized coordinate defined by

$$q = y - l_0 + \frac{(M+m)g}{k} \tag{3.110}$$

to rewrite the equation of motion in the form

$$(M+m)\ddot{q} + c\dot{q} + kq = me\omega^2 \sin\omega t, \tag{3.111}$$

or simply

$$\ddot{q} + 2\zeta\omega_n\dot{q} + \omega_n^2 q = \frac{me\omega^2}{M+m}e^{i\omega t}, \tag{3.112}$$

with the understanding that the physical response is the imaginary part of the solution. Note that the natural frequency and damping ratio are

$$\omega_n = \sqrt{k/(M+m)}, \quad \zeta = c/\left[2(M+m)\omega_n\right], \tag{3.113}$$

respectively.

3.5.5.3 Analysis

The steady-state response is

$$q(t) = Qe^{i\omega t}. \tag{3.114}$$

Substitute this into Eq. (3.112) and solve for the amplitude Q:

$$Q(\omega) = \frac{me\omega^2/(M+m)}{\omega_n^2 - \omega^2 + i2\zeta\omega_n\omega}. \tag{3.115}$$

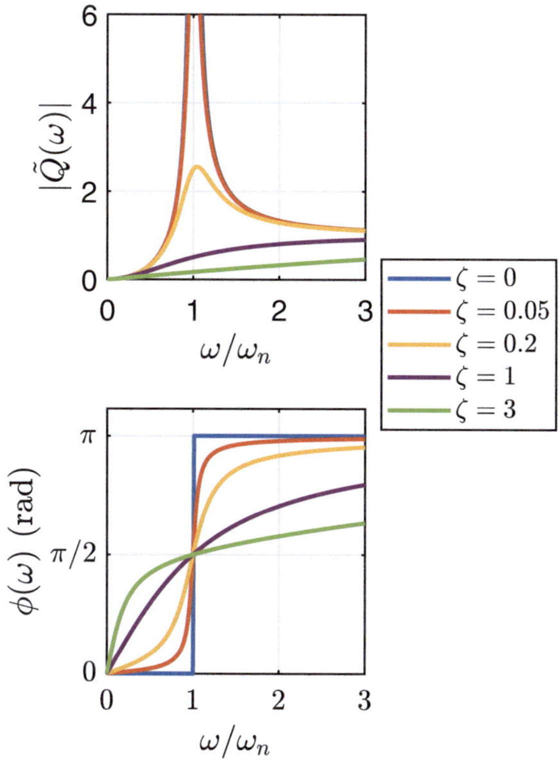

Fig. 3.31 Frequency response curve of the normalized amplitude $\tilde{Q}$

Figure 3.31 shows the frequency response curve of the normalized amplitude $\tilde{Q}$, which is defined by

$$\tilde{Q}(\omega) = \frac{(M+m)Q(\omega)}{me} = \frac{\left(\frac{\omega}{\omega_n}\right)^2}{1 - \left(\frac{\omega}{\omega_n}\right)^2 + i2\zeta\left(\frac{\omega}{\omega_n}\right)} = \left|\tilde{Q}\right| e^{-i\phi}. \tag{3.116}$$

This plot illustrates that the vibration amplitude remains small when the washing machine operates at lower speeds. The amplitude progressively grows as ω increases, peaking at the resonant frequency

$$\omega_r = \frac{\omega_n}{\sqrt{1-2\zeta^2}}. \tag{3.117}$$

This peak occurs for $\zeta < \sqrt{2}/2$. The term $\sqrt{1-2\zeta^2} < 1$ within this range, indicating that $\omega_r > \omega_n$ in a system with rotating unbalance. This stands in contrast to the behavior depicted in the other frequency response curves showcased in Figs. 3.10 and 3.26. When damping is small, the resonant frequency is $\omega_r \approx \omega_n$. This is called the *critical speed* of a rotating machine.

At high frequencies, or supercritical speeds, rotating machines exhibit a fascinating phenomenon known as *autocentering*, where the system's center of mass remains at the equilibrium position during rotation. This characteristic is important for the smooth operation of rotating machinery because vibrations are minimized.

3.6 Summary

The vibration of SDOF system excited by a harmonic force is governed by the equation of motion

$$m\ddot{q} + c\dot{q} + kq = F_0 e^{i\omega t},$$

where F_0 is the amplitude of the force, and ω is its frequency. The solution of this equation is

$$q(t) = q_h(t) + q_p(t),$$

where the homogeneous part q_h assumes one of the forms shown in Eq. (3.3), and the particular solution may be computed by the method of undetermined coefficients:

$$q_p(t) = Qe^{i\omega t}.$$

The complete solution $q(t)$ is used to satisfy the the ICs $q(0) = q_0$ and $\dot{q}(0) = \dot{q}_0$. In the presence of damping, q_h represents a transient on the solution, and q_p is the steady-state response.

The FRF is defined as the ratio between vibration output and force input according to

$$H(\omega) = \frac{Q}{F_0}.$$

This parameter is commonly expressed in polar form

$$H(\omega) = |H(\omega)|e^{-i\phi(\omega)},$$

where the absolute value is

$$|H(\omega)| = \frac{1/k}{\sqrt{\left[1-\left(\frac{\omega}{\omega_n}\right)^2\right]^2+\left[2\zeta\left(\frac{\omega}{\omega_n}\right)\right]^2}}$$

and the phase angle is given by

$$\phi(\omega) = \arctan\frac{2\zeta\left(\frac{\omega}{\omega_n}\right)}{1-\left(\frac{\omega}{\omega_n}\right)^2}.$$

The steady-state response

$$q_p(t) = |H(\omega)|F_0 e^{i(\omega t-\phi(\omega))}.$$

is magnified by $|H(\omega)|$ and delayed by $\phi(\omega)$. The effect of ω on the response can be readily inferred from the frequency response curve of $H(\omega)$ shown in Fig. 3.10. In particular, the peak of vibration amplitude occurs at the resonant frequency

$$\omega_r = \omega_n\sqrt{1-2\zeta^2}.$$

for $\zeta < 0.707$. For lightly damped systems, resonance occurs when $\omega_r = \omega_n$.

A system can also be excited by imposed harmonic base motion or by rotating unbalance. However, the mathematical apparatus discussed for the case of applied harmonic force can be seamlessly extended to treat these other cases.

3.7 Problems

3.1 When is the sum of two harmonic functions periodic?

3.2 Consider the case when an SDOF system is excited by a harmonic force. Obtain approximations for ω_r and $|H(\omega_r)|$ when $0 < \zeta << 1$.

3.3 Consider the mass-spring-damper system shown in Fig. 3.1.

(a) Show that the amplitude of the force transmitted to the fixed support, denoted as $|F_T|$, satisfies

$$\frac{|F_T|}{F_0} = \frac{\sqrt{1 + \left[2\zeta\left(\frac{\omega}{\omega_n}\right)\right]^2}}{\sqrt{\left[1 - \left(\frac{\omega}{\omega_n}\right)^2\right]^2 + \left[2\zeta\left(\frac{\omega}{\omega_n}\right)\right]^2}}. \tag{3.118}$$

(b) Evaluate $|F_T|/F_0$ at $\omega/\omega_n = \sqrt{2}$ and interpret the result.
(c) Determine the frequency at which $|F_T|$ is maximum.

3.4 Suppose the mass-spring-damper system shown in Fig. 3.1 is excited by the harmonic force $f(t) = F_0\cos(\omega t)$. Find the driving frequency ω at which the amplitude of velocity attains its maximum and determine the maximum amplitude of velocity.

3.5 Suppose the mass-spring-damper system shown in Fig. 3.1 is excited by the harmonic force $f(t) = F_0\cos(\omega t)$. Find the driving frequency ω at which the amplitude of acceleration attains its maximum and determine the maximum amplitude of acceleration.

3.6 Recall the suspension system shown in Fig. 3.24 and the associated transmissibility from Eq. (3.101). Show that the peak of $|G|$ occurs at the frequency

$$\omega_r = \omega_n \frac{\sqrt{-1 + \sqrt{1 + 8\zeta^2}}}{2\zeta}. \tag{3.119}$$

3.7 Derive the result from Eq. (3.117) and establish an expression for $\left|\tilde{Q}\right|_{max}$.

3.8 Show that the center of mass of the system "washing machine+laundry" from Fig. 3.29 lies at the static equilibrium position and does not move at high frequencies.

3.9 The rod with mass m and length L shown in Fig. 3.32 is being excited by an external moment $M(t)\mathbf{k}$. The position vectors of points A, B, a, and b are represented by

$$\mathbf{x}_A = -\frac{L}{5}\mathbf{e}_1, \tag{3.120}$$

$$\mathbf{x}_B = \frac{4L}{5}\mathbf{e}_1, \tag{3.121}$$

$$\mathbf{r}_a = -\frac{L}{5}\mathbf{i} - l_A\mathbf{j}, \tag{3.122}$$

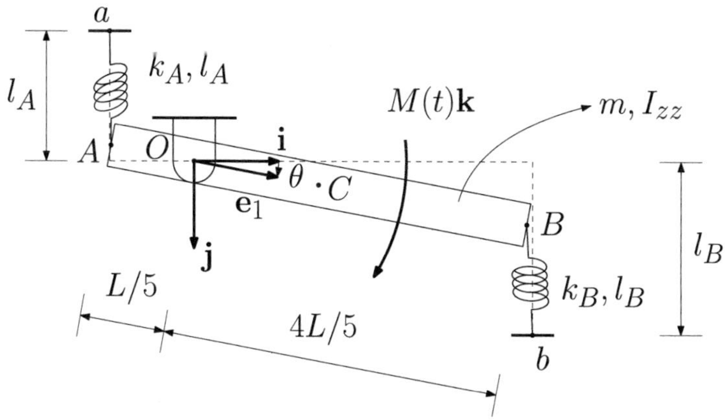

Fig. 3.32 Bar with mass m and length L that is free to rotate around the fixed axis passing through O. The center of mass of the bar is located at C. The gravitational force acts along the $\mathbf{j}$ direction

$$\mathbf{r}_b = \frac{4L}{5}\mathbf{i} + l_B\mathbf{j}, \tag{3.123}$$

respectively. Assume that θ varies slightly from $\theta = 0$ and that the applied moment is harmonic with amplitude M_0 and frequency ω. If $k_1 = 5 \times 10^4$ N/m, $k_2 = 5 \times 10^3$ N/m, $m = 20$ kg, $L = 1$ m, and $\omega = 20$ rad/s, determine the values of M_0 for which the steady-state amplitude will be less than 1°.

3.10 Consider the mechanical system shown in Fig. 3.33. The block with mass m_1 slides on a smooth surface when excited by the force $f(t)$. This block is connected, via an ideal cable, to a drum with a moment of inertia I_{zz}^O relative to the axis $\mathbf{k}$ passing through the fixed point O. Another block, with mass m_2, is also connected to this drum. Using the reference system shown, the positions of the blocks can be represented by

$$\mathbf{r}_1 = -r_1\mathbf{j} + x(t)\mathbf{i}, \tag{3.124}$$

$$\mathbf{r}_2 = r_2\mathbf{i} + y(t)\mathbf{j}, \tag{3.125}$$

where $x < 0$. Suppose that

$$\begin{aligned} &k = 3 \times 10^6, \text{N/m}, \quad m_1 = 35, \text{kg}, \quad m_2 = 25, \text{kg}, \quad I_{zz}^O = 2, \text{kg.m}^2 \\ &c = 3000, \text{Ns/m}, \quad r_1 = 11, \text{cm}, \quad r_2 = 25, \text{cm}, \end{aligned} \tag{3.126}$$

and that the applied force is harmonic with frequency $\omega = 100$ rad/s and amplitude $F_0 = 1000$ N. Determine the steady-state amplitude.

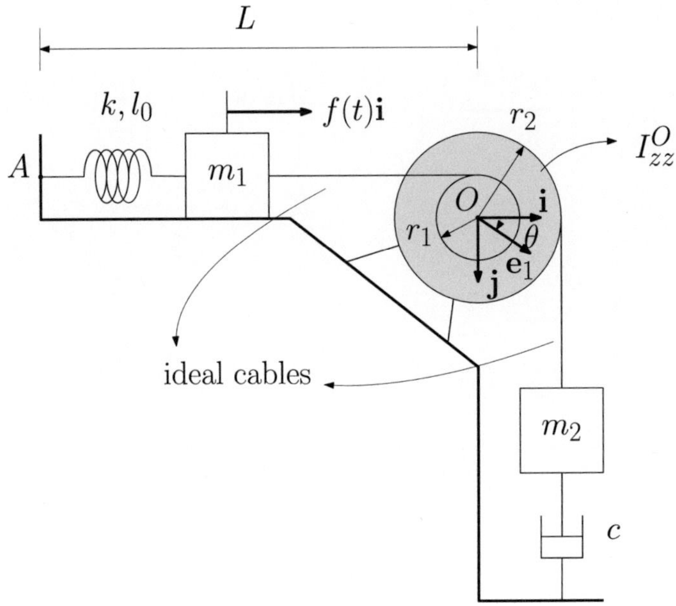

Fig. 3.33 Mechanical system formed by two blocks and a drum. Gravitational force acts along the **j** direction

Part II
Vibrations of Systems with Multiple Degrees of Freedom

Chapter 4
Equations of Motion of Systems with Multiple Degrees of Freedom

The objective of this chapter is to introduce the form of the equations of motion for systems with multiple degrees of freedom. This chapter first addresses these equations through two classical examples from mechanics. Next, the general form of the equations is derived in a broader context using the generalized Taylor series, highlighting the limitations and applicability of linear vibration theory. The chapter concludes with three solved examples that effectively illustrate the application of the main concepts developed herein. MATLAB codes are provided throughout to aid in visualizing the physics of the problems studied.

4.1 Systems with Multiple Degrees of Freedom

Part I of these notes focused on SDOF systems. These models simplify the dynamics to a single variable and can provide satisfactory results for pendulums or mass-spring-damper systems. However, they have limitations when it comes to accurately reproducing the behavior of more complex mechanical systems, particularly when interactions between multiple components or different types of motion are involved. In such cases, SDOF models are typically used during the preliminary design stages to provide a quick, albeit rough, assessment of system behavior.

Consider, for example, the motorcycle suspension system discussed in Sect. 3.5.4. That model only accounted for the vehicle's vertical vibrations and neglected pitch (tilting forward or backward). Pitch is noticeable when a motorcycle hits a bump, causing the front and rear suspensions to compress and extend differently. A model with more than one DOF would be necessary to account for this additional *mode of vibration*.

Supplementary Information The online version contains supplementary material available at (https://doi.org/10.1007/978-3-031-83995-5_4).

R. Gonçalves Salsa Junior, *Fundamentals of the Theory of Mechanical Vibrations*, https://doi.org/10.1007/978-3-031-83995-5_4

As another example, consider a bridge subjected to dynamic loads. A model with only one DOF would overlook interactions between different forms of vibration, such as bending, torsion, and longitudinal waves propagating through the structure. Such a simplified model would fail to predict catastrophic events like the collapse of the Tacoma Narrows Bridge, where strong winds induced high-frequency vertical forces that destabilized the torsional mode of vibration, ultimately leading to the bridge's failure.

Multi-degree-of-freedom (MDOF) models offer greater accuracy, but this comes at the cost of increased mathematical complexity. Unlike in the SDOF case, a system with n DOFs requires n generalized coordinates to uniquely define its configuration at any given time. These coordinates are governed by n interdependent ODEs, which can be derived using the Newton–Euler method. This involves constructing FBDs for each system component and accounting for interaction forces and moments. As the number of interconnected bodies increases, this process becomes increasingly laborious.

The objective of this chapter is to introduce the general form of the equations of motion for MDOF systems. A secondary objective is to review how to use the laws of motion to derive these equations while highlighting the similarities and differences compared to the SDOF case. The presentation is organized similarly to Chap. 1. Section 4.2 begins with an applied example involving a system with two particles, and a more complex system is studied in Sect. 4.3. The general form of the equations of motion is introduced in Sect. 4.4, followed by three additional examples provided in Sect. 4.5.

4.2 A System of Two Interconnected Particles

The simplest system with more than one DOF is shown in Fig. 4.1. It comprises two particles that are free to translate vertically. External forces $f_1(t)$ and $f_2(t)$ act on particles with masses m_1 and m_2, respectively. Each particle is attached to a fixed support by a damper-spring set. Additionally, the particles are interconnected by a viscous damper with coefficient c_2 and a linear spring with stiffness k_2 and unstretched length l_2. This connection *couples* the vibrations of the particles, causing the motion of one to influence the other. The purpose of this example is to illustrate how the same four steps from Sect. 1.2.2 can be employed to derive the equations of motion for this system.

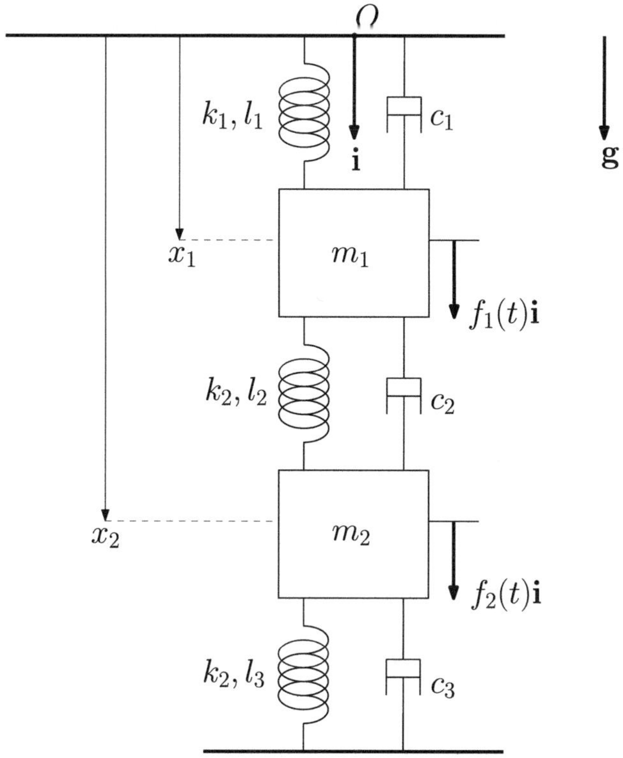

Fig. 4.1 A system with two particles free to translate vertically. The gravitational force acts along the **i** direction. The springs are unstretched when $x_1 = l_1$ and $x_2 = l_1 + l_2$

4.2.1 Kinematic Analysis

Using the Cartesian reference system depicted in Fig. 4.1, the position vectors of the particles with masses m_1 and m_2 are given by

$$\mathbf{r}_1 = x_1\mathbf{i} + y_1\mathbf{j} + z_1\mathbf{k}, \quad \mathbf{r}_2 = x_2\mathbf{i} + y_2\mathbf{j} + z_2\mathbf{k}, \tag{4.1}$$

respectively. These representations indicate that six coordinates are required to specify the positions of two particles moving without kinematic constraints. However, the assumption of rectilinear motion along the vertical direction introduces the following holonomic constraints:

$$\psi_1 = \mathbf{r}_1 \cdot \mathbf{j} = 0, \quad \psi_2 = \mathbf{r}_1 \cdot \mathbf{k} = 0, \quad \psi_3 = \mathbf{r}_2 \cdot \mathbf{j} = 0, \quad \psi_4 = \mathbf{r}_2 \cdot \mathbf{k} = 0. \tag{4.2}$$

It follows from Eq. (1.4) that the system has $n = 6 - 4 = 2$ DOFs. The Cartesian coordinates x_1 and x_2, as shown in Fig. 4.1, fully describe the system's dynamics.

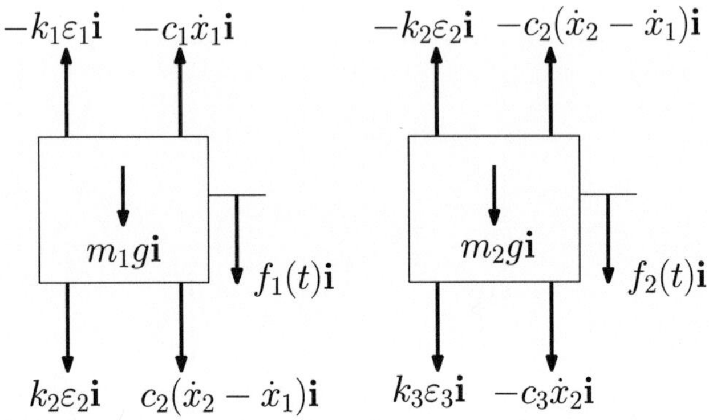

Fig. 4.2 FBDs of the particles

The position, velocity, and acceleration vectors for each particle can be represented by

$$\mathbf{r}_1 = x_1\mathbf{i}, \quad \mathbf{v}_1 = \dot{x}_1\mathbf{i}, \quad \mathbf{a}_1 = \ddot{x}_1\mathbf{i}, \tag{4.3}$$

$$\mathbf{r}_2 = x_2\mathbf{i}, \quad \mathbf{v}_2 = \dot{x}_2\mathbf{i}, \quad \mathbf{a}_2 = \ddot{x}_2\mathbf{i}. \tag{4.4}$$

4.2.2 *The Differential Equations of Motion*

The FBDs of each particle are shown in Fig. 4.2. The change of length of the topmost spring is

$$\epsilon_1 = x_1 - l_1. \tag{4.5}$$

The change in length of the spring connecting the particles, which depends on their relative displacement, is

$$\epsilon_2 = x_2 - x_1 - l_2. \tag{4.6}$$

Assuming that the spring in the bottom is unstretched when $x_2 = l_1 + l_2$, its change of length is

$$\epsilon_3 = l_1 + l_2 - x_2. \tag{4.7}$$

This spring is compressed ($\epsilon_3 < 0$) if $x_2 > l_1 + l_2$ or stretched ($\epsilon_3 > 0$) if $x_2 < l_1 + l_2$. All springs are unstretched when $x_1 = l_1$ and $x_2 = l_1 + l_2$.

Apply Newton's second law to each particle to obtain the following governing differential equations

$$m_1\ddot{x}_1 = f_1(t) + m_1 g - k_1\epsilon_1 + k_2\epsilon_2 - c_1\dot{x}_1 + c_2(\dot{x}_2 - \dot{x}_1), \tag{4.8}$$

$$m_2\ddot{x}_2 = f_2(t) + m_2 g - k_2\epsilon_2 + k_3\epsilon_3 - c_2(\dot{x}_2 - \dot{x}_1) - c_3\dot{x}_3. \tag{4.9}$$

The motion of the system is described by a system of two second-order linear ODEs, corresponding to its two DOFs. The intermediate spring-damper set *couples* these equations, meaning that the motion of one particle influences the motion of the other. Mathematically, this results in x_1, x_2, and their first derivatives appearing in both equations. Consequently, these differential equations cannot be solved individually. They must be solved simultaneously using the ICs $x_1(0)$, $x_2(0)$, $\dot{x}_1(0)$, and $\dot{x}_2(0)$.

4.2.3 Equilibrium Positions

The equilibrium positions of the system are constant solutions given by

$$x_1 = x_1^e, \quad x_2 = x_2^e. \tag{4.10}$$

Ignoring the external forces f_1 and f_2, substitute these constant solutions into Eqs. (4.8) and (4.9) to obtain

$$x_1^e = l_1 + \frac{[(k_2 + k_3)m_1 + k_2 m_2]g}{k_1 k_2 + k_1 k_3 + k_2 k_3}, \tag{4.11}$$

$$x_2^e = l_1 + l_2 + \frac{[k_2 m_1 + (k_1 + k_2)m_2]g}{k_1 k_2 + k_1 k_3 + k_2 k_3}. \tag{4.12}$$

Define new generalized coordinates q_1 and q_2 as

$$q_1 = x_1 - x_1^e, \quad q_2 = x_2 - x_2^e. \tag{4.13}$$

These new coordinates measure the positions of the particles from their static equilibrium positions. They can be used to rewrite Eqs. (4.8) and (4.9) according with

$$m_1\ddot{q}_1 + (c_1 + c_2)\dot{q}_1 - c_2\dot{q}_2 + (k_1 + k_2)q_1 - k_2 q_2 = f_1(t), \tag{4.14}$$

$$m_2\ddot{q}_2 - c_2\dot{q}_1 + (c_2 + c_3)\dot{q}_2 - k_2 q_1 + (k_2 + k_3)q_2 = f_2(t). \tag{4.15}$$

These equations can be used to describe the system vibrations when disturbed from equilibrium.

4.2.4 Arranging the Equations of Motion in the Matrix-Vector Form

The equations of motion (4.14) and (4.15) are usually arranged as a single matrix-vector equation. Define the vector of generalized coordinates as

$$\mathbf{q} = \begin{bmatrix} q_1 \\ q_2 \end{bmatrix}. \tag{4.16}$$

The aforementioned system of ODEs can be expressed as

$$\mathbf{M}\ddot{\mathbf{q}} + \mathbf{C}\dot{\mathbf{q}} + \mathbf{K}\mathbf{q} = \mathbf{f}(t), \tag{4.17}$$

where

$$\mathbf{f}(t) = \begin{bmatrix} f_1(t) \\ f_2(t) \end{bmatrix} \tag{4.18}$$

is the vector of external forces and the matrix coefficients are

$$\mathbf{M} = \begin{bmatrix} m_1 & 0 \\ 0 & m_2 \end{bmatrix}, \tag{4.19}$$

$$\mathbf{C} = \begin{bmatrix} c_1 + c_2 & -c_2 \\ -c_2 & c_2 + c_3 \end{bmatrix}, \tag{4.20}$$

$$\mathbf{K} = \begin{bmatrix} k_1 + k_2 & -k_2 \\ -k_2 & k_2 + k_3 \end{bmatrix}. \tag{4.21}$$

These matrices are referred to as the mass, damping, and stiffness matrices, respectively. Notably, they are symmetric:[1]

$$\mathbf{M} = \mathbf{M}^T, \quad \mathbf{C} = \mathbf{C}^T, \quad \mathbf{K} = \mathbf{K}^T. \tag{4.22}$$

The matrix-vector form (4.17) offers a compact and concise representation of systems of differential equations. It simplifies their presentation and enables effi-

[1] See Appendix C for a review of basic matrix properties and algebra.

cient application of operations, including differentiation, integration, transposition, and inversion. Its greatest strength, however, lies in its ability to reveal the vibratory characteristics of MDOF systems through inspection of the coefficient matrices. For example, the off-diagonal entries indicate whether the equations are coupled. Note how the differential equations (4.14) and (4.15) would be independent if all the coefficient matrices were diagonal. System stability can also be assessed based on properties of the coefficient matrices, such as symmetry and skew symmetry. Additionally, the matrix-vector equation integrates seamlessly with programming languages and numerical libraries used in computational simulations.

4.2.5 *Energy of the System*

The total mechanical energy of the system, which is the sum of the kinetic and potential energies of its components, is given by

$$E = \frac{1}{2}m_1\dot{x}_1^2 + \frac{1}{2}m_2\dot{x}_2^2 + \frac{1}{2}k_1\epsilon_1^2 + \frac{1}{2}k_2\epsilon_2^2 + \frac{1}{2}k_3\epsilon_3^2 - m_1gx_1 - m_2gx_2. \tag{4.23}$$

Use the change of coordinates $x_1 = q_1 + x_1^e$ and $x_2 = q_2 + x_2^e$ to express E equivalently as

$$E = E_q + U_e, \tag{4.24}$$

where U_e is the potential energy at the equilibrium position and

$$E_q = \frac{1}{2}m_1\dot{q}_1^2 + \frac{1}{2}m_2\dot{q}_2^2 + \frac{1}{2}k_1q_1^2 + \frac{1}{2}k_2(q_2 - q_1)^2 + \frac{1}{2}k_3q_2^2 \tag{4.25}$$

accounts for the fluctuation in energy. This expression can be rewritten in matrix-vector form as

$$E_q = \frac{1}{2}\dot{\mathbf{q}}^T\mathbf{M}\dot{\mathbf{q}} + \frac{1}{2}\mathbf{q}^T\mathbf{K}\mathbf{q}. \tag{4.26}$$

This form is preferred for its compactness and applicability to various mechanical systems.

The rate of change of energy is

$$\dot{E} = \dot{E}_q = \frac{1}{2}\ddot{\mathbf{q}}^T\mathbf{M}\dot{\mathbf{q}} + \frac{1}{2}\dot{\mathbf{q}}^T\mathbf{M}\ddot{\mathbf{q}} + \frac{1}{2}\dot{\mathbf{q}}^T\mathbf{K}\mathbf{q} + \frac{1}{2}\mathbf{q}^T\mathbf{K}\dot{\mathbf{q}}. \tag{4.27}$$

This can be simplified noting that the terms $\ddot{\mathbf{q}}^T \mathbf{M} \dot{\mathbf{q}}$ and $\mathbf{q}^T \mathbf{K} \dot{\mathbf{q}}$ are scalars and therefore unaffected by transposition:

$$\ddot{\mathbf{q}}^T \mathbf{M} \dot{\mathbf{q}} = \left(\ddot{\mathbf{q}}^T \mathbf{M} \dot{\mathbf{q}} \right)^T = \dot{\mathbf{q}}^T \mathbf{M} \ddot{\mathbf{q}}, \tag{4.28}$$

$$\mathbf{q}^T \mathbf{K} \dot{\mathbf{q}} = \left(\mathbf{q}^T \mathbf{K} \dot{\mathbf{q}} \right)^T = \dot{\mathbf{q}}^T \mathbf{K} \mathbf{q} \tag{4.29}$$

Thus, $\dot{E}$ becomes

$$\dot{E} = \dot{\mathbf{q}}^T (\mathbf{M} \ddot{\mathbf{q}} + \mathbf{K} \dot{\mathbf{q}}). \tag{4.30}$$

The equations of motion (4.17) imply that

$$\dot{E} = -\dot{\mathbf{q}}^T \mathbf{C} \dot{\mathbf{q}} + \dot{\mathbf{q}}^T \mathbf{f}(t). \tag{4.31}$$

In the case of free vibration, where $\mathbf{f} = \mathbf{0}$, the rate of change of energy is

$$\dot{E} = -\dot{\mathbf{q}}^T \mathbf{C} \dot{\mathbf{q}} = -\frac{1}{2} c_1 \dot{q}_1^2 - \frac{1}{2} c_2 (\dot{q}_2 - \dot{q}_1)^2 - \frac{1}{2} c_3 \dot{q}_2^2 \leq 0. \tag{4.32}$$

The time evolution of energy is shown in Fig. 4.3 for both damped ($\mathbf{C} \neq \mathbf{0}$) and undamped ($\mathbf{C} = \mathbf{0}$) cases. Energy is conserved for the undamped system. The corresponding responses $q_1(t)$ and $q_2(t)$ are shown in Fig. 4.4. As with SDOF systems, damping causes energy dissipation and amplitude decay. However, in the case of free vibration, the oscillations are not necessarily harmonic.

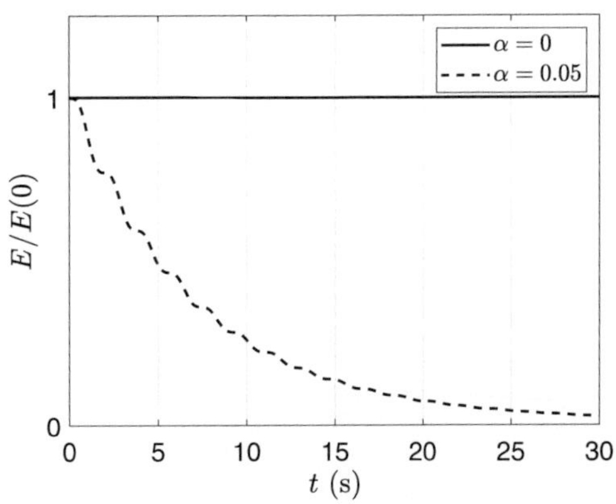

Fig. 4.3 Variation of energy for the system with two particles in free vibration ($f_1 = f_2 = 0$). Plot generated with $m_1 = m_2 = 1$ kg, $k_1 = k_2 = k_3 = 1$ N/m, $c_1 = c_2 = c_3 = \alpha$, and the ICs $[q_1(0) \quad q_2(0) \quad \dot{q}_1(0) \quad \dot{q}_2(0)]^T = [1 \quad -1 \quad 0 \quad 1]^T$

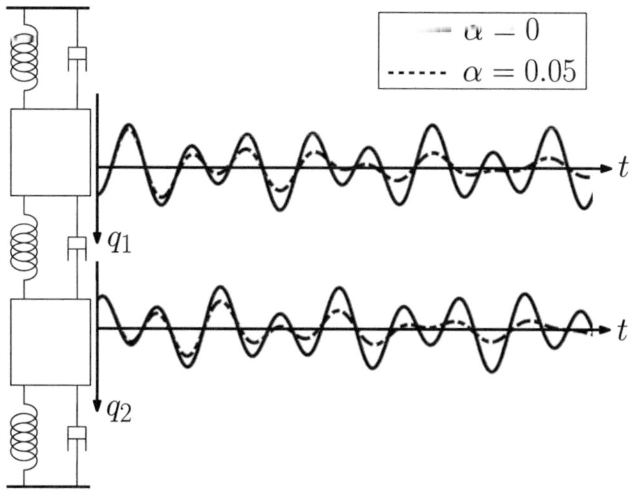

Fig. 4.4 Displacement of the particles in free vibration ($f_1 = f_2 = 0$). Plot produced with $m_1 = m_2 = 1\,\mathrm{kg}$, $k_1 = k_2 = k_3 = 1\,\mathrm{N/m}$, $c_1 = c_2 = c_3 = \alpha$, and the ICs $[q_1(0) \quad q_2(0) \quad \dot{q}_1(0) \quad \dot{q}_2(0)]^T = [1 \quad -1 \quad 0 \quad 1]^T$

4.3 The Carts-Pendulum System

The vibrations of the two-particle system presented in the previous section are governed by a system of two linear differential equations. This case is not representative of mechanics in general, as most systems are inherently nonlinear. Therefore, the aim of this section is to investigate the form of the equations of motion for systems with nonlinearities. This is accomplished through an illustrative example that highlights the differences and similarities compared to the case covered in the previous section. In particular, the equations of motion for the system shown in Fig. 4.5 are derived and analyzed. This system consists of two carts, with masses m_1 and m_2, which are free to translate along a smooth horizontal track. The nonlinearity of the system is due to a pendulum of length l and mass m_3 that is attached to the cart with mass m_2. In this model, the carts and the pendulum bob are represented as particles, while the cable is considered ideal.

4.3.1 Kinematic Analysis

The first step in deriving the equations of motion involves defining the coordinate systems used. The geometry of the problem suggests that the position vectors of the carts with masses m_1 and m_2, as well as the pendulum bob, can be represented using

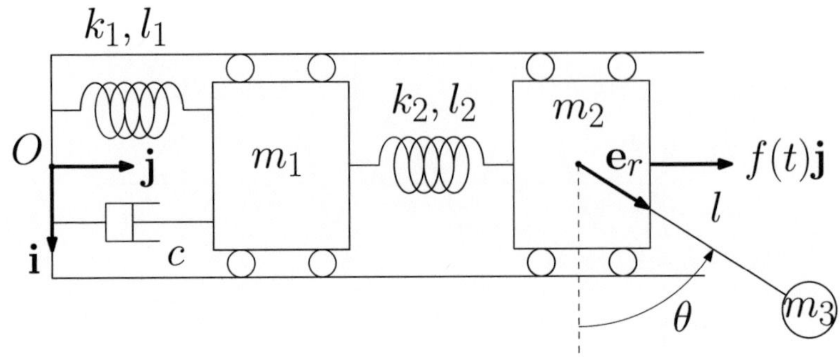

Fig. 4.5 A nonlinear system composed of three particles. The gravitational force acts along the **i** direction and an external force $f(t)$ excites one of the particles

Cartesian and cylindrical-polar coordinates, as follows:

$$\mathbf{r}_1 = x_1\mathbf{i} + y_1\mathbf{j} + z_1\mathbf{k}, \tag{4.33}$$

$$\mathbf{r}_2 = x_2\mathbf{i} + y_2\mathbf{j} + z_2\mathbf{k}, \tag{4.34}$$

$$\mathbf{r}_3 = \mathbf{r}_2 + r\mathbf{e}_r + z_3\mathbf{k}, \tag{4.35}$$

respectively. While nine coordinates are necessary to uniquely determine the configuration of the system, only three are needed because the motion of the particles must satisfy the following six holonomic constraints:

$$\psi_1 = \mathbf{r}_1 \cdot \mathbf{i} = x_1 = 0, \quad \psi_2 = \mathbf{r}_1 \cdot \mathbf{k} = z_1 = 0, \tag{4.36}$$

$$\psi_3 = \mathbf{r}_2 \cdot \mathbf{i} = x_2 = 0, \quad \psi_4 = \mathbf{r}_2 \cdot \mathbf{k} = z_2 = 0, \tag{4.37}$$

$$\psi_5 = (\mathbf{r}_3 - \mathbf{r}_2) \cdot \mathbf{e}_r - l = 0, \quad \psi_6 = (\mathbf{r}_3 - \mathbf{r}_2) \cdot \mathbf{k} = 0. \tag{4.38}$$

Therefore, the generalized coordinates y_1, y_2, and θ can be used to parameterize the motion of the particles by

$$\mathbf{r}_1 = y_1\mathbf{j}, \quad \dot{\mathbf{r}}_1 = \dot{y}_1\mathbf{j}, \quad \ddot{\mathbf{r}}_1 = \ddot{y}_1\mathbf{j}, \tag{4.39}$$

$$\mathbf{r}_2 = y_2\mathbf{j}, \quad \dot{\mathbf{r}}_2 = \dot{y}_2\mathbf{j}, \quad \ddot{\mathbf{r}}_2 = \ddot{y}_2\mathbf{j}, \tag{4.40}$$

$$\mathbf{r}_3 = \mathbf{r}_2 + l\mathbf{e}_r, \quad \dot{\mathbf{r}}_3 = \dot{\mathbf{r}}_2 + l\dot{\theta}\mathbf{e}_\theta, \quad \ddot{\mathbf{r}}_3 = \ddot{\mathbf{r}}_2 + l\ddot{\theta}\mathbf{e}_\theta - l\dot{\theta}^2\mathbf{e}_r. \tag{4.41}$$

4.3.2 Obtaining the Equations of Motion

The FBDs of all particles are illustrated in Fig. 4.6. The carts experience the reaction forces $-N_1\mathbf{i}$ and $-N_2\mathbf{i}$ from the smooth tracks. Additionally, the pendulum bob

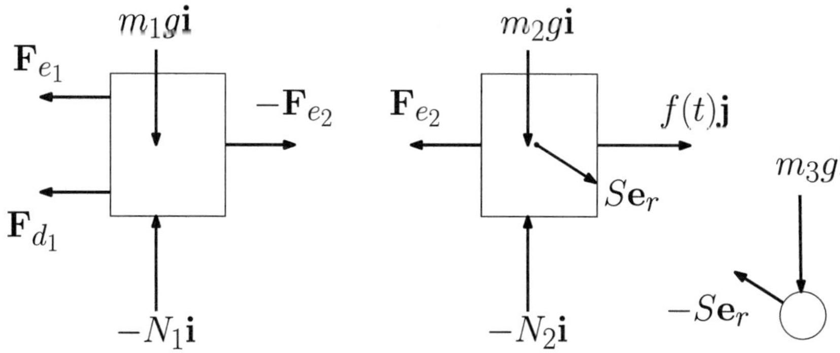

Fig. 4.6 FBDs of the carts and pendulum

and the cart with mass m_2 are acted by the the traction force S directed along the cable. The elastic forces of the springs are given by

$$\mathbf{F}_{e_1} = -k_1(y_1 - l_1)\mathbf{j}, \quad \mathbf{F}_{e_2} = -k_2(y_2 - y_1 - l_2)\mathbf{j}, \tag{4.42}$$

and the damping force on the first car is opposite to its velocity:

$$\mathbf{F}_{d_1} = -c\dot{y}_1\mathbf{j}. \tag{4.43}$$

Three differential equations are necessary to solve for the free coordinates y_1, y_2, and θ. To obtain these equations, start by applying Newton's second law to each particle:

$$m_1\ddot{y}_1 = -c\dot{y}_1 - k_1(y_1 - l_1) + k_2(y_2 - y_1 - l_2), \tag{4.44}$$

$$m_2\ddot{y}_2\mathbf{j} = f(t)\mathbf{j} + (m_2g - N_2)\mathbf{i} - k_2(y_2 - y_1 - l_2)\mathbf{j} + S\mathbf{e}_r, \tag{4.45}$$

$$m_3(\ddot{y}_2\mathbf{j} + l\ddot{\theta}\mathbf{e}_\theta - l\dot{\theta}^2\mathbf{e}_r) = m_3g\mathbf{i} - S\mathbf{e}_r. \tag{4.46}$$

The first one, Eq. (4.44), is already in the desired form. The remaining two independent differential equations have to be extracted from Eqs. (4.44)–(4.46). They are not as straightforward to visualize because Eqs. (4.45) and (4.46) contain the additional unknowns N_2 and S. To obtain the second ODE, add Eqs. (4.45) and (4.46) to obtain

$$\begin{aligned}(m_2 + m_3)\ddot{y}_2\mathbf{j} + m_3(l\ddot{\theta}\mathbf{e}_\theta - l\dot{\theta}^2\mathbf{e}_r) &= f(t)\mathbf{j}\\ &+ [(m_2 + m_3)g - N_2]\mathbf{i} - k_2(y_2 - y_1 - l_2)\mathbf{j}.\end{aligned} \tag{4.47}$$

Take the projection of this equation along the $\mathbf{j}$ direction to remove N_2:

$$(m_2 + m_3)\ddot{y}_2 + m_3l\ddot{\theta}\cos\theta - m_3l\dot{\theta}^2\sin\theta = f(t) - k_2(y_2 - y_1 - l_2). \tag{4.48}$$

To eliminate S and obtain the last equation, take the projection of Eq. (4.46) along $\mathbf{e}_\theta$:

$$m_3 \cos\theta \ddot{y}_2 + m_3 l \ddot{\theta} = -m_3 g \sin\theta. \tag{4.49}$$

Two of the equations in the system formed by (4.44), (4.48), and (4.49) are nonlinear. This system of three ODEs can be expressed in the following matrix-vector form:

$$\mathbf{M}(\theta)\begin{bmatrix} \ddot{y}_1 \\ \ddot{y}_2 \\ \ddot{\theta} \end{bmatrix} + \mathbf{g}(y_1,\, y_2,\, \theta,\, \dot{y}_1,\, \dot{\theta}) = \begin{bmatrix} 0 \\ f(t) \\ 0 \end{bmatrix}, \tag{4.50}$$

where the acceleration coefficient, $\mathbf{M}(\theta)$, is a symmetric matrix function given by

$$\mathbf{M}(\theta) = \begin{bmatrix} m_1 & 0 & 0 \\ 0 & m_2 + m_3 & m_3 l \cos\theta \\ 0 & m_3 l \cos\theta & m_3 l^2 \end{bmatrix}. \tag{4.51}$$

Note that that Eq. (4.49) has been multiplied by l to render the matrix coefficient $\mathbf{M}(\theta)$ symmetric.[2] The vector function $\mathbf{g}(y_1,\, y_2,\, \theta,\, \dot{y}_1,\, \dot{\theta})$ has entries

$$\mathbf{g}(y_1,\, y_2,\, \theta,\, \dot{y}_1,\, \dot{\theta}) = \begin{bmatrix} c\dot{y}_1 + k_1(y_1 - l_1) - k_2(y_2 - y_1 - l_2) \\ -m_3 l \dot{\theta}^2 \sin\theta + k_2(y_2 - y_1 - l_2) \\ m_3 g l \sin\theta \end{bmatrix}. \tag{4.52}$$

4.3.3 Equilibrium Positions

To understand how the system vibrates when disturbed from the static equilibrium position, ignore $f(t)$ and look for constant solutions

$$y_1 = y_1^e, \quad y_2 = y_2^e, \quad \theta = \theta_e. \tag{4.53}$$

[2] This adjustment would not be necessary if the equations were derived using the Lagrangian method. This topic is beyond the scope of these notes.

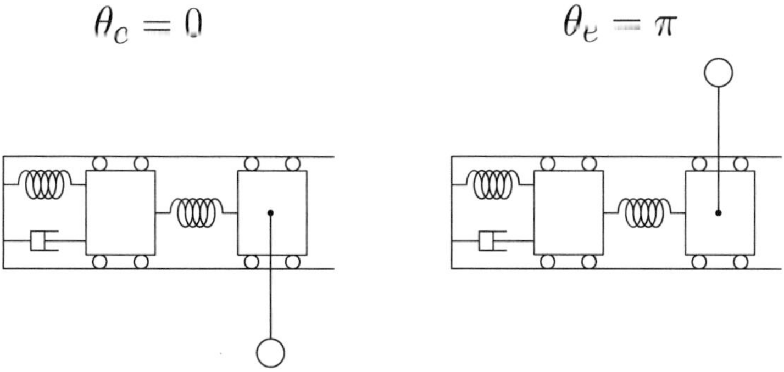

Fig. 4.7 Equilibrium positions of the carts-pendulum system

Substitute this into Eq. (4.50) to obtain the equilibrium equations

$$\begin{bmatrix} k_1(y_1^e - l_1) - k_2(y_2^e - y_1^e - l_2) \\ \\ k_2(y_2^e - y_1^e - l_2) \\ \\ m_3 g l \sin\theta_e \end{bmatrix} = \begin{bmatrix} 0 \\ \\ 0 \\ \\ 0 \end{bmatrix}. \tag{4.54}$$

The solutions of this system of equations are

$$y_1^e = l_1, \quad y_2^e = l_1 + l_2, \quad \theta_e = 0, \ \pm\pi, \ \pm 2\pi, \ldots . \tag{4.55}$$

The two physically distinct equilibrium positions are shown in Fig. 4.7. Intuitively, the equilibrium position with $\theta_e = 0$ is stable. This means that if the system experiences a small disturbance from the state $(y_1, \ y_2, \ \theta, \ \dot{y}_1, \ \dot{y}_2, \ \dot{\theta}) = (0, \ 0, \ 0, \ 0, \ 0, \ 0)$, its future states will remain confined within a small neighborhood around this equilibrium.

4.3.4 Simulating the Motion

Analytical solutions in closed form for the system of nonlinear differential equations in (4.50) are not possible. Numerical simulations must be employed to study the motion predicted by these equations. The numerical approach requires first rewriting Eq. (4.50) as

$$\begin{bmatrix} m_1 & 0 & 0 \\ 0 & m_2+m_3 & m_3 l\cos\theta \\ 0 & m_3 l\cos\theta & m_3 l^2 \end{bmatrix} \begin{bmatrix} \ddot{y}_1 \\ \ddot{y}_2 \\ \ddot{\theta} \end{bmatrix} + \begin{bmatrix} c\dot{y}_1 + (k_1+k_2)(y_1-y_1^e) - k_2(y_2-y_2^e) \\ -m_3 l\dot{\theta}^2\sin\theta + k_2(y_2-y_2^e) - k_2(y_1-y_1^e) \\ m_3 g l\sin\theta \end{bmatrix} = \begin{bmatrix} 0 \\ f(t) \\ 0 \end{bmatrix} \tag{4.56}$$

and subsequently introducing the variables

$$\begin{aligned} z_1 &= y_1 - y_1^e, \quad z_2 = y_2 - y_2^e, \quad z_3 = \theta, \\ z_4 &= \dot{y}_1, \quad z_5 = \dot{y}_2, \quad z_6 = \dot{\theta}, \end{aligned} \tag{4.57}$$

to convert the Eqs. (4.56) into the system of six first-order ODEs:

$$\frac{d}{dt}\begin{bmatrix} z_1 \\ z_2 \\ z_3 \\ z_4 \\ z_5 \\ z_6 \end{bmatrix} = \begin{bmatrix} z_4 \\ z_5 \\ z_6 \\ \mathbf{M}(z_3)^{-1}\left(-\mathbf{g}(z_1,z_2,z_3,z_4,z_6) + \begin{bmatrix} 0 \\ f(t) \\ 0 \end{bmatrix}\right) \end{bmatrix}. \tag{4.58}$$

The MATLAB code `Chapt4_code1`, provided in the *Electronic Supplementary Material*, numerically integrates Eq. (4.58) and generates the plots shown in Fig. 4.8 using the parameters:

$$m_1 = m_2 = 1\,\text{kg}, \quad m_3 = 0.1\,\text{kg}, \quad l = 0.5\,\text{m}, \quad k_1 = k_2 = 10\,\text{N/m}. \tag{4.59}$$

Figure 4.8a shows the system response when one of the carts is subjected to a harmonic force. The transient vibrations present in the initial stage decay due to damping. Afterward, the carts oscillate harmonically in phase and at the same

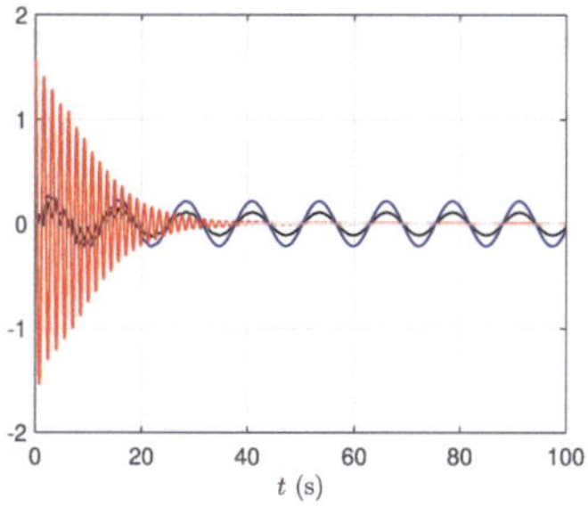

(a) Plot generated with $f(t) = \sin(0.5t)$, $\mathbf{z}(0) = [0 \quad 0 \quad \pi/2 \quad 0 \quad 0 \quad 0]^T$, and $c = 1$ Ns/m.

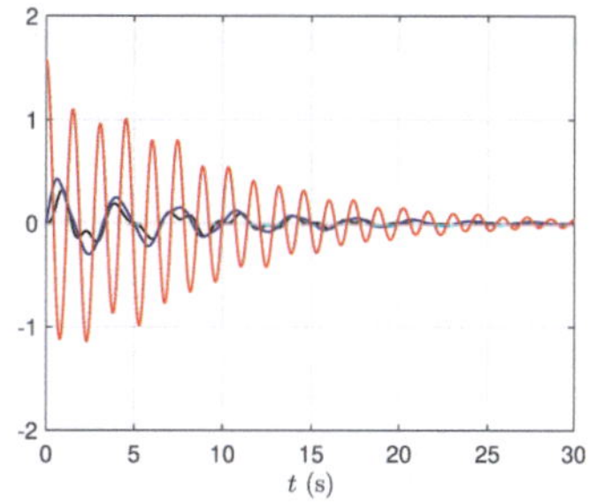

(b) Plot generated with $f = 0$, $\mathbf{z}(0) = [0 \quad 0 \quad \pi/2 \quad 0 \quad 1 \quad 0]^T$, and $c = 1$ Ns/m.

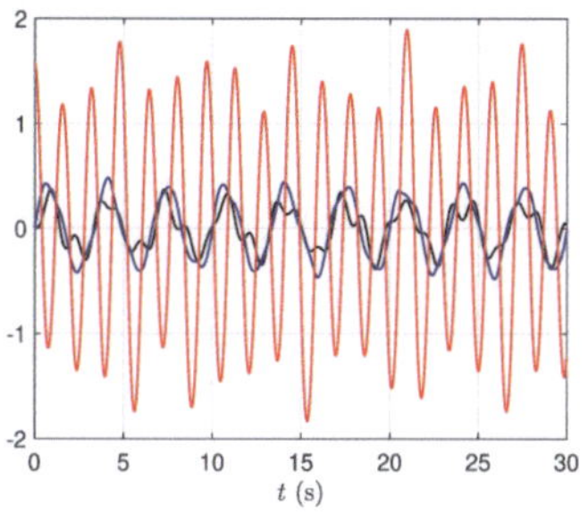

(c) Results generated with $f = 0$, $\mathbf{z}(0) = [0 \quad 0 \quad \pi/2 \quad 0 \quad 1 \quad 0]^T$, and $c = 0$.

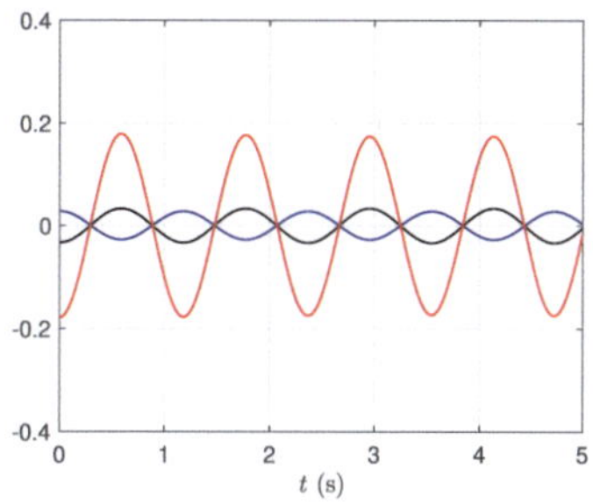

(d) Results generated with $f = 0$, $\mathbf{z}(0) = [-0.0331 \quad 0.0276 \quad -0.1795 \quad 0 \quad 0 \quad 0]^T$, and $c = 0$.

Fig. 4.8 Time evolution of $y_1 - y_1^e$ (—), $y_2 - y_2^e$ (—), and θ (—) obtained by numerically integrating Eq. (4.56)

frequency as the excitation, while the pendulum practically ceases to oscillate. Figure 4.8b–d depict the behavior during free vibration. In the presence of damping, the oscillations of all DOFs decay even though only one of carts is connected to a damper. The oscillations persist in the absence of damping. Figure 4.8d shows a vibration pattern, known as a natural mode, where all components oscillate harmonically at the same frequency.

4.3.5 *Linearization*

As is common in vibration analysis, Eq. (4.50) can be linearized to reveal the governing equations for small vibrations around the stable equilibrium. The entries in the vector of nonlinear functions

$$\mathbf{g}(y_1,\ y_2,\ \theta,\ \dot{y}_1,\ \dot{y}_2,\ \dot{\theta}) = \begin{bmatrix} g_1(y_1, y_2, \dot{y}_1) \\ g_2(y_1, y_2, \theta, \dot{\theta}) \\ g_3(\theta) \end{bmatrix} \tag{4.60}$$

can be expanded as a multivariate Taylor series around the stable equilibrium state $(y_1,\ y_2,\ \theta,\ \dot{y}_1,\ \dot{y}_2,\ \dot{\theta}) = (y_1^e,\ y_2^e,\ 0,\ 0,\ 0,\ 0)$ according to

$$\begin{aligned} g_1(y_1,\ y_2,\ \dot{y}_1) \approx g_1(y_1^e,\ y_2^e,\ 0) &+ \left.\frac{\partial g_1}{\partial \dot{y}_1}\right|_{(y_1^e,\, y_2^e,\, 0)} \times \dot{y}_1 \\ &+ \left.\frac{\partial g_1}{\partial y_1}\right|_{(y_1^e,\, y_2^e,\, 0)} \times (y_1 - y_1^e) \\ &+ \left.\frac{\partial g_1}{\partial y_2}\right|_{(y_1^e,\, y_2^e,\, 0)} \times (y_2 - y_2^e), \end{aligned} \tag{4.61}$$

$$\begin{aligned} g_2(y_1,\ y_2,\ \theta, \dot{\theta}) \approx g_2(y_1^e,\ y_2^e,\ 0,\ 0) &+ \left.\frac{\partial g_2}{\partial \dot{\theta}}\right|_{(y_1^e,\, y_2^e,\, 0,\, 0)} \times \dot{\theta} \\ &+ \left.\frac{\partial g_2}{\partial \theta}\right|_{(y_1^e,\, y_2^e,\, 0,\, 0)} \times \theta \\ &+ \left.\frac{\partial g_2}{\partial y_1}\right|_{(y_1^e,\, y_2^e,\, 0,\, 0)} \times (y_1 - y_1^e) \\ &+ \left.\frac{\partial g_2}{\partial y_2}\right|_{(y_1^e,\, y_2^e,\, 0,\, 0)} \times (y_2 - y_2^e), \end{aligned} \tag{4.62}$$

and

$$g_3(\theta) = g_3(0) + \left.\frac{\partial g_3}{\partial \theta}\right|_0 \times \theta. \tag{4.63}$$

Evaluate the partial derivatives at the equilibrium state to verify that the approximation for $\mathbf{g}$ is

$$\mathbf{g}(y_1,\ y_2,\ \theta,\ \dot{y}_1,\ \dot{\theta}) \approx \begin{bmatrix} c\dot{y}_1 + (k_1 + k_2)(y_1 - y_1^e) - k_2(y_2 - y_2^e) \\ -k_2(y_1 - y_1^e) + k_2(y_2 - y_2^e) \\ m_3 g l \theta \end{bmatrix}. \tag{4.64}$$

The other nonlinear terms in Eq. (4.50) appear in the acceleration coefficient $\mathbf{M}(\theta)$ To derive the governing linear differential equations of motion, the entries in this matrix must also be expanded as a Taylor series. The approximation of the entries of $\mathbf{M}(\theta)$ should include only the zeroth-order terms in the Taylor series, as retaining first-order terms would result in a nonlinear system of equations, with θ multiplying $\ddot{y}_2$. With this in mind, the matrix coefficient is approximated by

$$\mathbf{M}(\theta) \approx \mathbf{M}(0) = \begin{bmatrix} m_1 & 0 & 0 \\ 0 & m_2 + m_3 & m_3 l \\ 0 & m_3 l & m_3 l^2 \end{bmatrix}. \tag{4.65}$$

The approximations derived herein generate the following system of coupled linear ODEs:

$$\mathbf{M}\begin{bmatrix} \ddot{q}_1 \\ \ddot{q}_2 \\ \ddot{q}_3 \end{bmatrix} + \mathbf{C}\begin{bmatrix} \dot{q}_1 \\ \dot{q}_2 \\ \dot{q}_3 \end{bmatrix} + \mathbf{K}\begin{bmatrix} q_1 \\ q_2 \\ q_3 \end{bmatrix} = \begin{bmatrix} 0 \\ f(t) \\ 0 \end{bmatrix}, \tag{4.66}$$

where

$$q_1 = y_1 - y_1^e, \quad q_2 = y_2 - y_2^e, \quad q_3 = \theta \tag{4.67}$$

are generalized coordinates measured from the equilibrium position. The mass, damping, and stiffness matrices are given by

$$\mathbf{M} = \begin{bmatrix} m_1 & 0 & 0 \\ 0 & m_2 + m_3 & m_3 l \\ 0 & m_3 l & m_3 l^2 \end{bmatrix}, \tag{4.68}$$

$$\mathbf{C} = \begin{bmatrix} c & 0 & 0 \\ 0 & 0 & 0 \\ 0 & 0 & 0 \end{bmatrix}, \tag{4.69}$$

$$\mathbf{K} = \begin{bmatrix} k_1 + k_2 & -k_2 & 0 \\ -k_2 & k_2 & 0 \\ 0 & 0 & m_3 g l \end{bmatrix}, \tag{4.70}$$

respectively. The linearized equations of motion, as expressed in (4.66), take the same form as the matrix-vector equation (4.17) for a two-particle system.

4.3.6 Energy of the System

The total mechanical energy of this system is

$$\begin{aligned} E = &\frac{1}{2}m_1\dot{q}_1^2 + \frac{1}{2}(m_2 + m_3)\dot{q}_2^2 + \frac{1}{2}m_3 l^2\dot{q}_3^2 + m_3 l\dot{q}_3\dot{q}_2\cos(q_3) \\ &+ \frac{1}{2}k_1 q_1^2 + \frac{1}{2}k_2(q_2 - q_1)^2 - m_3 g l\cos(q_3). \end{aligned} \tag{4.71}$$

Under the small motion assumption, E can be approximated as a second-order multivariate Taylor series by

$$\begin{aligned} E \approx &\frac{1}{2}m_1\dot{q}_1^2 + \frac{1}{2}(m_2 + m_3)\dot{q}_2^2 + \frac{1}{2}m_3 l^2\dot{q}_3^2 + m_3 l\dot{q}_3\dot{q}_2 \\ &+ \frac{1}{2}k_1 q_1^2 + \frac{1}{2}k_2(q_2 - q_1)^2 - m_3 g l q_3^2 + U_e \end{aligned} \tag{4.72}$$

where $U_e = -m_3 g l$ is the potential energy of the system at the static equilibrium position. The approximation for energy can be expressed as

$$\begin{aligned} E = &\frac{1}{2}\begin{bmatrix} \dot{q}_1 & \dot{q}_2 & \dot{q}_3 \end{bmatrix}\mathbf{M}\begin{bmatrix} \dot{q}_1 \\ \dot{q}_2 \\ \dot{q}_3 \end{bmatrix} \\ &+ \frac{1}{2}\begin{bmatrix} q_1 & q_2 & q_3 \end{bmatrix}\mathbf{K}\begin{bmatrix} q_1 \\ q_2 \\ q_3 \end{bmatrix} + U_e. \end{aligned} \tag{4.73}$$

The rate of change of E is

$$\dot{E} = -\begin{bmatrix}\dot{q}_1 & \dot{q}_2 & \dot{q}_3\end{bmatrix}\mathbf{C}\begin{bmatrix}\dot{q}_1\\ \dot{q}_2\\ \dot{q}_3\end{bmatrix} + \begin{bmatrix}\dot{q}_1 & \dot{q}_2 & \dot{q}_3\end{bmatrix}\begin{bmatrix}0\\ f(t)\\ 0\end{bmatrix}. \tag{4.74}$$

Energy is conserved if the system is undamped and $f = 0$.

4.3.7 Alternative Derivation of the Equations of Motion

A faster route to arrive at Eq. (4.66) is to assume small motions from the outset. Taking the state $(y_1, y_2, \theta, \dot{y}_1, \dot{y}_2, \dot{\theta})$ to remain close to $(l_1, l_1 + l_2, 0, 0, 0, 0)$,[3] the following approximations may be introduced

$$\mathbf{e}_r \approx \mathbf{i} + \theta\mathbf{j}, \quad \mathbf{e}_\theta \approx -\theta\mathbf{i} + \mathbf{j}. \tag{4.75}$$

Consequently, the position and acceleration vectors of the carts and pendulum become

$$\mathbf{r}_1 = y_1\mathbf{j} = (q_1 + l_1)\mathbf{j}, \quad \ddot{\mathbf{r}}_1 = \ddot{q}_1\mathbf{j}, \tag{4.76}$$

$$\mathbf{r}_2 = y_2\mathbf{j} = (q_2 + l_1 + l_2)\mathbf{j}, \quad \ddot{\mathbf{r}}_2 = \ddot{q}_2\mathbf{j}, \tag{4.77}$$

$$\mathbf{r}_3 \approx \mathbf{r}_2 + l\mathbf{i} + lq_3\mathbf{j}, \quad \ddot{\mathbf{r}}_3 \approx (\ddot{q}_2 + l\ddot{q}_3)\mathbf{j}. \tag{4.78}$$

The FBDs are similar to those from Fig. 4.6, with the difference that the spring forces and traction force are now represented by

$$\mathbf{F}_{e_1} = -k_1 q_1\mathbf{j}, \quad \mathbf{F}_{e_2} = -k_2(q_2 - q_1)\mathbf{j}, \quad S\mathbf{e}_r \approx S\mathbf{i} + Sq_3\mathbf{j}. \tag{4.79}$$

Application of Newton's second law yields

$$m_1\ddot{q}_1 = -c\dot{q}_1 - k_1 q_1 + k_2(q_2 - q_1), \tag{4.80}$$

$$m_2\ddot{q}_2\mathbf{j} = f(t)\mathbf{j} + (m_2 g - N_2)\mathbf{i} - k_2(q_2 - q_1)\mathbf{j} + S(\mathbf{i} + q_3\mathbf{j}), \tag{4.81}$$

$$m_3(\ddot{q}_2 + l\ddot{q}_3)\mathbf{j} = m_3 g\mathbf{i} - S(\mathbf{i} + q_3\mathbf{j}). \tag{4.82}$$

Use the same steps from Sect. 4.3.2 to obtain Eq. (4.66).

[3] Or, equivalently, that $(q_1, q_2, q_3, \dot{q}_1, \dot{q}_2, \dot{q}_3)$ is close to $(0, 0, 0, 0, 0, 0)$.

4.4 General Form of the Equations of Motion

The analysis of SDOF systems is greatly facilitated by using a prototype system. Solving the mass-spring-damper equation is sufficient to understand the behavior of other mechanical systems with one DOF under the "small vibrations" regime.

At first glance, it may seem unlikely that a prototype system could be applied to understand the behavior of systems with multiple DOFs. This is because MDOF systems may exhibit various modes of vibration, such as translation, rotation, torsion, or even bending. However, consider the example of the carts-pendulum system from Sect. 4.3. The small vibrations of this system are governed by three linear differential equations that, when expressed in the matrix-vector form (4.66), exhibit similar structure as the equations of motion (4.17) for a linear system composed of two particles. These examples suggest that a prototype form of the equations of motion does exist for MDOF systems.

To derive the prototype form of the equations of motion, consider a mechanical system with n DOFs. Suppose the dynamics of this system is specified by the generalized coordinates $x_1, x_2, \ldots, x_n$, which can be arranged as the n-dimensional vector

$$\mathbf{x} = \begin{bmatrix} x_1 \\ x_2 \\ \vdots \\ x_n \end{bmatrix}. \tag{4.83}$$

Assume that the motion of such a system is governed by a system of n differential equations that can be arranged as

$$\mathbf{M}(\mathbf{x},\ t)\,\ddot{\mathbf{x}} + \mathbf{g}(\mathbf{x},\ \dot{\mathbf{x}},\ t) = \mathbf{f}(t), \tag{4.84}$$

where $\mathbf{M}(\mathbf{x},\ t)$ is a symmetric $n \times n$ matrix function, $\mathbf{g}(\mathbf{x},\ \dot{\mathbf{x}},\ t)$ is an n-dimensional vector function, and

$$\mathbf{f}(t) = \begin{bmatrix} f_1(t) \\ f_2(t) \\ \vdots \\ f_n(t) \end{bmatrix} \tag{4.85}$$

is a vector of externally applied loads that depend on time only. The entries in $\mathbf{M}(\mathbf{x},\ t)$ and $\mathbf{g}(\mathbf{x},\ \dot{\mathbf{x}},\ t)$ may contain nonlinear functions.

Since the effect of $\mathbf{f}(t)$ is to disturb the system from equilibrium, the equilibria may be obtained by ignoring $\mathbf{f}(t)$ and substituting the constant vector $\mathbf{x} = \mathbf{x}_e$ into (4.84):

$$\mathbf{g}(\mathbf{x}_e, \mathbf{0}, t) = \mathbf{0}. \tag{4.86}$$

The equilibrium equations is a system of n nonlinear algebraic equations for $\mathbf{x}_e$.

The equations that govern the small vibratory behavior can be obtained using the generalized Taylor expansion about the stable equilibrium $(\mathbf{x},\ \dot{\mathbf{x}}) = (\mathbf{x}_e,\ \mathbf{0})$. The vector of nonlinear functions is approximated by

$$\mathbf{g}(\mathbf{x},\ \dot{\mathbf{x}},\ t) = \begin{bmatrix} g_1(\mathbf{x}_e,\ \dot{\mathbf{x}},\ t) \\ g_2(\mathbf{x}_e,\ \dot{\mathbf{x}},\ t) \\ \vdots \\ g_n(\mathbf{x}_e,\ \dot{\mathbf{x}},\ t) \end{bmatrix}$$

$$\approx \begin{bmatrix} g_1(\mathbf{x}_e,\ \mathbf{0},\ t) + \sum\limits_{j=1}^{n} a_{1j}(t)\dot{x}_j + \sum\limits_{j=1}^{n} b_{1j}(t)(x_j - x_j^e) \\ g_2(\mathbf{x}_e,\ \mathbf{0},\ t) + \sum\limits_{j=1}^{n} a_{2j}(t)\dot{x}_j + \sum\limits_{j=1}^{n} b_{2j}(t)(x_j - x_j^e) \\ \vdots \\ g_n(\mathbf{x}_e,\ \mathbf{0},\ t) + \sum\limits_{j=1}^{n} a_{nj}(t)\dot{x}_j + \sum\limits_{j=1}^{n} b_{nj}(t)(x_j - x_j^e) \end{bmatrix}. \tag{4.87}$$

The equilibrium equation (4.86) implies that $g_j(\mathbf{x},\ \dot{\mathbf{x}},\ t) = 0$ for $j = 1, \ldots, n$. Define the time-dependent coefficients

$$a_{lj}(t) = \frac{\partial g_l}{\partial \dot{x}_j}(\mathbf{x}_e,\ \mathbf{0},\ t), \quad b_{lj}(t) = \frac{\partial g_l}{\partial x_j}(\mathbf{x}_e,\ \mathbf{0},\ t), \tag{4.88}$$

to express $\mathbf{g}(\mathbf{x},\ \dot{\mathbf{x}},\ t)$ as

$$\mathbf{g}(\mathbf{x},\ \dot{\mathbf{x}},\ t) \approx \mathbf{A}(t)\dot{\mathbf{x}} + \mathbf{B}(t)(\mathbf{x} - \mathbf{x}_e), \tag{4.89}$$

where the matrix coefficients are

$$\mathbf{A}(t) = \begin{bmatrix} a_{11}(t) & \cdots & a_{1n}(t) \\ \vdots & & \vdots \\ a_{n1}(t) & \cdots & a_{nn}(t) \end{bmatrix}, \tag{4.90}$$

$$\mathbf{B}(t) = \begin{bmatrix} b_{11}(t) & \cdots & b_{1n}(t) \\ \vdots & & \vdots \\ b_{n1}(t) & \cdots & b_{nn}(t) \end{bmatrix}. \tag{4.91}$$

The Taylor expansion of the acceleration coefficient includes only the zeroth-order term:

$$\mathbf{M}(\mathbf{x},\ t) = \mathbf{M}(\mathbf{x}_e,\ t) + \ldots \approx \mathbf{M}(t). \tag{4.92}$$

Since $\mathbf{M}(\mathbf{x},\ t)$ multiplies $\ddot{\mathbf{x}}$, the inclusion of higher-order terms in this expansion would introduce products between the coordinates and their second derivatives. Such terms cannot appear in a system of linear differential equations. The aforementioned Taylor expansions produce the following approximation for Eq. (4.84):

$$\mathbf{M}(t)\ddot{\mathbf{q}} + \mathbf{A}(t)\dot{\mathbf{q}} + \mathbf{B}(t)\mathbf{q} = \mathbf{f}(t), \tag{4.93}$$

where the change of coordinates $\mathbf{q} = \mathbf{x} - \mathbf{x}_e$ has been applied. This is the general form of the equations of motion for "small" vibrations. This linear system is time-invariant if and only if it possesses constant coefficients. Under the assumption of time invariance, these equations take the form

$$\mathbf{M}\ddot{\mathbf{q}} + \mathbf{A}\dot{\mathbf{q}} + \mathbf{B}\mathbf{q} = \mathbf{f}(t), \tag{4.94}$$

where $\mathbf{M}$, $\mathbf{A}$, and $\mathbf{B}$ are constant square matrices of order n. The mass matrix $\mathbf{M} = \mathbf{M}^T$ is symmetric, but $\mathbf{A}$ and $\mathbf{B}$ are arbitrary. However, a square matrix can always be decomposed as the sum of symmetric and skew-symmetric parts.[4] In that view,

[4] See Appendix C.

Table 4.1 Prototype form of the equations of motion

	$n = 1$	$n > 1$
Generalized coordinates measured from equilibrium	q	$\mathbf{q} = [q_1 \quad \dots \quad q_n]^T$
Nonlinear equations of motion	$\ddot{q} + g(q, \dot{q}, t) = f(t)$	$\mathbf{M}(\mathbf{q}, t)\ddot{\mathbf{q}} + \mathbf{g}(\mathbf{q}, \dot{\mathbf{q}}, t) = \mathbf{f}(t)$
Equations of motion for "small" vibrations	$m\ddot{q} + c\dot{q} + kq = f(t)$	$\mathbf{M}\ddot{\mathbf{q}} + \mathbf{C}\dot{\mathbf{q}} + \mathbf{K}\mathbf{q} = \mathbf{f}(t)$

the equations of motion of a linear time-invariant system may be expressed as

$$\mathbf{M}\ddot{\mathbf{q}} + (\mathbf{C} + \mathbf{G})\dot{\mathbf{q}} + (\mathbf{K} + \mathbf{H})\mathbf{q} = \mathbf{f}(t). \tag{4.95}$$

The coefficient matrix of velocity was decomposed as $\mathbf{A} = \mathbf{C} + \mathbf{G}$, where $\mathbf{C} = \mathbf{C}^T$ is a damping matrix and $\mathbf{G} = -\mathbf{G}^T$ is a gyroscopic matrix. Likewise, the coefficient matrix of displacement is $\mathbf{A} = \mathbf{K} + \mathbf{H}$, where $\mathbf{K} = \mathbf{K}^T$ is a stiffness matrix and $\mathbf{H} = -\mathbf{H}^T$ is a circulatory matrix. For the vast majority of systems, the gyroscopic and circulatory matrices are zero, resulting in the system of equations:

$$\mathbf{M}\ddot{\mathbf{q}} + \mathbf{C}\dot{\mathbf{q}} + \mathbf{K}\mathbf{q} = \mathbf{f}(t). \tag{4.96}$$

Much of the interest in linear vibrations and structural dynamics lies on damped systems, where the coefficients are the symmetric matrices **M**, **C**, and **K**. For this reason, Eq. (4.96) is considered the prototype equations of motion for MDOF systems. The derivation of the prototype form of the equations of motion is summarized in Table 4.1, which also compares it to the SDOF case. This comparison shows that equations share a close resemblance. The equations of motion for a system with $n > 1$ DOFs has matrix coefficients of order n. An SDOF system can be interpreted as the limiting case when $n = 1$ and the matrices reduce to scalars.

4.5 Illustrative Examples

This section presents three examples that illustrate the steps involved in deriving the equations of motion for MDOF systems and the subsequent linearization of these equations.

4.5.1 An Autoparametric System

A mass-spring-damper system, as depicted in Fig. 4.9, experiences resonance when excited by a harmonic force with frequency equal to its natural frequency. One strategy to reduce or completely suppress these violent vibrations involves attaching a mechanical appendage, such as the pendulum shown in that figure. This approach is considered a *passive control* strategy because it relies solely on the inherent

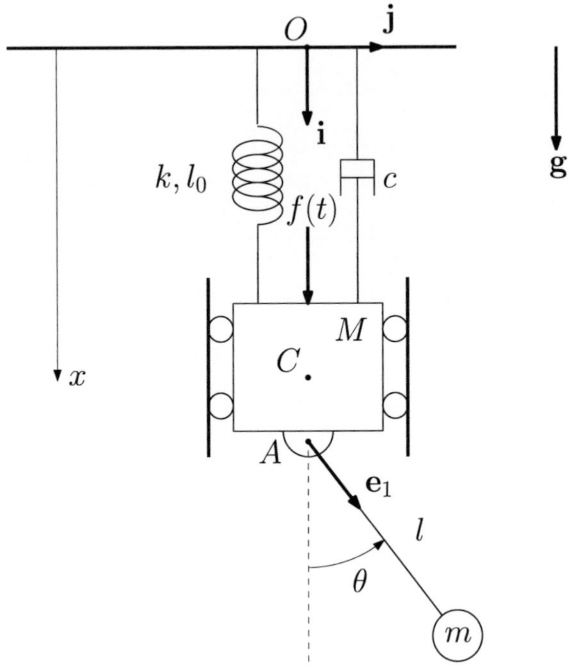

Fig. 4.9 A pendulum with mass m and length l is attached to a mass-spring-damper system that is driven by a harmonic force $f(t) = F_0 \cos(\omega t)$. The gravitational forces act along $\mathbf{i}$

characteristics of the mechanical system to achieve vibration reduction. The goal of this example is to derive the equations of motion for this system and use numerical simulation to analyze the effectiveness of the pendulum in reducing the vibrations of the mass-spring-damper system.

4.5.1.1 Kinematic Analysis

The cart shown in Fig. 4.9 is modeled as a rigid body with mass M and center of mass at C. It is connected to a fixed support by a viscous damper with coefficient c and a linear spring with stiffness k and unstretched length l_0. The rigid body can translate along a smooth track and is subjected to a force $f(t)$. A rigid pendulum of length l and concentrated mass m is attached to the translating body via the pin joint at A. A Cartesian reference system $\{\mathbf{i},\ \mathbf{j},\ \mathbf{k} = \mathbf{i} \times \mathbf{j}\}$ is fixed at the origin O, while the corotational basis $\{\mathbf{e}_1,\ \mathbf{e}_2,\ \mathbf{e}_3 = \mathbf{k} = \mathbf{e}_1 \times \mathbf{e}_2\}$ is attached to the rigid pendulum at A.

The first step in dynamic analysis is to determine the number of DOFs that the system has. If both rigid bodies were free to move in the plane of motion, the system would have six DOFs. However, the presence of kinematic constraints significantly reduces this number. Two constraints are related to the translational motion of the cart. Firstly, its angular velocity must be zero:

$$\boldsymbol{\omega}_1 \cdot \mathbf{k} = 0. \tag{4.97}$$

This holonomic constraint implies that the body is not rotating. Additionally, the position vector of the cart's center of mass, $\mathbf{x}_1$, must satisfy

$$\psi_2 = \mathbf{x}_1 \cdot \mathbf{j} = 0, \tag{4.98}$$

since the translation is rectilinear. The remaining two constraints are related to the connection of the two bodies at the pin joint A. If $\mathbf{x}_2$ is the position of the center of mass of the rigid pendulum, this connection can be expressed mathematically by:

$$\psi_3 = \mathbf{x}_2 \cdot \mathbf{i} - \mathbf{c}_0 \cdot \mathbf{i} - l\mathbf{e}_1 \cdot \mathbf{i} - \mathbf{x}_1 \cdot \mathbf{i} = 0, \tag{4.99}$$

$$\psi_4 = \mathbf{x}_2 \cdot \mathbf{j} - \mathbf{c}_0 \cdot \mathbf{j} - l\mathbf{e}_1 \cdot \mathbf{j} - \mathbf{x}_1 \cdot \mathbf{j} = 0, \tag{4.100}$$

where $\mathbf{c}_0$ is a constant vector. The system operates under four holonomic constraints, resulting in $n = 6 - 4 = 2$ DOFs. The chosen generalized coordinates, shown in Fig. 4.9, are x and θ. The position, velocity, and acceleration vectors of C have the representation

$$\mathbf{x}_1 = x\mathbf{i}, \quad \dot{\mathbf{x}}_1 = \dot{x}\mathbf{i}, \quad \ddot{\mathbf{x}}_1 = \ddot{x}\mathbf{i}, \tag{4.101}$$

respectively. The position vector of the pendulum bob is

$$\mathbf{x}_2 = \mathbf{x}_1 + \mathbf{c}_0 + l\mathbf{e}_1. \tag{4.102}$$

Differentiating this expression twice yields:

$$\ddot{\mathbf{x}}_2 = \ddot{x}\mathbf{i} + l\ddot{\theta}\mathbf{e}_2 - l\dot{\theta}^2\mathbf{e}_1. \tag{4.103}$$

4.5.1.2 Equations of Motion

The FBDs of the cart and pendulum are shown in Fig. 4.10. Vectors $\mathbf{N} = N\mathbf{j}$ and $\mathbf{M}_R = M_R\mathbf{k}$ are the force and moment from the tracks that maintain the rectilinear translation of the body, respectively. The constraint force $\mathbf{R}$ appears due to the pin joint connection at A. It is assumed this connection is not ideal and that a counter moment $-c_\theta\dot{\theta}\mathbf{k}$ due to friction is present. There are additional unknowns ($\mathbf{N}$, $\mathbf{M}_R$, $\mathbf{R}$) that need to be eliminated to obtain two equations involving only x and θ.

Apply Euler's laws to the rigid pendulum to obtain:

$$\mathbf{R} = m\ddot{x}\mathbf{i} - mg\mathbf{i} + ml\ddot{\theta}\mathbf{e}_2 - ml\dot{\theta}^2\mathbf{e}_1, \tag{4.104}$$

$$0 = -c_\theta\dot{\theta} - l\mathbf{e}_2 \cdot \mathbf{R}. \tag{4.105}$$

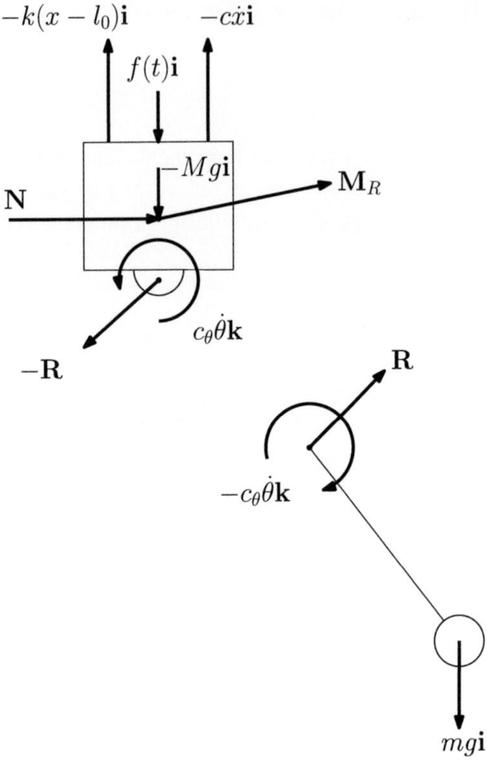

Fig. 4.10 FBDs of the translating cart and the rigid pendulum

The first differential equation that contains only x and θ as unknowns can be derived by substituting Eq. (4.104) into Eq. (4.105):

$$-ml\ddot{x}\sin\theta + ml^2\ddot{\theta} + c_\theta\dot{\theta} + mgl\sin\theta = 0. \tag{4.106}$$

To obtain the remaining differential equation, apply Euler's first law to the cart:

$$M\ddot{x}\mathbf{i} = \mathbf{N} - \mathbf{R} + F_0\cos(\omega t)\mathbf{i} - k(x-l_0)\mathbf{i} - c\dot{x}\mathbf{i} + Mg\mathbf{i}. \tag{4.107}$$

Finally, substitute Eq. (4.104) into Eq. (4.107) and take the projection of the resulting equation along $\mathbf{i}$ to arrive at the second ODE:

$$\begin{aligned}(M+m)\ddot{x} - ml\ddot{\theta}\sin\theta - ml\dot{\theta}^2\cos\theta + c\dot{x}\\ + k(x-l_0) - (M+m)g = F_0\cos(\omega t).\end{aligned} \tag{4.108}$$

The equations of motion (4.106) and (4.108) can be organized in the following vector-matrix form:

$$\begin{bmatrix} M+m & -ml\sin\theta \\ -ml\sin\theta & ml^2 \end{bmatrix}\begin{bmatrix}\ddot{x} \\ \ddot{\theta}\end{bmatrix} + \begin{bmatrix} -ml\dot{\theta}^2\cos\theta + c\dot{x} + k(x-l_0) - (M+m)g \\ c_\theta\dot{\theta} + mgl\sin\theta \end{bmatrix} = \begin{bmatrix} F_0\cos(\omega t) \\ 0 \end{bmatrix}. \tag{4.109}$$

The 2 × 2 acceleration coefficient is symmetric, and both this matrix and the vector function contain nonlinear entries that depend on θ.

4.5.1.3 Linearization

The static equilibrium position can be obtained by ignoring $f(t)$ and seeking constant solutions given by:

$$x = x_e, \quad \theta = \theta_e. \tag{4.110}$$

Substitute these solutions into Eq. (4.109) to obtain

$$\begin{bmatrix} k(x_e - l_0) - (M+m)g \\ mgl\sin\theta_e \end{bmatrix} = \begin{bmatrix} 0 \\ 0 \end{bmatrix}. \tag{4.111}$$

The equilibrium equations are independent because the only terms that coupled these equations depended on $\dot{\theta}$. Solving these equations yields:

$$x_e = l_0 + \frac{(M+m)g}{k}, \quad \theta_e = 0, \pm\pi, \pm 2\pi, \ldots. \tag{4.112}$$

The value x_e is length of the spring that supports the total weight of the system. The physical interpretation of the multiple values for θ_e is shown in Fig. 4.11.

Assuming that any state $(x, \theta, \dot{x}, \dot{\theta})$ stays confined in a small neighborhood around the stable equilibrium $(x_e, 0, 0, 0)$ when the external force $f(t)$ acts on the system, the following first-order approximations may be introduced

$$\begin{bmatrix} M+m & -ml\sin\theta \\ -ml\sin\theta & ml^2 \end{bmatrix} \approx \begin{bmatrix} M+m & 0 \\ 0 & ml^2 \end{bmatrix}, \tag{4.113}$$

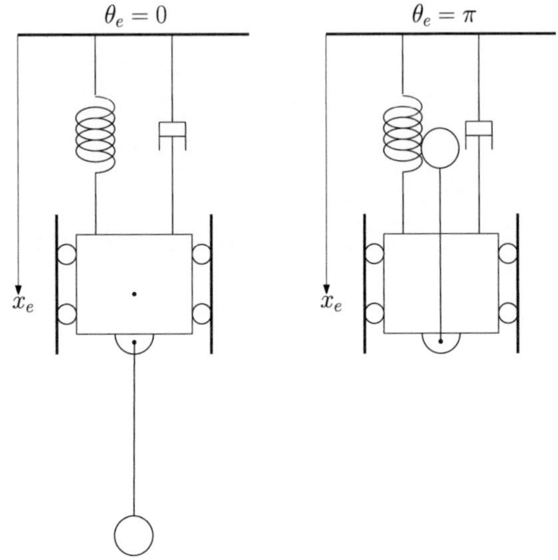

Fig. 4.11 Equilibrium positions of the system

$$\begin{bmatrix} -ml\dot{\theta}^2\cos\theta + c\dot{x} + k(x-l_0) - (M+m)g \\ c_\theta\dot{\theta} + mgl\sin\theta \end{bmatrix} \approx \begin{bmatrix} c\dot{x} + k(x-x_e) \\ c_\theta\dot{\theta} + mgl\theta \end{bmatrix}. \tag{4.114}$$

The system of nonlinear equations (4.109) is linearized into

$$\begin{aligned} &\begin{bmatrix} M+m & 0 \\ 0 & ml^2 \end{bmatrix}\begin{bmatrix} \ddot{q}_1 \\ \ddot{q}_2 \end{bmatrix} + \begin{bmatrix} c & 0 \\ 0 & c_\theta \end{bmatrix}\begin{bmatrix} \dot{q}_1 \\ \dot{q}_2 \end{bmatrix} \\ &\quad + \begin{bmatrix} k & 0 \\ 0 & mgl \end{bmatrix}\begin{bmatrix} q_1 \\ q_2 \end{bmatrix} = \begin{bmatrix} F_0\cos(\omega t) \\ 0 \end{bmatrix}, \end{aligned} \tag{4.115}$$

where $q_1 = x - x_e$ and $q_2 = \theta$. The equations are independent because the coefficient matrices are diagonal. In these equations, the mass-spring-damper system undergoes forced vibrations, while the pendulum experiences damped free vibrations. The equations are decoupled through linearization because coupling in Eq. (4.109) is caused solely by nonlinear terms. Generally, linearization does not remove coupling.

4.5.1.4 Alternative Derivation of the Equations of Motion

If the objective is only to obtain Eq. (4.115), it can be assumed from the outset that $(x, \theta, \dot{x}, \dot{\theta})$ is close to $(x_e, 0, 0, 0)$. This assumption implies that $\mathbf{e}_1 \approx \mathbf{i} + \theta\mathbf{j}$ and

$\mathbf{e}_2 \approx \theta\mathbf{i} + \mathbf{j}$. Consequently, the position and acceleration vectors of the pendulum bob are approximated as:

$$\mathbf{x}_2 \approx x\mathbf{i} + l\mathbf{i} + l\theta\mathbf{j}, \quad \ddot{\mathbf{x}}_2 \approx \ddot{x}\mathbf{i} + l\ddot{\theta}\mathbf{j}, \tag{4.116}$$

respectively. The FBDs remain the same as those shown in Fig. 4.10. Applying Euler's first law to the rigid pendulum gives:

$$\mathbf{R} = m\ddot{x}\mathbf{i} - mg\mathbf{i} + ml\ddot{\theta}\mathbf{j}. \tag{4.117}$$

Use this expression in conjunction with Euler's second law to derive one of the equations of motion:

$$ml^2\ddot{q}_2 + c_\theta\dot{q}_2 + mglq_2 = 0, \tag{4.118}$$

where $q_2 = \theta$. The other equation is obtained by first applying Euler's first law to the cart and subsequently substituting the aforementioned expression for $\mathbf{R}$ into the resulting equation. Then, take the projection along $\mathbf{i}$ to get

$$(M+m)\ddot{x} + c\dot{x} + k(x - l_0) - (M+m)g = F_0\cos(\omega t). \tag{4.119}$$

Substitute $q_1 = x - x_e$ into the aforementioned equation to arrive at the equation of motion

$$(M+m)\ddot{q}_1 + c\dot{q}_1 + kq_1 = F_0\cos(\omega t). \tag{4.120}$$

4.5.1.5 Analysis of the Response

The system motion can now be analyzed using the known equations of motion. Suppose the mass-spring-damper system is defined by the parameters

$$M = 5\,\text{kg}, \quad c = 1.9686\,\text{Ns/m}, \quad k = 436\,\text{N/m}, \tag{4.121}$$

and that it is excited by a force with amplitude $F_0 = 4.36$ N and frequency $\omega = 8.8589$ rad/s. To reduce the vibration, a pendulum with mass $m = 0.5556$ kg was attached to the cart. The MATLAB code `Chapt4_code2`, from the *Electronic Supplementary Material*, numerically integrates Eqs. (4.109) and generates the plots shown in Fig. 4.12. The ICs in the code are

$$q_1(0) = 0.005, \quad \dot{q}_1(0) = 0; \quad q_2(0) = 0.02, \quad \dot{q}_2 = 0. \tag{4.122}$$

First, consider the cases shown in Fig. 4.12a and b, where the pendulum has a length of $l = 50$ cm. In the initial moments, the analytical solutions of the linearized system of Eqs. (4.115) closely follow the numerical solutions of the nonlinear

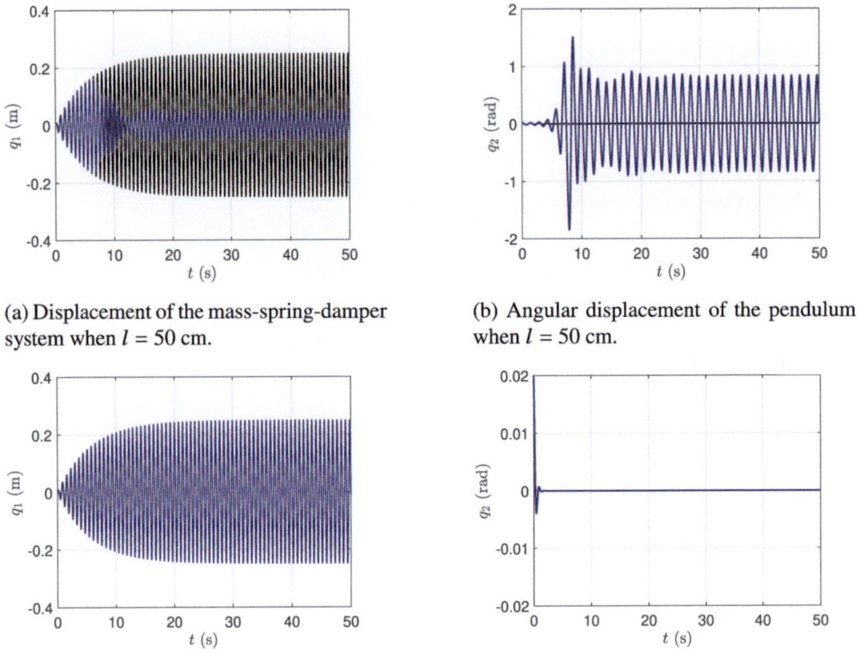

(a) Displacement of the mass-spring-damper system when $l = 50$ cm.

(b) Angular displacement of the pendulum when $l = 50$ cm.

(c) Displacement of the mass-spring-damper system when $l = 15.4$ cm.

(d) Angular displacement of the pendulum when $l = 15.4$ cm.

Fig. 4.12 Numerical solutions of Eq. (4.109) (—) compared to the analytical solutions of the linearized counterpart (4.115) (—)

system (4.109). This explains why the mass-spring-damper motion does not affect the pendulum motion during this period. Since the system is excited at the natural frequency $\omega_1 = \sqrt{k/(M+m)}$, the amplitude of vertical vibrations increases as time progresses. The nonlinear terms in the equations of motion (4.109) become significant in the system's dynamics after the cart's amplitude reaches a certain value, and the pendulum begins to oscillate profusely. The mass-spring-damper and pendulum motions are no longer independent, and the solutions predicted by the linearized equations no longer match the system's behavior. In fact, the solutions of Eq. (4.109) show that the destabilization of the pendulum motion reduces the cart's vibration, eventually reaching a steady-state vibration amplitude of 1/4 of that predicted by the solutions of Eq. (4.115).

The destabilization of the pendulum is a nonlinear phenomenon known as autoparametric resonance, which cannot be predicted by the linearized equations (4.115). Autoparametric resonance occurs in systems consisting of two or more parts that can influence each other to the point where one destabilizes the other. In this example, autoparametric resonance occurs when the pendulum is *tuned* such that its natural frequency $\omega_2 = \sqrt{g/l}$ is half the excitation frequency. This phenomenon does not occur if pendulums with different lengths are chosen.

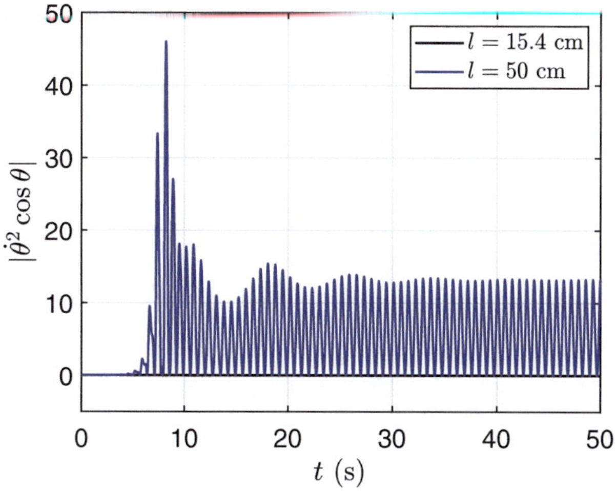

Fig. 4.13 Plot of the nonlinear term $\dot{\theta}^2 \cos\theta$ constructed with the numerical solutions of (4.109)

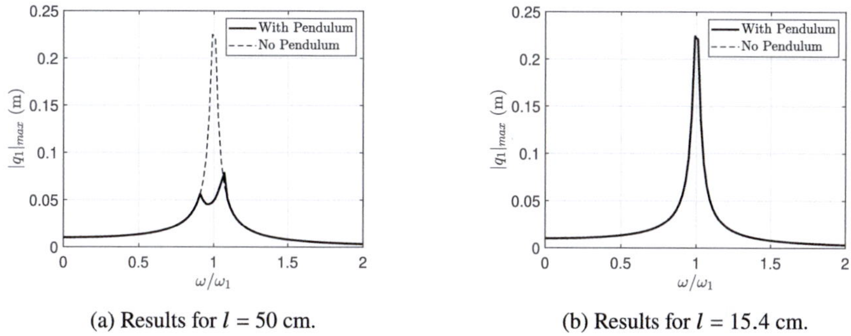

Fig. 4.14 Steady-state amplitude of the cart

If $l = 15.4$ cm, the pendulum is not destabilized. The nonlinear terms that couple the equations of motion (4.109) never become significant with this configuration, as illustrated by the comparison in Fig. 4.13. Consequently, the solutions of Eq. (4.115) match those of Eq. (4.109). This behavior is showcased in Figs. 4.12c and d.

The steady-state amplitude of the primary system is plotted against the excitation frequency in Fig. 4.14. Away from the natural frequency $\omega_1 = \sqrt{k/(M+m)}$, the amplitude of the primary system is the same as that without the pendulum. Vibration reduction occurs near ω_1: the pendulum absorbs and then dissipates much of the energy to reduce the cart.

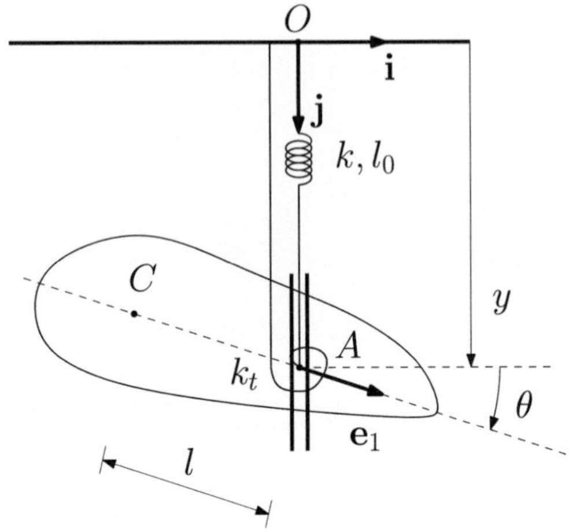

Fig. 4.15 Model of an airfoil. The gravitational force acts along **j**

4.5.2 *The Equations of Motion of an Airfoil*

An airfoil has the cross-sectional shape designed to create a pressure difference between the upper and lower surfaces during airflow, generating lift perpendicular to the direction of the flow. It is a fundamental component of wings, blades, and other aerodynamic surfaces used in aircraft, helicopters, wind turbines, and other applications requiring lift or thrust generation. The objective of this example is to derive the equations of motion for the simplified airfoil shown in Fig. 4.15. It is modeled as a rigid body with mass m and center of mass located at C. The body also has a moment of inertia I_{zz} about the axis parallel to $\mathbf{k}$ that passes through C. Point A of the body is constrained to move vertically inside a smooth track. This point is connected to a fixed support by both a torsional spring with stiffness k_t and a linear spring with stiffness k and unstretched length l_0. With reference to Fig. 4.15, a Cartesian reference system $\{\mathbf{i},\ \mathbf{j},\ \mathbf{k} = \mathbf{i} \times \mathbf{j}\}$ is placed at the origin O, which lies on the fixed support. A corotational basis $\{\mathbf{e}_1,\ \mathbf{e}_2,\ \mathbf{e}_3 = \mathbf{k} = \mathbf{e}_1 \times \mathbf{e}_2\}$ tracks the rotation of the airfoil.

4.5.2.1 Kinematic Analysis

The motion of the airfoil must satisfy the requirement that point A moves inside the vertical track:

$$\psi = \mathbf{x}_A \cdot \mathbf{i} = 0. \tag{4.123}$$

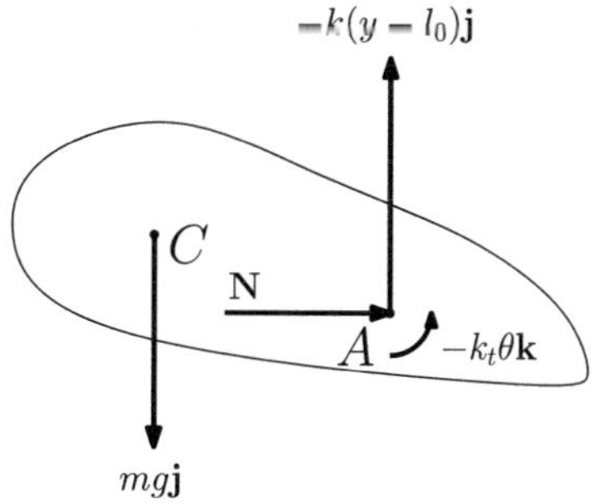

Fig. 4.16 FBD of the airfoil

This is the only constraint, meaning the system has $n = 3 - 1 = 2$ DOFs. With reference to Fig. 4.15, the coordinates y and θ are used to parameterize the position vectors of A and C according to

$$\mathbf{x}_A = y\mathbf{j}, \quad \mathbf{x}_C = \mathbf{x}_A - l\mathbf{e}_1, \tag{4.124}$$

respectively. It can be verified that the acceleration of C is

$$\ddot{\mathbf{x}}_C = \ddot{y}\mathbf{j} - l\ddot{\theta}\mathbf{e}_2 + l\dot{\theta}^2\mathbf{e}_1. \tag{4.125}$$

4.5.2.2 Equations of Motion

The FBD of the airfoil is shown in Fig. 4.16. The constraint force $\mathbf{N} = N\mathbf{i}$ is necessary to keep A moving inside the vertical track. It is an additional unknown that needs to be eliminated. Use Euler's laws of motion to obtain the equations

$$N\mathbf{i} - k(y - l_0)\mathbf{j} + mg\mathbf{j} = m\ddot{y}\mathbf{j} - ml\ddot{\theta}\mathbf{e}_2 + ml\dot{\theta}^2\mathbf{e}_1, \tag{4.126}$$

$$-k_t\theta - kl(y - l_0)\cos\theta - Nl\sin\theta = I_{zz}\ddot{\theta}. \tag{4.127}$$

The first equation of motion can be extracted by taking the projection of Eq. (4.126) along $\mathbf{j}$:

$$m\ddot{y} - ml\ddot{\theta}\cos\theta + ml\dot{\theta}^2\sin\theta + k(y - l_0) - mg = 0. \tag{4.128}$$

Additional steps are necessary to obtain the second equation of motion. First, take the projection of (4.126) along $\mathbf{i}$ to find the constraint force N:

$$N = ml\ddot{\theta}\sin\theta + ml\dot{\theta}^2\cos\theta. \tag{4.129}$$

Additionally, the relationship

$$-k(y - l_0) = m\ddot{y} - ml\ddot{\theta}\cos\theta + ml\dot{\theta}^2\sin\theta - mg \tag{4.130}$$

follows from Eq. (4.128). Substitute Eqs. (4.129) and (4.130) into (4.127) to obtain the second equation of motion:

$$-ml\ddot{y}\cos\theta + (I_{zz} + ml^2)\ddot{\theta} + k_t\theta + mgl\cos\theta = 0. \tag{4.131}$$

The two equations of motion can be expressed in the standard matrix-vector form:

$$\begin{bmatrix} m & -ml\cos\theta \\ -ml\cos\theta & I_{zz} + ml^2 \end{bmatrix} \begin{bmatrix} \ddot{y} \\ \ddot{\theta} \end{bmatrix} + \begin{bmatrix} ml\dot{\theta}^2\sin\theta + k(y - l_0) - mg \\ k_t\theta + mgl\cos\theta \end{bmatrix} = \begin{bmatrix} 0 \\ 0 \end{bmatrix}. \tag{4.132}$$

4.5.2.3 Linearization

The equilibrium positions y_e and θ_e are solutions of the equilibrium equations

$$\begin{bmatrix} k(y_e - l_0) - mg \\ k_t\theta_e + mgl\cos\theta_e \end{bmatrix} = \begin{bmatrix} 0 \\ 0 \end{bmatrix}. \tag{4.133}$$

The equilibrium coordinate y_e is

$$y_e = l_0 + \frac{mg}{k}. \tag{4.134}$$

At equilibrium, the airfoil is not horizontal but rotated by an angle θ_e that satisfies the nonlinear equation

$$\theta_e = -\frac{mgl}{k_t}\cos\theta_e. \tag{4.135}$$

This transcendental equation can only be solved numerically.

A multivariate Taylor series expansion around the equilibrium state $(y, \theta, \dot{y}, \dot{\theta}) = (y_e, \theta_e, 0, 0)$ is used to linearize the equations of motion (4.132) as

$$\begin{bmatrix} m & -ml\cos\theta_e \\ -ml\cos\theta_e & I_{zz} + ml^2 \end{bmatrix} \begin{bmatrix} \ddot{q}_1 \\ \ddot{q}_2 \end{bmatrix} + \begin{bmatrix} k & 0 \\ 0 & k_t + mgl\sin\theta_e \end{bmatrix} \begin{bmatrix} q_1 \\ q_2 \end{bmatrix} = \begin{bmatrix} 0 \\ 0 \end{bmatrix}. \tag{4.136}$$

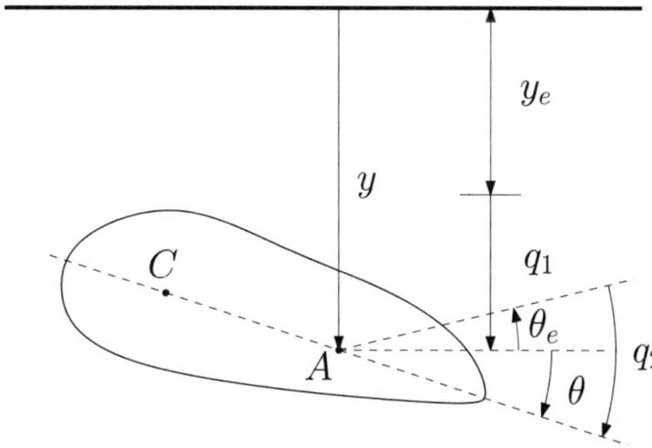

Fig. 4.17 Generalized coordinates defined from the static equilibrium position

where $q_1 = y - y_e$ and $q_2 = \theta - \theta_e$ are the generalized coordinates measured from the equilibrium position shown in Fig. 4.17. The airfoil is elastically decoupled because only the stiffness matrix is diagonal.

In real systems, such as an airplane wing, the equilibrium angle θ_e can be very small. Therefore, the additional approximations $\cos\theta_e \approx 1$ and $\sin\theta_e \approx 0$ can be substituted into (4.136) to simplify it as

$$\begin{bmatrix} m & -ml \\ -ml & I_{zz} + ml^2 \end{bmatrix} \begin{bmatrix} \ddot{q}_1 \\ \ddot{q}_2 \end{bmatrix} + \begin{bmatrix} k & 0 \\ 0 & k_t \end{bmatrix} \begin{bmatrix} q_1 \\ q_2 \end{bmatrix} = \begin{bmatrix} 0 \\ 0 \end{bmatrix}. \tag{4.137}$$

4.5.2.4 Alternative Derivation of the Equations of Motion

The derivation of the linearized equations of motion (4.137) can be streamlined by introducing suitable approximations from the outset. Assume that the states $(y, \theta, \dot{y}, \dot{\theta})$ of the system remain in a small neighborhood around $(y_e, \theta_e, 0, 0)$ when this equilibrium state is perturbed. From this, it follows that

$$\mathbf{e}_1 \approx \mathbf{i} + q_2\mathbf{j}, \tag{4.138}$$

$$\mathbf{e}_2 \approx -q_2\mathbf{i} + \mathbf{j}, \tag{4.139}$$

$$\mathbf{x}_C \approx -l\mathbf{i} + (q_1 + y_e - lq_2)\mathbf{j}, \tag{4.140}$$

$$\ddot{\mathbf{x}}_C \approx (\ddot{q}_1 - l\ddot{q}_2)\mathbf{j}. \tag{4.141}$$

It was assumed that θ_e is very small. Referring to the FBD shown in Fig. 4.16, it can be verified that application of Euler's laws yields

$$m(\ddot{q}_1 - l\ddot{q}_2) = -k(q_1 + y_e - l_0) + mg, \tag{4.142}$$

$$N = 0, \tag{4.143}$$

$$I_{zz}\ddot{q}_2 = -k_t(q_2 + \theta_e) - kl(q_1 + y_e - l_0). \tag{4.144}$$

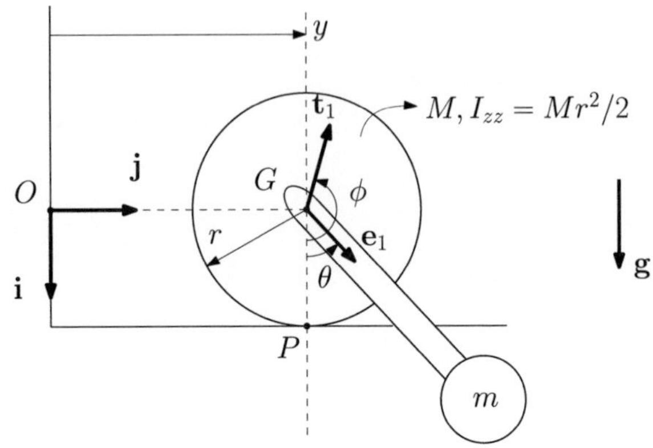

Fig. 4.18 A system composed of a cylinder connected to a pendulum. The force of gravity acts along **i**

The system of equations in (4.137) can be derived by following the same steps as in Sect. 4.5.2.2.

4.5.3 An Unrestrained System

The objective of this last example is to derive the equations of motion of the system shown in Fig. 4.18. The system consists of a homogeneous cylinder with mass M and radius r that rolls without slipping on a flat surface. A rigid pendulum with length L and concentrated mass m is attached to the cylinder via an ideal pin joint located at the cylinder's center of mass C.

4.5.3.1 Kinematic Analysis

A fixed origin is placed at O together with the orthonormal Cartesian vectors $\{\mathbf{i}, \mathbf{j}, \mathbf{k}\}$. The corotational basis $\{\mathbf{t}_1 = \cos\phi\mathbf{i} + \sin\phi\mathbf{j},\ \mathbf{t}_2 = -\sin\phi\mathbf{i} + \cos\phi\mathbf{j},\ \mathbf{t}_3 = \mathbf{k} = \mathbf{t}_1 \times \mathbf{t}_2\}$ tracks the angular motion of the cylinder, while $\{\mathbf{e}_1 = \cos\theta\mathbf{i} + \sin\theta\mathbf{j},\ \mathbf{e}_2 = -\sin\theta\mathbf{i} + \cos\theta\mathbf{j},\ \mathbf{e}_3 = \mathbf{k} = \mathbf{e}_1 \times \mathbf{e}_2\}$ tracks that of the pendulum. The position, velocity, and acceleration vectors of the mass center C of the cylinder can be represented by

$$\mathbf{x}_1 = y\mathbf{j}, \quad \dot{\mathbf{x}}_1 = \dot{y}\mathbf{j}, \quad \ddot{\mathbf{x}}_1 = \ddot{y}\mathbf{j}, \tag{4.145}$$

respectively. Similarly, the motion of the pendulum can be parameterized by

$$\mathbf{x}_2 = \mathbf{x}_1 + L\mathbf{e}_1, \quad \dot{\mathbf{x}}_2 = \dot{\mathbf{x}}_1 + L\dot{\theta}\mathbf{e}_2, \quad \ddot{\mathbf{x}}_2 = \ddot{\mathbf{x}}_1 + L\ddot{\theta}\mathbf{e}_2 - L\dot{\theta}^2\mathbf{e}_1. \tag{4.146}$$

Three coordinates are used to parameterize the vectors in Eqs. (4.145) and (4.146), but the system has only two DOFs. The motion of the center of mass of the cylinder satisfies the constraint

$$\psi_1 = \mathbf{x}_1 \cdot \mathbf{i} = 0. \tag{4.147}$$

In addition, the pin joint connecting the cylinder and the pendulum at C introduces the constraints

$$\psi_2 = (\mathbf{x}_2 - \mathbf{x}_1 - L\mathbf{e}_1) \cdot \mathbf{i} = 0, \tag{4.148}$$

$$\psi_3 = (\mathbf{x}_2 - \mathbf{x}_1 - L\mathbf{e}_1) \cdot \mathbf{j} = 0. \tag{4.149}$$

The last kinematic constraint concerns the rolling condition, whereas the velocity of the instantaneous point of contact P between the cylinder and the horizontal surface satisfies

$$\mathbf{v}_P = \mathbf{0}. \tag{4.150}$$

This constraint implies that y, ϕ, and their derivatives are related by

$$\Delta y = -r\Delta\phi, \quad \dot{y} = -r\dot{\phi}, \quad \ddot{y} = -r\ddot{\phi}. \tag{4.151}$$

Therefore, the motion of this system can be completely characterized by ϕ and θ only.

4.5.3.2 Equations of Motion

The FBDs for each rigid body is shown in Fig. 4.19. The normal force $\mathbf{N}$ and static friction force $\mathbf{F}_f$ act at P. There is an unknown force $\mathbf{R}$ at C due to the pin joint connection. Apply Euler's laws to the cylinder to obtain

$$\mathbf{N} + \mathbf{F}_f = -Mg\mathbf{i} + \mathbf{R} - Mr\ddot{\phi}\mathbf{j}, \tag{4.152}$$

$$r\mathbf{j} \cdot (\mathbf{N} + \mathbf{F}_f) = \frac{1}{2}Mr^2\ddot{\phi}. \tag{4.153}$$

Proceeding similarly for the rigid pendulum:

$$\mathbf{R} = -mg\mathbf{i} - mr\ddot{\phi}\mathbf{j} + mL\ddot{\theta}\mathbf{e}_2 - mL\dot{\theta}^2\mathbf{e}_1, \tag{4.154}$$

$$\mathbf{e}_2 \cdot \mathbf{R} = 0. \tag{4.155}$$

To obtain the first equation of motion, start by substituting Eq. (4.154) into (4.152):

$$\mathbf{N} + \mathbf{F}_f = -(M+m)g\mathbf{i} - (M+m)r\ddot{\phi}\mathbf{j} + mL\ddot{\theta}\mathbf{e}_2 - mL\dot{\theta}^2\mathbf{e}_1. \tag{4.156}$$

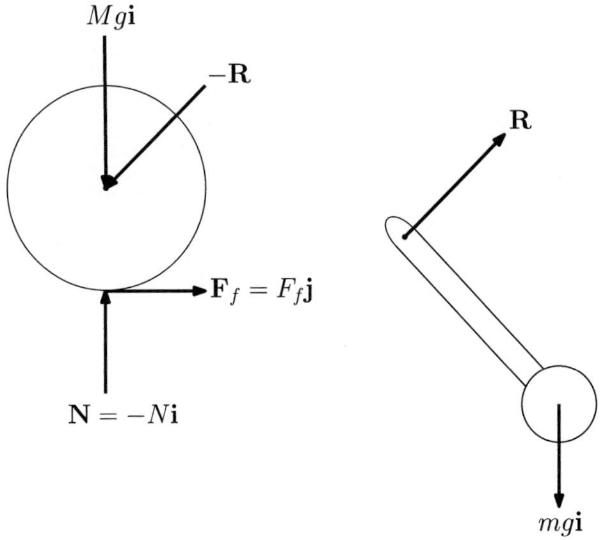

Fig. 4.19 FBDs of the cylinder and pendulum

This expression is substituted into Eq. (4.152) to reveal the first equation of motion:

$$\frac{1}{2}(3M + 2m)r^2\ddot{\phi} - mrL\ddot{\theta}\cos\theta + mrL\dot{\theta}^2\sin\theta = 0. \tag{4.157}$$

The second equation is obtained by taking the inner product of Eq. (4.154) with the vector $L\mathbf{e}_2$ while considering the result from Eq. (4.155):

$$-mrL\ddot{\phi}\cos\theta + mL^2\ddot{\theta} + mgL\sin\theta = 0. \tag{4.158}$$

Finally, Eqs. (4.157) and (4.158) can be expressed as

$$\begin{bmatrix} \frac{1}{2}(3M + 2m)r^2 & -mrL\cos\theta \\ -mrL\cos\theta & mL^2 \end{bmatrix} \begin{bmatrix} \ddot{\phi} \\ \ddot{\theta} \end{bmatrix} + \begin{bmatrix} mrL\dot{\theta}^2\sin\theta \\ mgL\sin\theta \end{bmatrix} = \begin{bmatrix} 0 \\ 0 \end{bmatrix}. \tag{4.159}$$

4.5.3.3 Numerical Integration

The equations of motion (4.159) form a system of two nonlinear ODEs that can only be solved numerically. The MATLAB code `Chapt4_code3` from the *Electronic Supplementary Material* numerically integrates these equations using the parameters

$$M = 8\,\text{kg}, \quad m = 1\,\text{kg}, \quad L = 0.7\,\text{m}, \quad r = 0.2\,\text{m}. \tag{4.160}$$

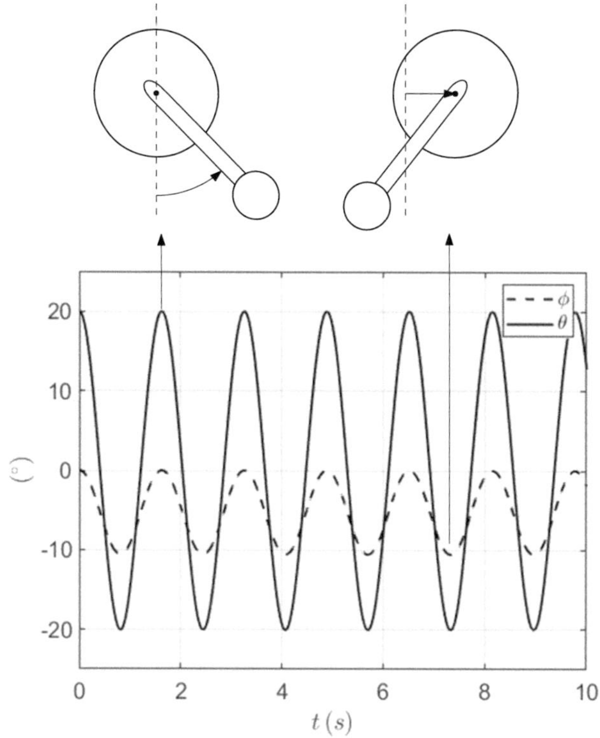

Fig. 4.20 Solution of Eq. (4.159) for the ICs $[\phi(0) \quad \theta(0) \quad \dot{\phi}(0) \quad \dot{\theta}(0)] = [0 \quad \pi/9 \quad 0 \quad 0]$

One type of motion for this system is showcased in Fig. 4.20, where the cylinder and the pendulum oscillate back and forth. Figure 4.21 shows another type of motion allowed for the system. If only the cylinder is given an initial push, it oscillates continually with constant angular velocity, and the pendulum is carried without oscillating. This type of motion, called *rigid body mode*, is possible because there are no restraints that keep the system in place.

4.5.3.4 Linearization

The constant solutions $\phi = \phi_e$ and $\theta = \theta_e$ are substituted into Eq. (4.159) so the following equilibrium equation is obtained:

$$\sin\theta_e = 0. \tag{4.161}$$

The constant angle ϕ_e can take any value and θ_e can be any integer multiple of π. The two physically distinct equilibrium configurations are depicted in Fig. 4.22.

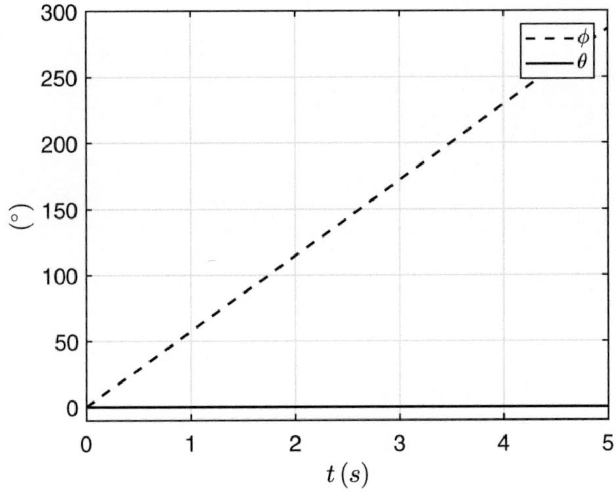

Fig. 4.21 Solution of Eq. (4.159) for the ICs $[\phi(0) \quad \theta(0) \quad \dot{\phi}(0) \quad \dot{\theta}(0)] = [0 \quad 0 \quad 1 \quad 0]$

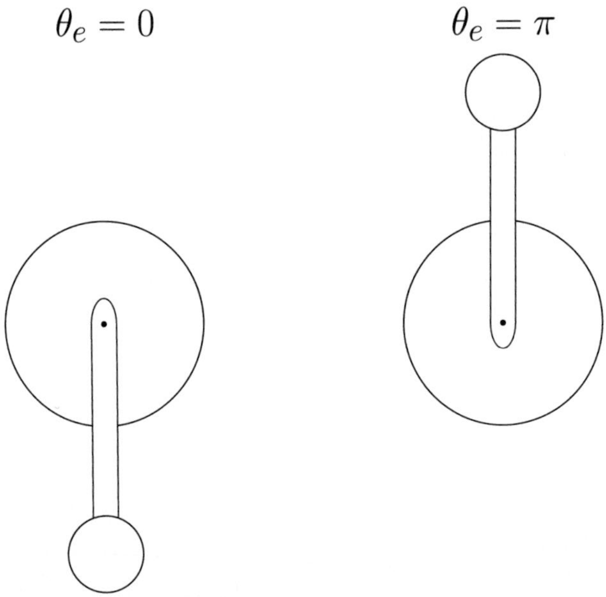

Fig. 4.22 Equilibrium positions

Assuming that the state $(\phi, \theta, \dot{\phi}, \dot{\theta})$ remains close to (0, 0, 0, 0) during the motion, Eq. (4.159) is linearized according with

$$\begin{bmatrix} \frac{1}{2}(3M+2m)r^2 & -mrL \\ -mrL & mL^2 \end{bmatrix} \begin{bmatrix} \ddot{\phi} \\ \ddot{\theta} \end{bmatrix} + \begin{bmatrix} 0 & 0 \\ 0 & mgL \end{bmatrix} \begin{bmatrix} \phi \\ \theta \end{bmatrix} = \begin{bmatrix} 0 \\ 0 \end{bmatrix}. \tag{4.162}$$

The mass and stiffness matrices are symmetric, but **K** has another interesting property. Recall that the potential energy associated with the linearized system is

$$U = \frac{1}{2} \begin{bmatrix} \phi & \theta \end{bmatrix} \mathbf{K} \begin{bmatrix} \phi \\ \theta \end{bmatrix} = \frac{1}{2} mgL\theta^2. \tag{4.163}$$

The potential energy is $U = 0$ for the rigid body mode allowed for this system ($\phi \neq 0$ but with $\theta = 0$). This result is translated mathematically into the stiffness matrix **K** being positive semidefinite, i.e., there are nonzero vectors **x** such that

$$\mathbf{x}^T\mathbf{K}\mathbf{x} = 0. \tag{4.164}$$

4.6 Summary

Suppose the motion of a mechanical system with n DOFs is described by n generalized coordinates $x_1, x_2, \ldots, x_n$. In general, these coordinates satisfy the system of nonlinear second-order ODEs

$$\mathbf{M}(\mathbf{x}, t)\ddot{\mathbf{x}} + \mathbf{g}(\mathbf{x}, \dot{\mathbf{x}}, t) = \mathbf{f}(t),$$

where

$$\mathbf{x} = \begin{bmatrix} x_1 \\ x_2 \\ \vdots \\ x_n \end{bmatrix}$$

is the vector of generalized coordinates,

$$\mathbf{f}(t) = \begin{bmatrix} f_1(t) \\ f_2(t) \\ \vdots \\ f_n(t) \end{bmatrix}$$

is the vector of externally applied loads, $\mathbf{g}(\mathbf{x},\, \dot{\mathbf{x}},\, t)$ is an n-dimensional vector function and $\mathbf{M}(\mathbf{x},\, t)$ is a square matrix of order n. In the analysis of small vibrations, it is customary to linearize the equations of motion around a stable equilibrium $\mathbf{x} = \mathbf{x}_e$. The generalized Taylor expansion around this equilibrium point often leads to the approximation

$$\mathbf{M}\ddot{\mathbf{q}} + \mathbf{C}\dot{\mathbf{q}} + \mathbf{K}\mathbf{q} = \mathbf{f}(t), \tag{4.165}$$

where

$$\mathbf{q} = \begin{bmatrix} q_1 \\ q_2 \\ \vdots \\ q_n \end{bmatrix}$$

is the vector of generalized coordinates measured from the equilibirum position. The coefficients **M**, **C**, and **K** are constant and symmetric matrices of order n called mass, damping, and stiffness matrices, respectively. Under broad conditions, this is the prototype equations of motion for MDOF systems.

4.7 Problems

4.1 Check that the system of differential equations (4.96) satisfies the principle of superposition.

4.2 As shown in Fig. 4.23 a particle with mass m is attached to a point O by an elastic spring with stiffness k and unstretched length l_0. The particle's motion takes

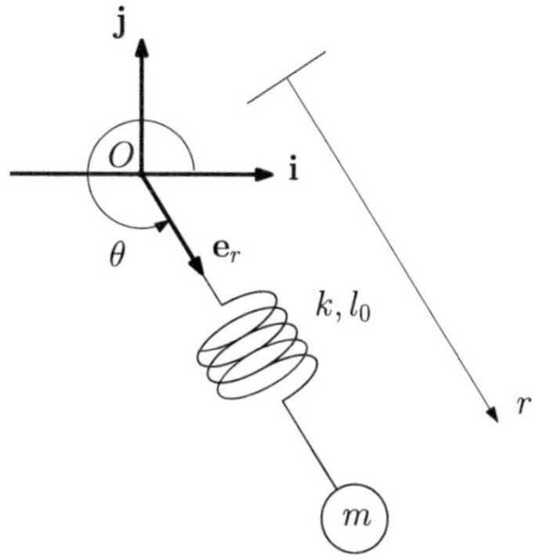

Fig. 4.23 Particle connected to the fixed origin O by a linear spring with stiffness k and unstretched length l_0. The force of gravity acts along $-\mathbf{j}$

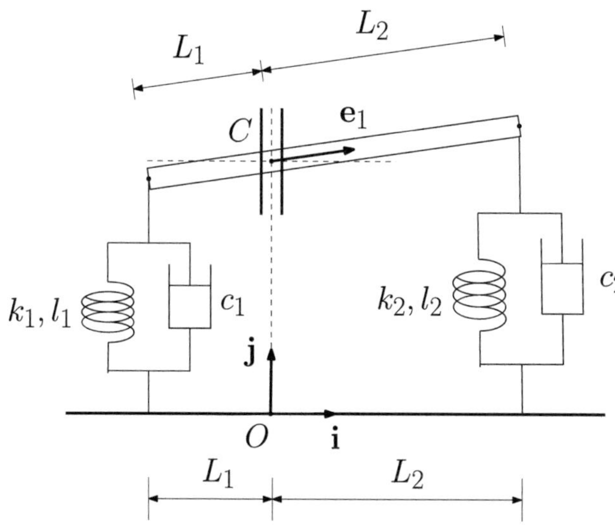

Fig. 4.24 Homogeneous bar. The gravitational force acts along $-\mathbf{j}$

place in the $\mathbf{i}-\mathbf{j}$ plane. Find the nonlinear equations of motion and then linearize them around the stable equilibrium.

4.3 The homogeneous bar shown in Fig. 4.24 has mass m and moment of inertia I_{zz} with respect to the $\mathbf{k}$ axis that passes through the center of mass C. The center of mass C is constrained to move inside a smooth vertical track. Its position vector can be represented by

$$\mathbf{x}_C = x\mathbf{j}. \tag{4.166}$$

Assuming the bar executes small vibrations, use Euler's laws to find the equations of motion for this system.

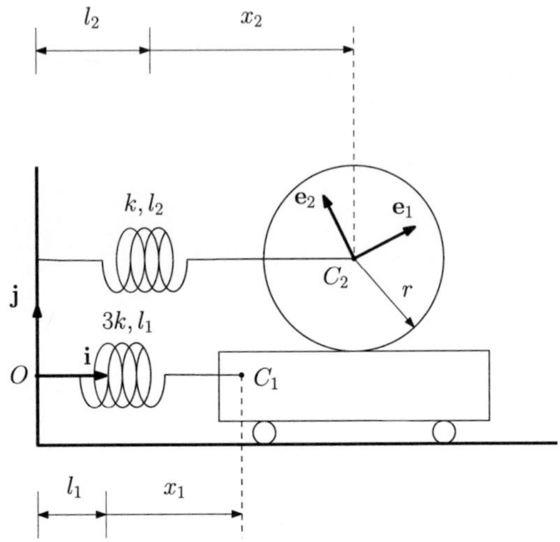

Fig. 4.25 Cylinder and cart system. The gravitational force acts along $-\mathbf{j}$

4.4 Figure 4.25 shows a mechanical system where a cylinder rolls without slipping on a cart that translates on a smooth surface. The cylinder has a radius r, mass m, and the position vector of its center of mass C_2 is

$$\mathbf{x}_2 = (l_2 + x_2)\mathbf{i} + y_0\mathbf{j}, \tag{4.167}$$

where y_0 is constant. The cart has a mass of $2m$ and the position vector of its mass center C_1 is

$$\mathbf{x}_1 = (l_1 + x_1)\mathbf{i}. \tag{4.168}$$

Find the equations of motion for this system.

4.5 A rigid bar has mass m_3 concentrated at its free end. As shown in Fig. 4.26, this bar is connected to two particles by means of linear springs. The positions of the points of interest can be represented by

$$\mathbf{r}_1 = a_1\mathbf{i} - y_1(t)\mathbf{j}, \tag{4.169}$$

$$\mathbf{r}_2 = (a_1 + a_2 + a_3)\mathbf{i} - y_2(t)\mathbf{j}, \tag{4.170}$$

$$\bar{\mathbf{x}}_3 = (a_1 + a_2 + a_3 + a_4)\mathbf{e}_1, \tag{4.171}$$

$$\mathbf{x}_A = a_1\mathbf{e}_1, \tag{4.172}$$

$$\mathbf{x}_B = (a_1 + a_2)\mathbf{e}_1, \tag{4.173}$$

$$\mathbf{x}_C = (a_1 + a_2 + a_3)\mathbf{e}_1, \tag{4.174}$$

$$\mathbf{r}_D = (a_1 + a_2)\mathbf{i} + l_3\mathbf{j}, \tag{4.175}$$

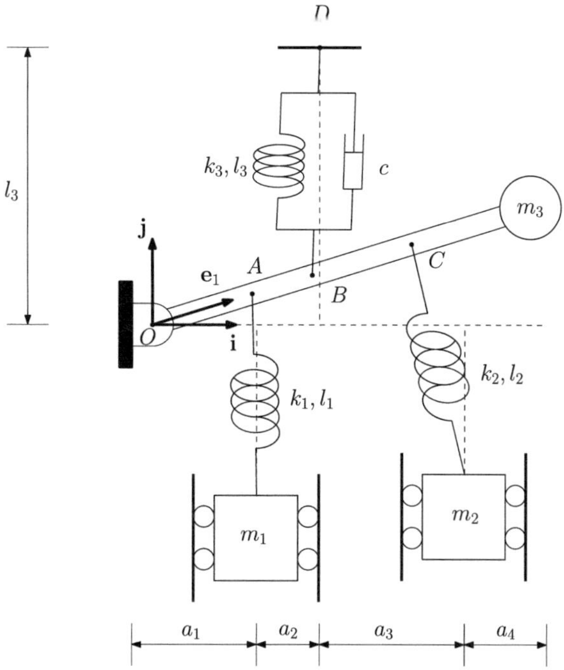

Fig. 4.26 System consisting of a bar connected to two blocks. The gravitational force acts along $-\mathbf{j}$

where $y_1, y_2 > 0$. Assuming that the system is under the small vibrations regime, find the equations of motion.

4.6 The model shown in Fig. 4.27 is used to analyze the vibrations experienced by a building due to earthquakes. The earthquakes are represented by the base motion $y(t)$. The building is a block with mass M and mass center C_1 connected to a bar with mass m and mass center C_2. A torsional spring with stiffness k_t is present at the pin joint connection at A. The position vector of C_1 has the representation

$$\mathbf{r}_1 = x\mathbf{j}, \tag{4.176}$$

while that of C_2 is

$$\bar{\mathbf{x}}_2 = \mathbf{r}_1 + \mathbf{c}_0 + l\mathbf{e}_1, \tag{4.177}$$

where $\mathbf{c}_0$ is a constant vector. Find the equations of motion for small vibrations assuming that θ varies slightly around $\theta = 0$.

4.7 Consider the double pendulum shown in Fig. 4.28. It is formed by two particles connected to ideal cables. The position vectors of the particles can be represented by

$$\mathbf{r}_1 = l_1\mathbf{e}_{r_1}(\theta_1), \tag{4.178}$$

$$\mathbf{r}_2 = \mathbf{r}_1 + l_2\mathbf{e}_{r_2}(\theta_2). \tag{4.179}$$

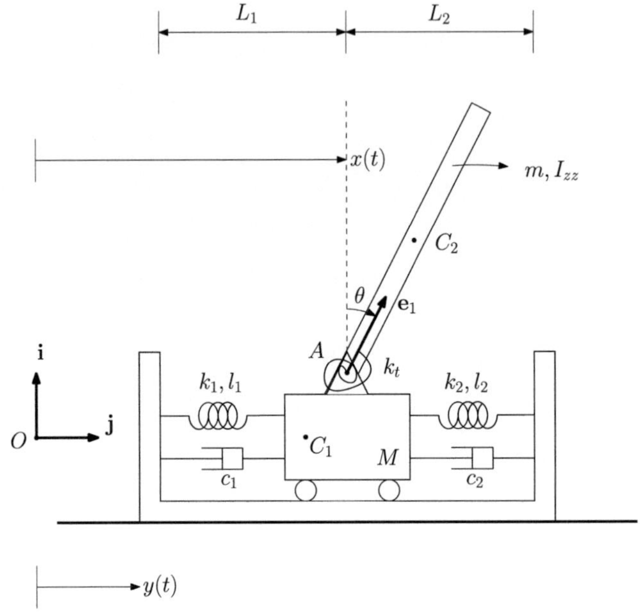

Fig. 4.27 Model for a building excited by an earthquake. The gravitational force acts along $-\mathbf{i}$

Fig. 4.28 Double pendulum. Gravitational forces act along $\mathbf{i}$

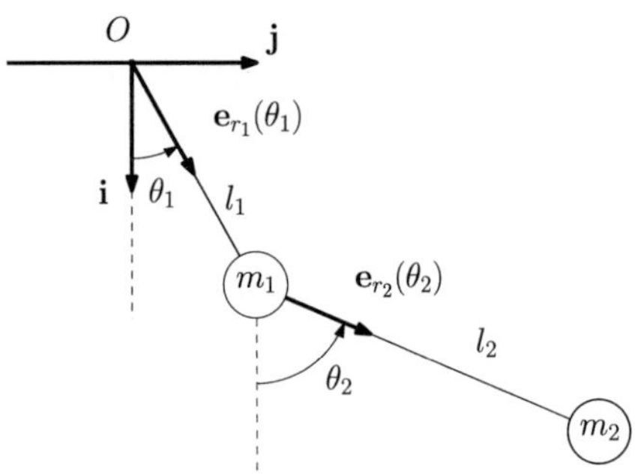

Find the nonlinear equations of motion. Use these equations to determine the equilibria and linearize the equations around the stable equilibrium position.

4.8 As shown in Fig. 4.29, a rigid bar of length $2l$, and masses $2m$ concentrated at the ends, is free to rotate around the axis $\mathbf{k}$ that passes through point O. One end of the bar is connected to a block with mass m. The positions of points A and B are

$$\mathbf{r}_A = l\mathbf{i} - l_0\mathbf{j}, \quad \mathbf{r}_B = -l\mathbf{i} + 2l_0\mathbf{j}. \tag{4.180}$$

The position of the block can be parameterized by

$$\mathbf{r} = -l\mathbf{i} + y\mathbf{j}. \tag{4.181}$$

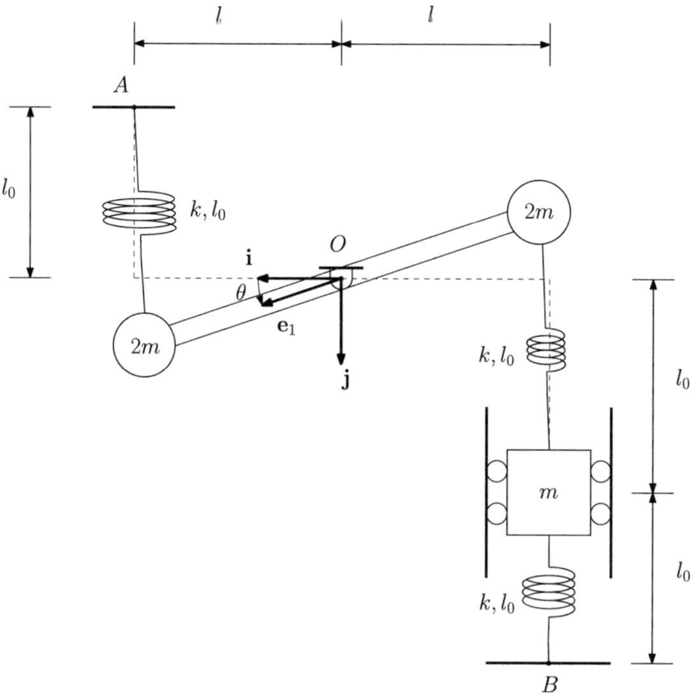

Fig. 4.29 System formed by a bar connected to a block. The gravitational force acts along **j**

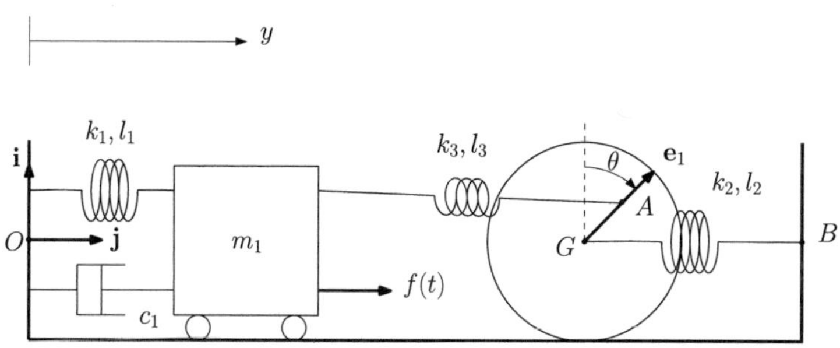

Fig. 4.30 Cylinder-block system. The gravitational force acts along $-\mathbf{i}$

Find the equations of motion for small vibrations around the equilibrium position.

4.9 The system shown in Fig. 4.30 consists of a uniform cylinder with radius r and mass m_2, which rolls without slipping. This cylinder is connected to a block with mass m_1 through a linear spring. The positions of the points of interest are parameterized by

$$\mathbf{r} = y\mathbf{j}, \tag{4.182}$$

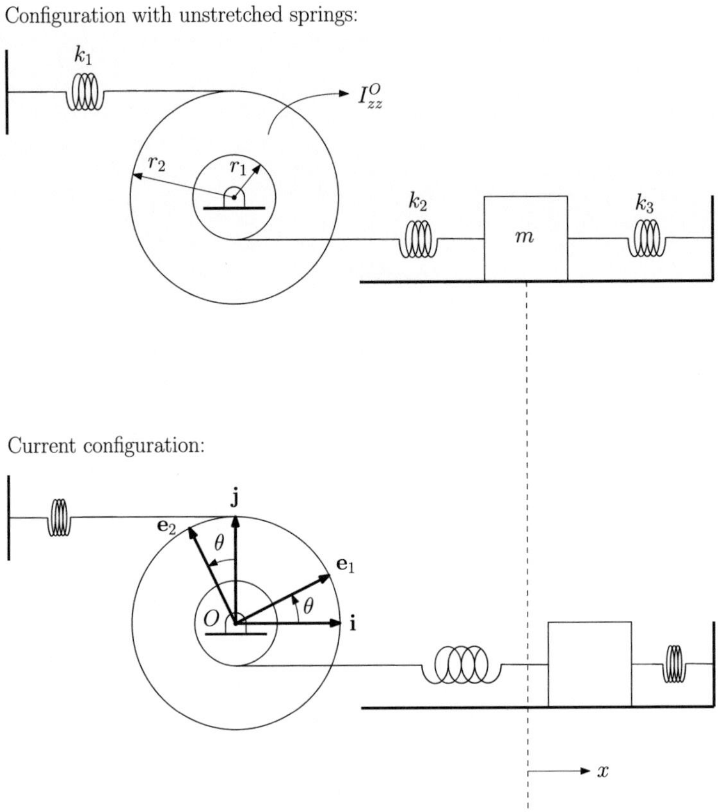

Fig. 4.31 Drum connected to a block that slides on a smooth surface. The gravitational force acts along $-\mathbf{j}$

$$\bar{\mathbf{x}} = x\mathbf{j}, \tag{4.183}$$

$$\mathbf{x}_A = \bar{\mathbf{x}} + a\mathbf{e}_1, \tag{4.184}$$

$$\mathbf{r}_B = (l_1 + l_2 + l_3)\mathbf{j}, \tag{4.185}$$

where a is a constant. Find the equations of motion.

4.10 The mechanical system shown in Fig. 4.31 consists of a drum connected to a block by means of a linear spring. In the position where the springs are relaxed, $\theta = 0$ and $x = 0$. Parameterize the position of the block by

$$\mathbf{r} = -r_1\mathbf{j} + (x - l_2)\mathbf{i}, \tag{4.186}$$

where l_2 is the relaxed length of the spring with stiffness k_2. Determine the equations of motion using coordinates measured from the equilibrium position.

Chapter 5
Free and Forced Vibrations of Systems with Multiple Degrees of Freedom

This chapter analyzes the free and forced vibrations of systems with multiple degrees of freedom. It covers both damped and undamped free vibrations, introducing key concepts such as natural frequencies, mode shapes, and complex modes. For forced vibrations, the frequency response function matrix is used to discuss the system response to varying excitation frequencies. The chapter concludes with an overview of modal analysis. The physics of free and forced vibrations are initially illustrated with examples, followed by the general theory. MATLAB codes are provided to solve generalized and quadratic eigenproblems.

5.1 Introduction

Recall from Chap. 4 that the dynamics of an n-DOF mechanical systems are completely described by n generalized coordinates $q_1, q_2, \ldots, q_n$. In the "small" vibrations regime, these coordinates satisfy a system of n second-order ODEs, generally cast in the compact matrix-vector form:

$$\mathbf{M}\ddot{\mathbf{q}} + \mathbf{C}\dot{\mathbf{q}} + \mathbf{K}\mathbf{q} = \mathbf{f}(t), \tag{5.1}$$

Supplementary Information The online version contains supplementary material available at (https://doi.org/10.1007/978-3-031-83995-5_5).

R. Gonçalves Salsa Junior, *Fundamentals of the Theory of Mechanical Vibrations*,
https://doi.org/10.1007/978-3-031-83995-5_5

where

$$\mathbf{q} = \begin{bmatrix} q_1 \\ q_2 \\ \vdots \\ q_n \end{bmatrix} \tag{5.2}$$

is the vector of generalized coordinates, and

$$\mathbf{f}(t) = \begin{bmatrix} f_1(t) \\ f_2(t) \\ \vdots \\ f_n(t) \end{bmatrix} \tag{5.3}$$

is the vector of externally applied loads. The $n \times n$ symmetric and constant matrices **M**, **C**, and **K** are called the mass, damping, and stiffness matrices.

The analysis of MDOF systems requires solving the prototype equations of motion (5.1). This process is clearly more complex than solving the equation of motion for SDOF systems, as it involves working with coupled differential equations. Nonetheless, many of the physical concepts and terminology from SDOF systems remain applicable. For instance, free vibration analysis still refers to problems where $\mathbf{f} = \mathbf{0}$, and the response is determined solely by the ICs $\mathbf{q}(0) = \mathbf{q}_0$ and $\dot{\mathbf{q}}(0) = \dot{\mathbf{q}}_0$. Forced vibration involves solving the system of differential equations (5.1) with $\mathbf{f} \neq \mathbf{0}$. This task is greatly facilitated by the fact that (5.1) is a system of *linear* differential equations. According to the principle of superposition, the response for forced vibrations is

$$\mathbf{q}(t) = \mathbf{q}_h(t) + \mathbf{q}_p(t). \tag{5.4}$$

The general homogeneous solution, $\mathbf{q}_h$, satisfies

$$\mathbf{M}\ddot{\mathbf{q}}_h + \mathbf{C}\dot{\mathbf{q}}_h + \mathbf{K}\mathbf{q}_h = \mathbf{0}, \tag{5.5}$$

while the particular solution $\mathbf{q}_p(t)$, which is independent of ICs, satisfies

$$\mathbf{M}\ddot{\mathbf{q}}_p + \mathbf{C}\dot{\mathbf{q}}_p + \mathbf{K}\mathbf{q}_p = \mathbf{f}(t). \tag{5.6}$$

The ICs $\mathbf{q}(0) = \mathbf{q}_0$ and $\dot{\mathbf{q}}(0) = \dot{\mathbf{q}}_0$ must be satisfied by the complete solution $\mathbf{q}$ to determine the arbitrary constants of integration contained in the homogeneous part.

The objective of this chapter is to investigate the solutions of the prototype equations of motion (5.1). The theoretical discussions rely on matrix algebra and eigenvalue problems. To enhance understanding and avoid heavy mathematics from the outset, the material is introduced through examples that illustrate the underlying physics before addressing the general cases. The chapter is organized as follows. The responses of undamped and damped systems in free vibrations are covered in Sects. 5.2 and 5.3, respectively. Forced vibrations problems are discussed in Sect. 5.4. MATLAB codes that numerically solve the associated eigenvalues problems are provided. Additionally, a technique known as modal analysis, which transforms MDOF systems into independent SDOF systems, is introduced in Sect. 5.5. The key results from this chapter are summarized in Sect. 5.6.

5.2 Free Vibration of Conservative Systems

The first case to be treated is that of the free vibration of undamped MDOF systems. Specifically, the goal is to solve

$$\mathbf{M}\ddot{\mathbf{q}} + \mathbf{K}\mathbf{q} = \mathbf{0}, \tag{5.7}$$

for a given set of ICs $\mathbf{q}(0) = \mathbf{q}_0$ and $\dot{\mathbf{q}}(0) = \dot{\mathbf{q}}_0$. Solving a system of differential equations is more complex than solving a single ODE, so the following examples introduce the physics of free vibration before the general case is presented.

5.2.1 *A System with Two Particles*

Consider the system with two particles shown in Fig. 5.1. This is a realization of the system from Sect. 4.2 with $m_1 = m_2 = m$, $c_1 = c_2 = c_3 = 0$, and $k_1 = k_2 = k_3 = k$. The equations of motion using the generalized coordinates q_1 and q_2 measured from the static equilibrium position take the form

$$\begin{bmatrix} m & 0 \\ 0 & m \end{bmatrix} \begin{bmatrix} \ddot{q}_1 \\ \ddot{q}_2 \end{bmatrix} + \begin{bmatrix} 2k & -k \\ -k & 2k \end{bmatrix} \begin{bmatrix} q_1 \\ q_2 \end{bmatrix} = \begin{bmatrix} 0 \\ 0 \end{bmatrix}. \tag{5.8}$$

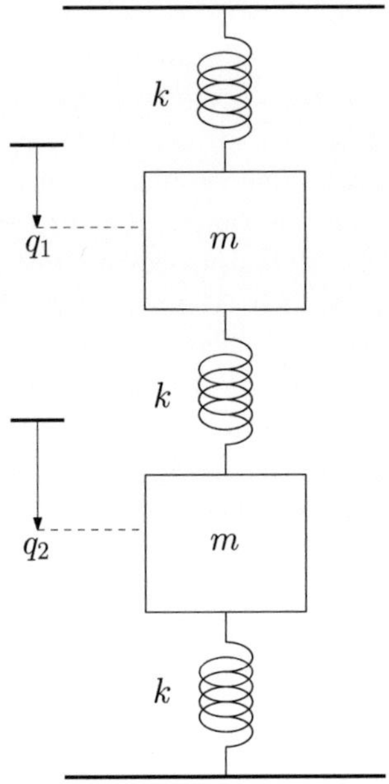

Fig. 5.1 A system with two particles

It is assumed the system is subjected to the ICs

$$\mathbf{q}(0) = \begin{bmatrix} q_{1_0} \\ \\ q_{2_0} \end{bmatrix}, \quad \dot{\mathbf{q}}(0) = \begin{bmatrix} \dot{q}_{1_0} \\ \\ \dot{q}_{2_0} \end{bmatrix}. \tag{5.9}$$

5.2.1.1 The Natural Frequencies

How can Eq. (5.8) be solved? Recall that the free vibration of an undamped SDOF system consists of harmonic oscillation with frequency equal to the natural frequency, which depends only on the system's inertia and stiffness. A parallel can be drawn by interpreting SDOF systems as special cases of MDOF systems when $n = 1$. Under this interpretation, one can seek a solution to Eq. (5.8) that corresponds to a response where both particles oscillate harmonically with the same natural frequency ω:

$$\mathbf{q}(t) = \mathbf{u}e^{i\omega t}. \tag{5.10}$$

The vector of constant amplitudes

$$\mathbf{u} = \begin{bmatrix} u_1 \\ \\ u_2 \end{bmatrix} \tag{5.11}$$

is also known as *mode shape* or *normal mode*. The mode shape $\mathbf{u}$ and natural frequency ω that are allowed as solutions can be determined by differentiating Eq. (5.10) twice and substituting the result into Eq. (5.8). This yields

$$(\mathbf{K} - \omega^2\mathbf{M})\mathbf{u} = \mathbf{0}. \tag{5.12}$$

Note that this equation can be thought of as a linear system of algebraic equations for the unknowns in $\mathbf{u}$. If the square matrix $\mathbf{K} - \omega^2\mathbf{M}$ is invertible, then the unique solution of Eq. (5.12) is $\mathbf{u} = \mathbf{0}$. This is the undesired trivial solution. Therefore, the assumed solution (5.10) allows only the values of ω that make the matrix $\mathbf{K} - \omega^2\mathbf{M}$ singular. A necessary and sufficient condition for this matrix to be singular is[1]

$$\det(\mathbf{K} - \omega^2\mathbf{M}) = 0. \tag{5.13}$$

Compute the determinant to obtain the *characteristic polynomial*

$$(\omega^2)^2 - \frac{4k}{m}(\omega^2) + \frac{3k^2}{m^2} = 0. \tag{5.14}$$

This is a polynomial of degree 4 in ω. As can be noted, the four roots occur in two pairs of the form $\pm\,\omega$. The negative roots have no physical meaning since a natural frequency is always $\omega \geq 0$, and mathematically, they merely represent the complex conjugates of (5.10). It is more practical to consider the characteristic polynomial as quadratic in ω^2 and compute the two natural frequencies as:

$$\omega_1 = \sqrt{\frac{k}{m}}, \quad \omega_2 = \sqrt{\frac{3k}{m}}. \tag{5.15}$$

This system has two DOFs and, therefore, has two natural frequencies that depend only on the system's inertia and stiffness. By convention, the natural frequencies are arranged in order of increasing magnitude such that $\omega_1 \leq \omega_2$. The lowest frequency ω_1 is called the fundamental frequency.

[1] Appendix C contains a concise review of matrix algebra.

5.2.1.2 Determination of the Mode Shapes

Once the natural frequencies are determined, it is necessary to find the vectors $\mathbf{u} \neq \mathbf{0}$ that are allowed as solutions. Substitute the first natural frequency into (5.12) to obtain

$$\begin{bmatrix} k & -k \\ -k & k \end{bmatrix} \begin{bmatrix} u_{11} \\ u_{21} \end{bmatrix} = \begin{bmatrix} 0 \\ 0 \end{bmatrix}. \tag{5.16}$$

This is a system of linear algebraic equations for the entries u_{11} and u_{21} of the mode shape $\mathbf{u}_1$, associated with the first natural frequency. It can be verified that any vector $\mathbf{u}_1$ with $u_{11} = u_{21} \neq 0$ is a solution to this system. The result is that the mode shape can only be determined up to a multiplicative constant. This is expected, as the singular matrix $\mathbf{K} - \omega_1^2\mathbf{M}$ causes the equations in (5.16) to be linearly dependent, resulting in infinitely many solutions. The process of choosing any particular solution is called *normalization*. For example, one normalization scheme involves choosing a unit first entry, i.e., $u_{11} = 1$:

$$\mathbf{u}_1 = \begin{bmatrix} 1 \\ 1 \end{bmatrix}. \tag{5.17}$$

Normalization has no physical significance and is just a matter of convenience.

The second mode shape can be obtained by substituting ω_2 into (5.12):

$$\begin{bmatrix} -k & -k \\ -k & -k \end{bmatrix} \begin{bmatrix} u_{12} \\ u_{22} \end{bmatrix} = \begin{bmatrix} 0 \\ 0 \end{bmatrix}. \tag{5.18}$$

There are infinitely many solutions for which the entries in $\mathbf{u}_2$ satisfy $u_{12} = -u_{22} \neq 0$. Choose the same normalization used for the first mode to obtain

$$\mathbf{u}_2 = \begin{bmatrix} 1 \\ -1 \end{bmatrix}. \tag{5.19}$$

5.2.1.3 The General Response

Since the system under analysis is linear, the general response is the superposition of all the solutions found in the previous analysis.[2] It takes the form

$$\begin{aligned}\mathbf{q}(t) = A_1 \begin{bmatrix} 1 \\ 1 \end{bmatrix} e^{i\omega_1 t} + B_1 \begin{bmatrix} 1 \\ 1 \end{bmatrix} e^{-i\omega_1 t} \\ + A_2 \begin{bmatrix} 1 \\ -1 \end{bmatrix} e^{i\omega_2 t} + B_2 \begin{bmatrix} 1 \\ -1 \end{bmatrix} e^{-i\omega_2 t},\end{aligned} \tag{5.20}$$

where the integration constants A_1, B_1, A_2, and B_2 are determined from the ICs. Using Euler's identity, the response can be recast as

$$\begin{aligned}\begin{bmatrix} q_1(t) \\ q_2(t) \end{bmatrix} = \begin{bmatrix} 1 \\ 1 \end{bmatrix} (a_1 \cos \omega_1 t + b_1 \sin \omega_1 t) \\ + \begin{bmatrix} 1 \\ -1 \end{bmatrix} (a_2 \cos \omega_2 t + b_2 \sin \omega_2 t),\end{aligned} \tag{5.21}$$

where a_1, b_1, a_2, and b_2 are also determined from the ICs. For illustration purposes, assume that $q_{1_0} = q_0$ and $q_{2_0} = \dot{q}_{1_0} = \dot{q}_{2_0} = 0$. It can be verified that

$$\begin{bmatrix} q_1(t) \\ q_2(t) \end{bmatrix} = \frac{q_0}{2} \begin{bmatrix} \cos\left(\sqrt{\frac{k}{m}}t\right) + \cos\left(\sqrt{\frac{3k}{m}}t\right) \\ \cos\left(\sqrt{\frac{k}{m}}t\right) - \cos\left(\sqrt{\frac{3k}{m}}t\right) \end{bmatrix}. \tag{5.22}$$

The responses predicted by this equation are illustrated in Fig. 5.2. While the free response of undamped SDOF systems is harmonic, the responses shown for q_1 and q_2 are neither harmonic nor periodic. This stark contrast in behavior is due to the superposition of two natural frequencies in Eq. (5.22).

[2] The solutions $\pm\,\omega$ share the same vector $\mathbf{u}$ because ω is squared in Eq. (5.12).

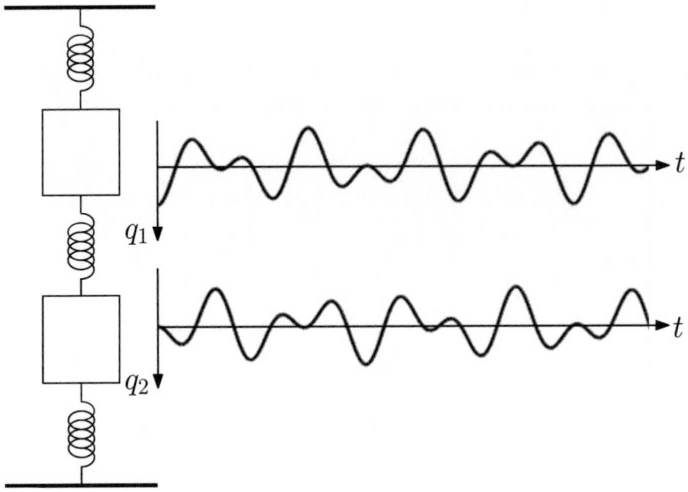

Fig. 5.2 Free vibration of the system with two particles for the ICs $q_{1_0} = q_0$ and $q_{2_0} = \dot{q}_{1_0} = \dot{q}_{2_0} = 0$

5.2.1.4 Motion as Modal Superposition

Note that the harmonic functions that share the same frequency in Eq. (5.21) can combined according to

$$a_1 \cos \omega_1 t + b_1 \sin \omega_1 t = X_1 \cos(\omega_1 t - \phi_1) \tag{5.23}$$

and

$$a_2 \cos \omega_2 t + b_2 \sin \omega_2 t = X_2 \cos(\omega_2 t - \phi_2). \tag{5.24}$$

The constants X_1, X_2, ϕ_1, and ϕ_2 can be determined in terms of a_1, b_1, a_2, and b_2 using the methodology presented in Chap. 2. The expressions aforementioned can be used to rewrite the general response of the system as

$$\mathbf{q}(t) = X_1 \mathbf{s}_1(t - \tau_1) + X_2 \mathbf{s}_2(t - \tau_2), \tag{5.25}$$

where the phase angles ϕ_1 and ϕ_2 have been replaced by the time delays defined by $\omega_1 \tau_1 = \phi_1$ and $\omega_2 \tau_2 = \phi_2$. The result from Eq. (5.25) is remarkable because it implies that, no matter how complicated the system response is, it can always be decomposed as the superposition of the two special motions given by

$$\mathbf{s}_1(t) = \begin{bmatrix} 1 \\ 1 \end{bmatrix} \cos\left(\sqrt{\frac{k}{m}} t\right), \quad \mathbf{s}_2(t) = \begin{bmatrix} 1 \\ -1 \end{bmatrix} \cos\left(\sqrt{\frac{3k}{m}} t\right). \tag{5.26}$$

Fig. 5.3 Plots for q_1 (—)and q_2 (- -)for each mode

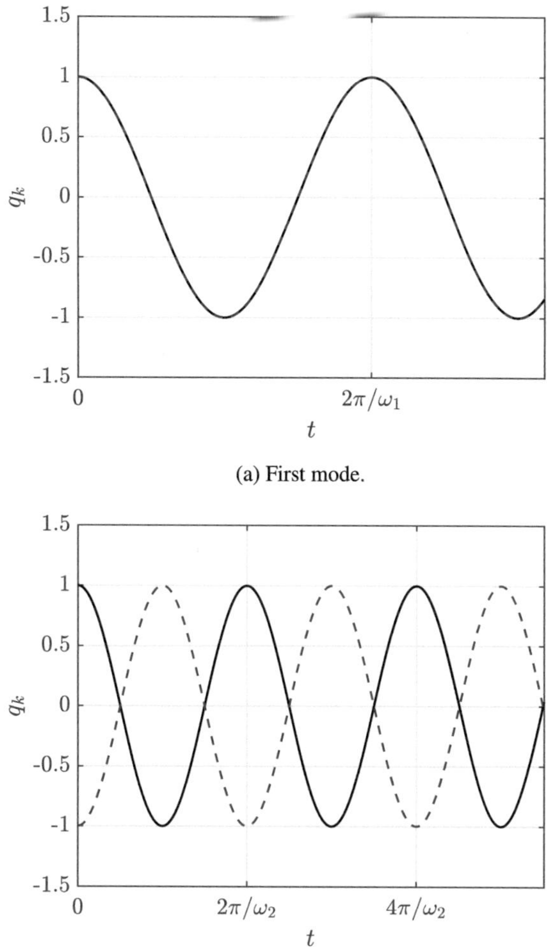

(a) First mode.

(b) Second mode.

For example, these two special motions contribute equally to the response in Eq. (5.22):

$$\mathbf{q}(t) = \frac{q_0}{2}\mathbf{s}_1(t) + \frac{q_0}{2}\mathbf{s}_2(t). \tag{5.27}$$

Motions like $\mathbf{s}_1(t)$ and $\mathbf{s}_2(t)$ are called normal or natural *modes* of vibration. In each mode, as illustrated in Fig. 5.3, the two particles perform simple harmonic motion with the same natural frequency, passing through their equilibrium positions at the same instant of time. The amplitude ratio of the particles and their relative phase lag are determined, for a specific mode, by the entries of the corresponding mode shape vector. For example, the ratio

$$\frac{u_{21}}{u_{11}} = 1 \tag{5.28}$$

indicates that the two particles have the same vibration amplitude and move in phase in the first mode. In contrast, the second mode is defined by

$$\frac{u_{22}}{u_{12}} = -1. \tag{5.29}$$

Therefore, the particles oscillate with the same amplitude but out of phase in this mode.

Although the general response of the system is a combination of the normal modes, each mode can be individually excited. It is easy to verify that the system vibrates in the first mode if it is excited with the ICs

$$\mathbf{q}(0) = \begin{bmatrix} 1 \\ 1 \end{bmatrix}, \quad \dot{\mathbf{q}}(0) = \mathbf{0}, \tag{5.30}$$

or in the second mode if

$$\mathbf{q}(0) = \begin{bmatrix} 1 \\ -1 \end{bmatrix}, \quad \dot{\mathbf{q}}(0) = \mathbf{0}. \tag{5.31}$$

5.2.2 A System with Three Particles

The previous example introduced the physics of undamped free vibration. It is still necessary to verify how those ideas extend to systems with more DOFs. To do that, consider as an example the system composed of three particles shown in Fig. 5.4. The equations of motion with respect to the generalized coordinates q_1, q_2, and q_3 measured from the equilibrium position take the form

$$\begin{bmatrix} m & 0 & 0 \\ 0 & m & 0 \\ 0 & 0 & m \end{bmatrix} \begin{bmatrix} \ddot{q}_1 \\ \ddot{q}_2 \\ \ddot{q}_3 \end{bmatrix} + \begin{bmatrix} 2k & -k & 0 \\ -k & 2k & -k \\ 0 & -k & 2k \end{bmatrix} \begin{bmatrix} q_1 \\ q_2 \\ q_3 \end{bmatrix} = \begin{bmatrix} 0 \\ 0 \\ 0 \end{bmatrix}. \tag{5.32}$$

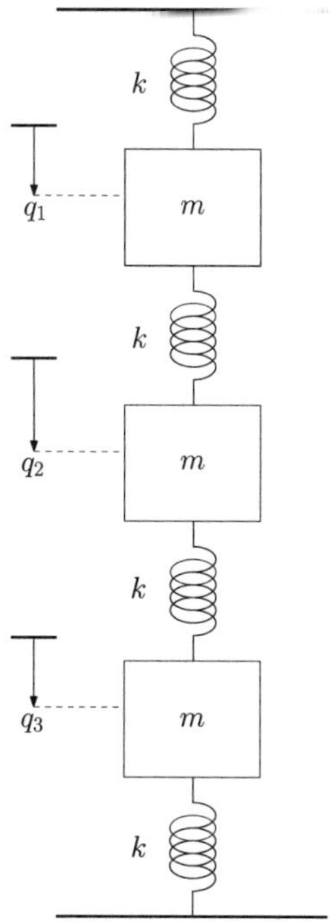

Fig. 5.4 A system of three particles particles with mass m interconnected by springs with stiffness k

5.2.2.1 Natural Frequencies and Mode Shapes

Following the same steps from Sect. 5.2.1, assume that the solution is oscillatory:

$$\mathbf{q}(t) = \mathbf{u}e^{i\omega t}, \tag{5.33}$$

where the vector of constant amplitudes is

$$\mathbf{u} = \begin{bmatrix} u_1 \\ u_2 \\ u_3 \end{bmatrix}. \tag{5.34}$$

Differentiate this expression two times and substitute into Eq. (5.32) to obtain

$$\left(\begin{bmatrix} 2k & -k & 0 \\ -k & 2k & -k \\ 0 & -k & 2k \end{bmatrix} - \omega^2 \begin{bmatrix} m & 0 & 0 \\ 0 & m & 0 \\ 0 & 0 & m \end{bmatrix} \right) \begin{bmatrix} u_1 \\ u_2 \\ u_3 \end{bmatrix} = \begin{bmatrix} 0 \\ 0 \\ 0 \end{bmatrix}. \tag{5.35}$$

The trivial solution can be avoided by finding the values of ω that satisfy $\det(\mathbf{K} - \omega^2\mathbf{M}) = 0$. The determinant results in the characteristic polynomial

$$(\omega^2)^3 - \frac{6k}{m}(\omega^2)^2 + \frac{10k^2}{m^2}(\omega^2) - \frac{4k^3}{m^3} = 0. \tag{5.36}$$

This is a cubic polynomial for ω^2. It can be verified that the three natural frequencies are

$$\omega_1 = \sqrt{2-\sqrt{2}}\sqrt{\frac{k}{m}}, \quad \omega_2 = \sqrt{2}\sqrt{\frac{k}{m}}, \quad \omega_3 = \sqrt{2+\sqrt{2}}\sqrt{\frac{k}{m}}. \tag{5.37}$$

This system has three natural frequencies because it has three DOFs. The mode shape associated with the first natural frequency can be found by solving the system of linear equations that results from substituting ω_1 into Eq. (5.35):

$$\begin{bmatrix} \sqrt{2}k & -k & 0 \\ -k & \sqrt{2}k & -k \\ 0 & -k & \sqrt{2}k \end{bmatrix} \begin{bmatrix} u_{11} \\ u_{21} \\ u_{31} \end{bmatrix} = \begin{bmatrix} 0 \\ 0 \\ 0 \end{bmatrix}. \tag{5.38}$$

This system has infinitely many solutions that satisfy

$$u_{11} = u_{31}, \tag{5.39}$$

$$u_{21} = \sqrt{2}u_{31}. \tag{5.40}$$

Choose a normalization with $u_{11} = 1$ to obtain

$$\mathbf{u}_1 = \begin{bmatrix} 1 \\ \sqrt{2} \\ 1 \end{bmatrix}. \tag{5.41}$$

The second mode shape vector is found by solving the following system of equations:

$$\begin{bmatrix} 0 & -k & 0 \\ -k & 0 & -k \\ 0 & -k & 0 \end{bmatrix} \begin{bmatrix} u_{12} \\ u_{22} \\ u_{32} \end{bmatrix} = \begin{bmatrix} 0 \\ 0 \\ 0 \end{bmatrix}. \tag{5.42}$$

The solutions must satisfy

$$u_{12} = -u_{32}, \tag{5.43}$$

$$u_{22} = 0. \tag{5.44}$$

Using the same normalization as before, the second mode shape vector becomes

$$\mathbf{u}_2 = \begin{bmatrix} 1 \\ 0 \\ -1 \end{bmatrix}. \tag{5.45}$$

The system of equations that can be solved for the third mode shape is

$$\begin{bmatrix} -\sqrt{2}k & -k & 0 \\ -k & -\sqrt{2}k & -k \\ 0 & -k & -\sqrt{2}k \end{bmatrix} \begin{bmatrix} u_{13} \\ u_{23} \\ u_{33} \end{bmatrix} = \begin{bmatrix} 0 \\ 0 \\ 0 \end{bmatrix}. \tag{5.46}$$

The entries of $\mathbf{u}_3$ satisfy

$$u_{13} = u_{33}, \tag{5.47}$$

$$u_{23} = -\sqrt{2}u_{33}. \tag{5.48}$$

One choice for this vector is

$$\mathbf{u}_3 = \begin{bmatrix} 1 \\ -\sqrt{2} \\ 1 \end{bmatrix}. \tag{5.49}$$

5.2.2.2 General Free Response

The general response in free vibration is given by

$$\begin{aligned}\mathbf{q}(t) = {} & A_1 \begin{bmatrix} 1 \\ \sqrt{2} \\ 1 \end{bmatrix} e^{i\omega_1 t} + B_1 \begin{bmatrix} 1 \\ \sqrt{2} \\ 1 \end{bmatrix} e^{-i\omega_1 t} + A_2 \begin{bmatrix} 1 \\ 0 \\ -1 \end{bmatrix} e^{i\omega_2 t} \\ & + B_2 \begin{bmatrix} 1 \\ 0 \\ -1 \end{bmatrix} e^{-i\omega_2 t} + A_3 \begin{bmatrix} 1 \\ -\sqrt{2} \\ 1 \end{bmatrix} e^{i\omega_3 t} + B_3 \begin{bmatrix} 1 \\ -\sqrt{2} \\ 1 \end{bmatrix} e^{-i\omega_3 t} .\end{aligned} \tag{5.50}$$

Use Euler's identity to express this vector in the form

$$\begin{aligned}\mathbf{q}(t) = {} & \begin{bmatrix} 1 \\ \sqrt{2} \\ 1 \end{bmatrix} (a_1 \cos \omega_1 t + b_1 \sin \omega_1 t) \\ & + \begin{bmatrix} 1 \\ 0 \\ -1 \end{bmatrix} (a_2 \cos \omega_2 t + b_2 \sin \omega_2 t) \\ & + \begin{bmatrix} 1 \\ -\sqrt{2} \\ 1 \end{bmatrix} (a_3 \cos \omega_3 t + b_3 \sin \omega_3 t) .\end{aligned} \tag{5.51}$$

The integration constants need to be determined from the ICs. For example, if

$$\mathbf{q}(0) = \begin{bmatrix} 1 \\ 0 \\ 0 \end{bmatrix}, \quad \dot{\mathbf{q}}(0) = \begin{bmatrix} 0 \\ 1 \\ 0 \end{bmatrix}, \tag{5.52}$$

it can be shown that

$$\mathbf{q}(t) = \frac{1}{4}\begin{bmatrix} 1 \\ \sqrt{2} \\ 1 \end{bmatrix}\left(\cos\omega_1 t + \frac{\sqrt{2}}{\omega_1}\sin\omega_1 t\right) + \frac{1}{2}\begin{bmatrix} 1 \\ 0 \\ -1 \end{bmatrix}\cos\omega_2 t$$
$$+ \frac{1}{4}\begin{bmatrix} 1 \\ -\sqrt{2} \\ 1 \end{bmatrix}\left(\cos\omega_3 t - \frac{\sqrt{2}}{\omega_3}\sin\omega_3 t\right). \tag{5.53}$$

The plots in Fig. 5.5 show the responses predicted by Eq. (5.53).

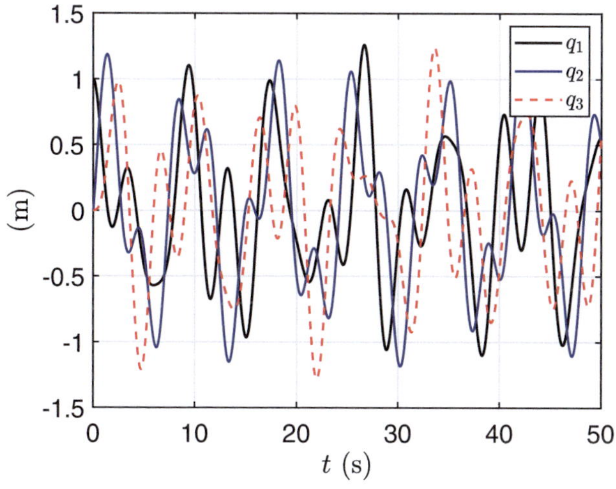

Fig. 5.5 Response of the system with three particles from Fig. 5.4. Plot constructed with $m = 1$ kg, $k = 1$ N/m, and the ICs $\mathbf{q}(0) = [1 \quad 0 \quad 0]^T$ (m) and $\dot{\mathbf{q}}(0) = [0 \quad 1 \quad 0]^T$ (m/s)

5.2.2.3 Normal Modes

This system can vibrate in three normal modes determined by the mode shapes $\mathbf{u}_1$, $\mathbf{u}_2$, and $\mathbf{u}_3$. In the first mode

$$\mathbf{s}_1(t) = \begin{bmatrix} 1 \\ \sqrt{2} \\ 1 \end{bmatrix} \cos\left(\sqrt{2-\sqrt{2}}\sqrt{\frac{k}{m}}t\right), \tag{5.54}$$

the particles oscillate harmonically with frequency ω_1 and in phase. In addition, the amplitude of vibration of the intermediate particle is $\sqrt{2}$ times greater than the other two. This is evidenced by the plot from Fig. 5.6a.

The second mode is

$$\mathbf{s}_2(t) = \begin{bmatrix} 1 \\ 0 \\ -1 \end{bmatrix} \cos\left(\sqrt{2}\sqrt{\frac{k}{m}}t\right). \tag{5.55}$$

In this mode, the top and bottom particles oscillate with the same amplitude but move out phase due to the sign difference in the first and third entries of $\mathbf{u}_2$. This means that the elastic forces exerted on the intermediate particle cancel out, and it stays at rest ($u_{22} = 0$), as illustrated in Fig. 5.6b. A non-clamped position in a vibrating system at which the displacement is zero at all times is called *node*.

In the third mode,

$$\mathbf{s}_3(t) = \begin{bmatrix} 1 \\ -\sqrt{2} \\ 1 \end{bmatrix} \cos\left(\sqrt{2}\sqrt{\frac{k}{m}}t\right), \tag{5.56}$$

the top and bottom particles oscillate in phase, while the intermediate particle moves out of phase with respect to them. In addition, this particle has greater amplitude of vibration, as demonstrated in Fig. 5.6c.

The free undamped response (5.51) can be expressed as the superposition of these three normal modes. For example, the response shown in Fig. 5.5 is

$$\mathbf{q}(t) = 0.5253\mathbf{s}_1(t - 1.4042) + 0.5\mathbf{s}_2(t) + 0.3148\mathbf{s}_3(t - 3.0469). \tag{5.57}$$

Fig. 5.6 Modes for the system with three particles from Fig. 5.4. Here, coordinates q_1, q_2, and q_3 are represented by curves (—), (—), and (- -), respectively

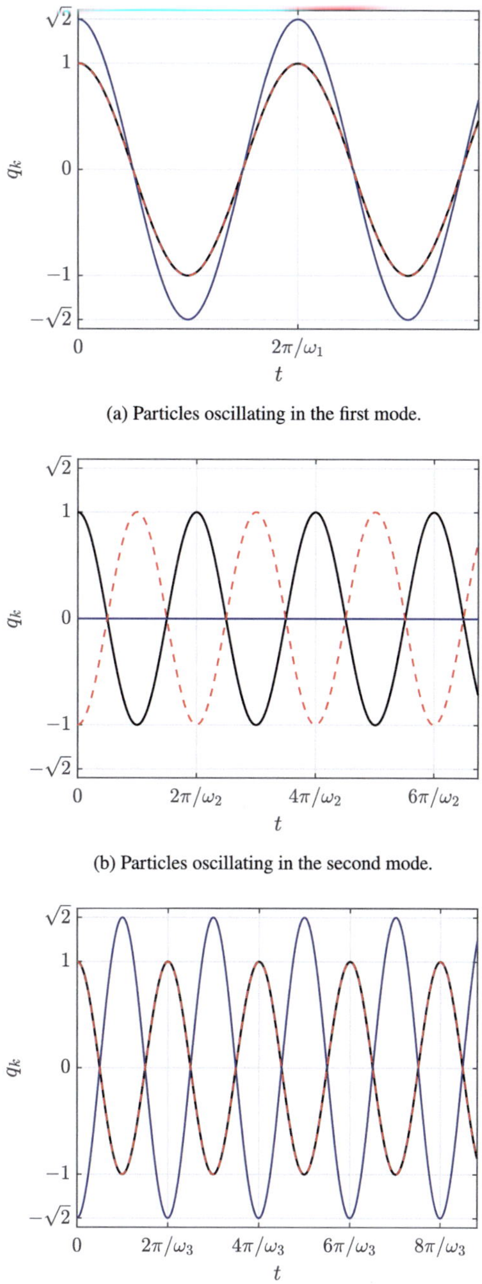

(a) Particles oscillating in the first mode.

(b) Particles oscillating in the second mode.

(c) Particles oscillating in the third mode.

5.2.3 *Computing the Response of Undamped Systems*

The previous examples indicate that there is a general procedure for computing the free response of undamped MDOF systems. The steps are outlined as follows.

5.2.3.1 Natural Frequencies

The response of n-DOFs system governed by Eq. (5.7) is postulated to be

$$\mathbf{q} = \mathbf{u}e^{i\omega t}, \tag{5.58}$$

where the mode shape $\mathbf{u} \neq \mathbf{0}$ is an n-dimensional vector, and ω is a natural frequency. Substituting the assumed solution into the equations of motion yields

$$(\mathbf{K} - \omega^2\mathbf{M})\mathbf{u} = \mathbf{0}. \tag{5.59}$$

The natural frequencies are roots of the characteristic polynomial

$$\det(\mathbf{K} - \omega^2\mathbf{M}) = 0, \tag{5.60}$$

which is of degree n in ω^2 when the determinant is expanded. There are n roots ω^2 that, by convention, are arranged in order of increasing magnitude: $0 \leq \omega_1 \leq \omega_2 \leq \ldots \leq \omega_n$.

5.2.3.2 Mode Shapes

The mode shapes associated with each natural frequency can be computed by solving the linear system of Eqs. (5.59). Each mode shape is determined up to a multiplicative constant. That is, if $\mathbf{u}$ is a mode shape associated with a natural frequency ω, then $\alpha\mathbf{u}$ is also a mode shape for any constant $\alpha \neq 0$. The process of choosing a particular scale factor α is called normalization. Normalization has no physical significance but serves as a convenient way to standardize the mode shapes.

An undamped linear dynamical system possesses n normal modes. The j-th normal mode ($j = 1, \ldots, n$) is the motion

$$\mathbf{s}_j(t) = \mathbf{u}_j \cos\omega_j t, \tag{5.61}$$

which represents a vibration pattern where all components of the system undergo simple harmonic motion with the same natural frequency ω_j, passing through their equilibrium positions at the same instant of time. The amplitude ratio between any two components remains constant over time, and any two components are either in phase or 180° out of phase. Each normal mode can be excited independently. For example, the mode $\mathbf{s}_j(t)$ can be excited with the ICs $\mathbf{q}(0) = \mathbf{u}_j$ and $\dot{\mathbf{q}}(0) = \mathbf{0}$.

5.2.3.3 System Response

The general response in free vibration is given by

$$\mathbf{q}(t) = \sum_{j=1}^{n} \mathbf{u}_j (a_j \cos \omega_j t + b_j \sin \omega_j t), \tag{5.62}$$

where the constants a_j, b_j ($j = 1, \dots, n$) are determined from the ICs. The response can also be represented as the superposition of the normal modes:

$$\mathbf{q}(t) = \sum_{j=1}^{n} X_j \mathbf{s}_j (t - \tau_j), \tag{5.63}$$

where the modal participation factors X_j and the time delays τ_j are also determined from the ICs. In general, the response of each DOF is neither harmonic nor periodic. An overview of the solution procedure for undamped free vibration is presented in Table 5.1 together with a comparison between SDOF and MDOF systems. An undamped SDOF system in free vibration can be viewed as a special case of Eq. (5.7) with $n = 1$: the system oscillates harmonically at its single natural frequency.

5.2.3.4 Computer Implementation

The determination of natural frequencies requires finding the roots of the characteristic polynomial (5.59). This can be challenging to implement in practice for systems with many DOFs because there is no general explicit formula to find the roots of polynomials with degree $n > 4$. Therefore, an alternative method is necessary to

Table 5.1 Solution procedure for undamped free vibration

	$n = 1$	$n > 1$
Equations of motion	$m\ddot{q} + kq = 0$	$\mathbf{M}\ddot{\mathbf{q}} + \mathbf{K}\mathbf{q} = \mathbf{0}$
Assumed solution	$q = ue^{i\omega_n t}$	$\mathbf{q} = \mathbf{u}e^{i\omega t}$
Characteristic polynomial	$k - m\omega_n^2 = 0$	$\det(\mathbf{K} - \omega^2 \mathbf{M}) = 0$
Natural frequencies	$\omega_n = \sqrt{k/m}$	$0 \leq \omega_1 \leq \ldots \leq \omega_n$
Mode shapes	$u = 1$	$(\mathbf{K} - \omega_j^2 \mathbf{M})\mathbf{u}_j = \mathbf{0}$
System response	$q(t) = A \cos(\omega_n t - \phi)$	$\mathbf{q}(t) = \sum_{j=1}^{n} A_j \mathbf{u}_j \cos(\omega_j t - \phi_j)$

compute the natural frequencies. The first step in finding this alternative method involves rearranging Eq. (5.59) as

$$\mathbf{M}^{-1}\mathbf{K}\mathbf{u} = \omega^2\mathbf{u}. \tag{5.64}$$

This expression represents the typical eigenvalue–eigenvector problem:[3]

$$\mathbf{A}\mathbf{x} = \lambda\mathbf{x}, \tag{5.65}$$

where the eigenvalues λ of the matrix $\mathbf{A} = \mathbf{M}^{-1}\mathbf{K}$ are the squares of the natural frequencies and the eigenvectors $\mathbf{x}$ are the mode shape vectors. Thus, determining the natural frequencies and mode shapes is equivalent to solving an eigenvalue problem. There are several well-known numerical methods that can be used to find eigenvalues and eigenvectors of a matrix, but some efficient ones for this particular case take advantage of the symmetry of $\mathbf{M}$ and $\mathbf{K}$ to solve the generalized eigenproblem (5.59). In MATLAB, the following command can be used:

```
% Solve eigenvalue problem:
[U,OM]  = eig(K,M);
```

This function returns the eigenvectors and eigenvalues in the form

$$\mathbf{U} = \begin{bmatrix}\mathbf{u}_1 & \mathbf{u}_2 & \cdots & \mathbf{u}_n\end{bmatrix}, \quad \mathbf{\Omega} = \begin{bmatrix}\omega_1^2 & & \\ & \ddots & \\ & & \omega_n^2\end{bmatrix}. \tag{5.66}$$

Matrix $\mathbf{U}$ contains the mode shapes in its columns and is called modal matrix. Usually, MATLAB employs a normalization scheme that does not produce eigenvectors with unit Euclidean norm:

$$\mathbf{u}^T\mathbf{u} \neq 1. \tag{5.67}$$

The other output, $\mathbf{\Omega}$, is a diagonal matrix called spectral matrix. The diagonal entries are the square of the natural frequencies.

[3] See Appendix C.

5.2.4 *Illustrative Examples*

Three additional examples are provided in this section to further clarify the application of the theory of undamped free vibration.

5.2.4.1 The Carts-Pendulum System

Recall the carts-pendulum system from Sect. 4.3, which has three DOFs. The objective of this example is to find the three natural frequencies and associated mode shapes. To achieve this, set the damping coefficient to $c = 0$ and use the data provided in the aforementioned section. The equations of motion for free vibration then become:

$$\begin{bmatrix} 1 & 0 & 0 \\ 0 & 1.1 & 0.05 \\ 0 & 0.05 & 0.025 \end{bmatrix} \begin{bmatrix} \ddot{q}_1 \\ \ddot{q}_2 \\ \ddot{q}_3 \end{bmatrix} + \begin{bmatrix} 20 & -10 & 0 \\ -10 & 10 & 0 \\ 0 & 0 & 0.4905 \end{bmatrix} \begin{bmatrix} q_1 \\ q_2 \\ q_3 \end{bmatrix} = \mathbf{0}. \tag{5.68}$$

Use the MATLAB code provided in Sect. 5.2.3.4 to compute the modal and spectral matrices:

$$\mathbf{U} = \begin{bmatrix} 0.4925 & -0.5191 & 0.6986 \\ 0.8122 & -0.0143 & -0.5832 \\ 0.3538 & 5.4336 & 3.7883 \end{bmatrix}, \tag{5.69}$$

$$\boldsymbol{\Omega} = \begin{bmatrix} 3.5088 & 0 & 0 \\ 0 & 19.7240 & 0 \\ 0 & 0 & 28.3491 \end{bmatrix}. \tag{5.70}$$

The natural frequencies can be extracted from the diagonal entries in $\boldsymbol{\Omega}$: $\omega_1 = 1.8732$ rad/s, $\omega_2 = 4.4412$ rad/s, and $\omega_3 = 5.3244$ rad/s. The three corresponding mode shapes occupy the columns of $\mathbf{U}$. The normalization used by MATLAB did not produce eigenvectors with unit Euclidean norm:

$$\mathbf{u}_1^T\mathbf{u}_1 = 1.0273, \tag{5.71}$$

$$\mathbf{u}_2^T\mathbf{u}_2 = 29.7933, \tag{5.72}$$

$$\mathbf{u}_3^T\mathbf{u}_3 = 15.1793. \tag{5.73}$$

It can also be verified that the mode shapes do not form an orthogonal set:

$$\mathbf{u}_1^T\mathbf{u}_2 = 1.6549, \tag{5.74}$$

$$\mathbf{u}_1^T\mathbf{u}_3 = 1.2105, \tag{5.75}$$

$$\mathbf{u}_2^T\mathbf{u}_3 = 20.2297. \tag{5.76}$$

The lack of orthogonality is generally the rule. However, as will be demonstrated in Sect. 5.5, these vectors are orthogonal when using the inner product weighted by the mass or stiffness matrices:

$$\mathbf{u}_1^T\mathbf{M}\mathbf{u}_2 = \mathbf{u}_1^T\mathbf{M}\mathbf{u}_3 = \mathbf{u}_2^T\mathbf{M}\mathbf{u}_3 = 0, \tag{5.77}$$

$$\mathbf{u}_1^T\mathbf{K}\mathbf{u}_2 = \mathbf{u}_1^T\mathbf{K}\mathbf{u}_3 = \mathbf{u}_2^T\mathbf{K}\mathbf{u}_3 = 0. \tag{5.78}$$

The natural modes are shown in Fig. 5.7. In the first mode, all DOFs oscillate in phase, with the carts vibrating with greater amplitude than the pendulum. The carts are in phase in the second mode, but the pendulum moved out of phase with respect to them. In the third mode, the cart that is connected to the pendulum oscillates out of phase with respect to the other two components of the system. A node is present between the two carts due to their phase opposition.

5.2.4.2 Two Coupled Pendulums

As shown in Fig. 5.8, two identical pendulums are pin-jointed at their endpoints and can move freely in a vertical plane. Each pendulum is modeled as a massless rigid rod of length L with a concentrated mass m at the free end. The pendulums would oscillate independently if not for the spring, which has stiffness k and unstretched length l_0, connecting them at points A and B distanced d from the pin joints. The objective is to determine the response of the system when it is excited by ICs.

Kinematic Analysis

Relative to the fixed origin O, the position vector of the mass center of the pendulum on the left is

$$\mathbf{x}_1 = L\mathbf{e}_1, \tag{5.79}$$

where $\{\mathbf{e}_1 = \cos\theta\mathbf{i} + \sin\theta\mathbf{j},\ \mathbf{e}_2 = -\sin\theta\mathbf{i} + \cos\theta\mathbf{j},\ \mathbf{e}_3 = \mathbf{k} = \mathbf{e}_1 \times \mathbf{e}_2\}$ is the corotational basis that tracks its oscillation. The position vector of the center of

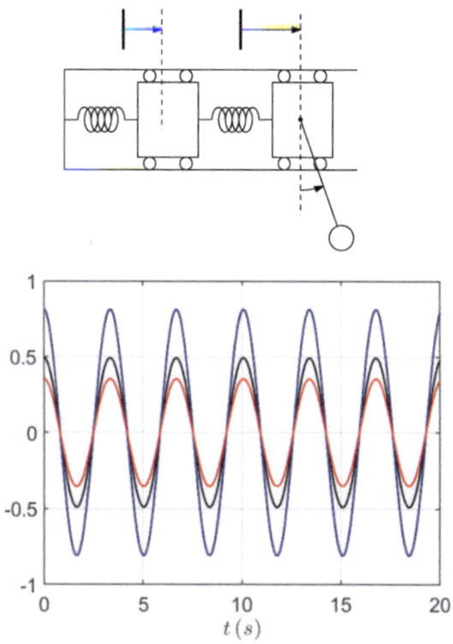

(a) First mode $\mathbf{s}_1(t) = \mathbf{u}_1 \cos \omega_1 t$.

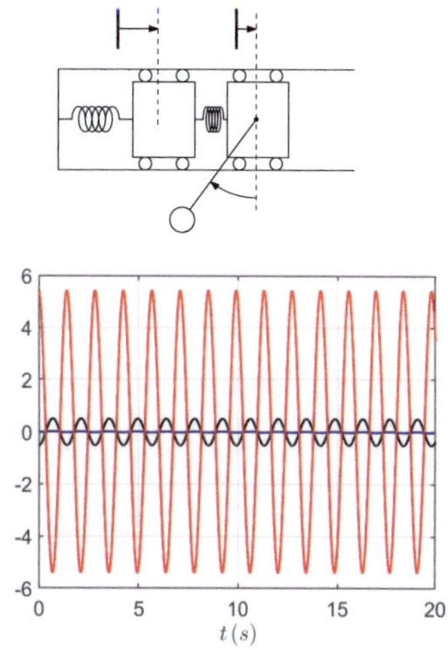

(b) Second mode $\mathbf{s}_2(t) = \mathbf{u}_2 \cos \omega_2 t$.

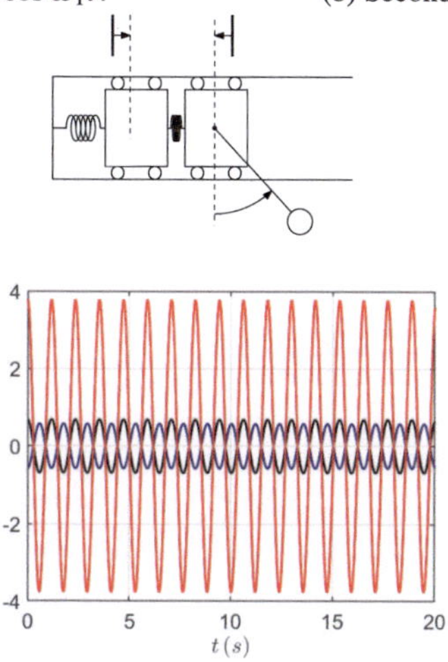

(c) Third mode $\mathbf{s}_3(t) = \mathbf{u}_3 \cos \omega_3 t$.

Fig. 5.7 Normal modes of the carts-pendulum system. A representative configuration for each component of the system is shown above each plot. The generalized coordinates q_1, q_2, and q_3 are represented by the curves (—) , (—) , and (—) , respectively

mass of the other pendulum is

$$\mathbf{x}_2 = l_0\mathbf{j} + L\mathbf{t}_1. \tag{5.80}$$

The basis $\{\mathbf{t}_1 = \cos\gamma\mathbf{i} + \sin\gamma\mathbf{j},\ \mathbf{t}_2 = -\sin\gamma\mathbf{i} + \cos\gamma\mathbf{j},\ \mathbf{t}_3 = \mathbf{k} = \mathbf{t}_1 \times \mathbf{t}_2\}$ corotates with the second pendulum. Differentiate these expressions with respect to time to obtain representations for the velocity vectors:

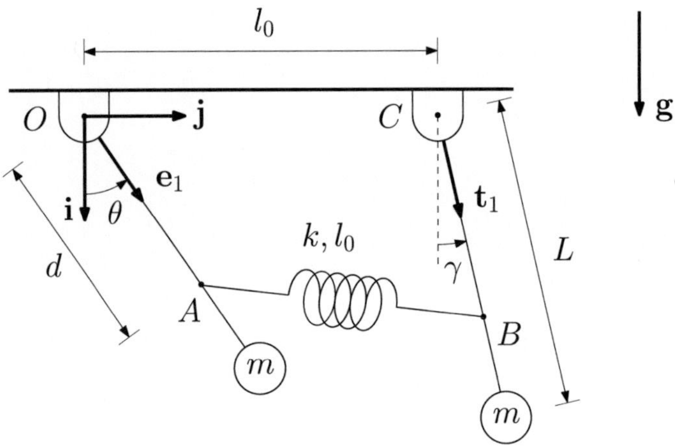

Fig. 5.8 Two pendulums connected by a linear spring. The gravitational force acts along **i**

$$\dot{\mathbf{x}}_1 = L\dot{\theta}\mathbf{e}_2, \quad \dot{\mathbf{x}}_2 = L\dot{\gamma}\mathbf{t}_2. \tag{5.81}$$

The angular momenta of the pendulums with respect to the pin-jointed ends are

$$\mathbf{H}_1^0 = \mathbf{x}_1 \times m\dot{\mathbf{x}}_1 = mL^2\dot{\theta}\mathbf{k}, \tag{5.82}$$

$$\mathbf{H}_2^C = (\mathbf{x}_2 - l_0\mathbf{j}) \times m\dot{\mathbf{x}}_2 = mL^2\dot{\gamma}\mathbf{k}. \tag{5.83}$$

Equations of Motion

The FBDs of the pendulums are depicted in Fig. 5.9. The weight of each pendulum, $mg\mathbf{i}$, acts on the respective mass center. The reaction forces $\mathbf{R}_O$ and $\mathbf{R}_C$ due to the pin joint connections are unknown. Since the objective is to obtain linearized equations, an approximation for the elastic force may be obtained using the small angles assumption, under which the position vectors of points A and B become

$$\mathbf{x}_A = d\mathbf{e}_1 \approx d(\mathbf{i} + \theta\mathbf{j}), \quad \mathbf{x}_B = l_0\mathbf{j} + d\mathbf{t}_1 \approx l_0\mathbf{j} + d(\mathbf{i} + \gamma\mathbf{j}), \tag{5.84}$$

respectively. These expressions imply that points A and B move only along the $\mathbf{j}$ direction when θ and γ are small. The elastic force then becomes

$$\mathbf{F}_e = -kd(\theta - \gamma)\mathbf{j}. \tag{5.85}$$

Apply Euler's second law to each pendulum and arrange the resulting equations as

$$\begin{bmatrix} mL^2 & 0 \\ 0 & mL^2 \end{bmatrix} \begin{bmatrix} \ddot{\theta} \\ \ddot{\gamma} \end{bmatrix} + \begin{bmatrix} mgL + kd^2 & -kd^2 \\ -kd^2 & mgL + kd^2 \end{bmatrix} \begin{bmatrix} \theta \\ \gamma \end{bmatrix} = \begin{bmatrix} 0 \\ 0 \end{bmatrix}. \tag{5.86}$$

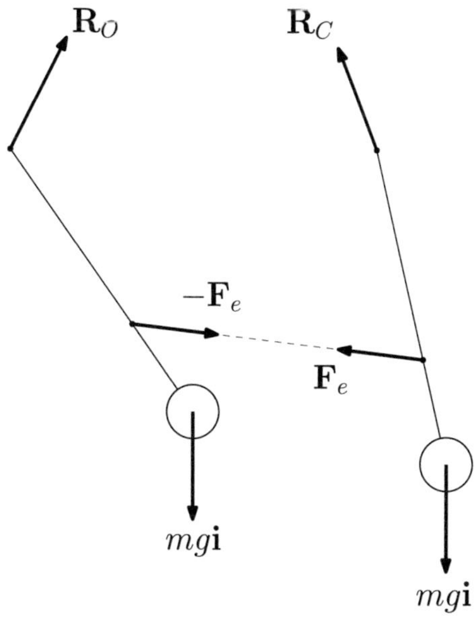

Fig. 5.9 FBDs of the pendulums

The stiffness matrix is diagonal if $k = 0$. When this happens, the equations of motion become uncoupled and both pendulums undergo independent simple harmonic motion with frequency $\omega_n = \sqrt{g/L}$.

Natural Frequencies and Mode Shapes

The characteristic polynomial can be written as

$$\left[\frac{g}{L} - (\omega)^2\right]\left[(\omega)^2 - \frac{g}{L} - \frac{2kd^2}{mL^2}\right] = 0. \tag{5.87}$$

The two natural frequencies are

$$\omega_1 = \sqrt{\frac{g}{L}}, \quad \omega_2 = \sqrt{\frac{g}{L} + \frac{2kd^2}{mL^2}}. \tag{5.88}$$

The first mode shape is the solution of the linear system

$$\begin{bmatrix} 1 & -1 \\ 0 & 0 \end{bmatrix} \mathbf{u}_1 = \mathbf{0}. \tag{5.89}$$

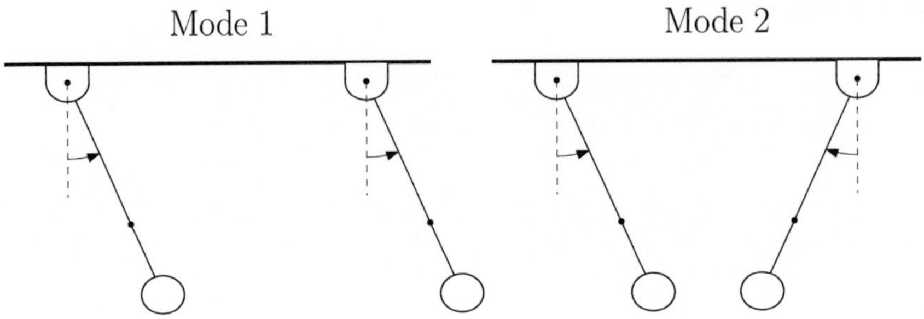

Fig. 5.10 Modes of the coupled pendulums

One solution is

$$\mathbf{u}_1 = \begin{bmatrix} 1 \\ 1 \end{bmatrix}. \tag{5.90}$$

The first mode is illustrated in Fig. 5.10. The pendulums oscillate in phase and with the same amplitude. It can be checked that the second mode shape is

$$\mathbf{u}_2 = \begin{bmatrix} 1 \\ -1 \end{bmatrix}. \tag{5.91}$$

As shown in Fig. 5.10, the pendulums oscillate out of phase and with the same amplitude in this mode.

Analysis of the Response

The general response is given by the superposition of the modes:

$$\begin{aligned} \begin{bmatrix} \theta(t) \\ \gamma(t) \end{bmatrix} &= \begin{bmatrix} 1 \\ 1 \end{bmatrix} (a_1 \cos \omega_1 t + b_1 \sin \omega_1 t) \\ &+ \begin{bmatrix} 1 \\ -1 \end{bmatrix} (a_2 \cos \omega_2 t + b_2 \sin \omega_2 t). \end{aligned} \tag{5.92}$$

Using the arbitrary ICs

$$\begin{bmatrix} \theta(0) \\ \gamma(0) \end{bmatrix} = \begin{bmatrix} \theta_0 \\ 0 \end{bmatrix}, \quad \begin{bmatrix} \dot{\theta}(0) \\ \dot{\gamma}(0) \end{bmatrix} = \mathbf{0}, \tag{5.93}$$

the response of the system becomes

$$\begin{bmatrix} \theta(t) \\ \gamma(t) \end{bmatrix} = \theta_0 \begin{bmatrix} \cos\left(\dfrac{\omega_1+\omega_2}{2}t\right)\cos\left(\dfrac{\omega_1-\omega_2}{2}t\right) \\ \sin\left(\dfrac{\omega_1+\omega_2}{2}t\right)\sin\left(\dfrac{\omega_1-\omega_2}{2}t\right) \end{bmatrix}. \tag{5.94}$$

In general, the oscillations of the pendulums are not harmonic nor periodic, as shown in Fig. 5.11. When the spring stiffness k is sufficiently small, the natural frequencies become $\omega_1 \approx \omega_2$. In this case, each pendulum's vibration is a superposition of two harmonics with close frequencies, resulting in beating. This phenomenon is

Fig. 5.11 Free vibration of the coupled pendulums. Plots produced with $m = 1$ kg, $L = d = 0.5$ m, and $\theta_0 = 10°$

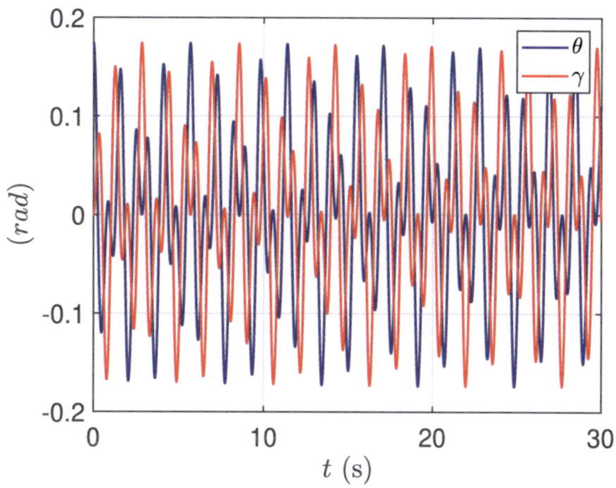

(a) Response for k = 20 N/m.

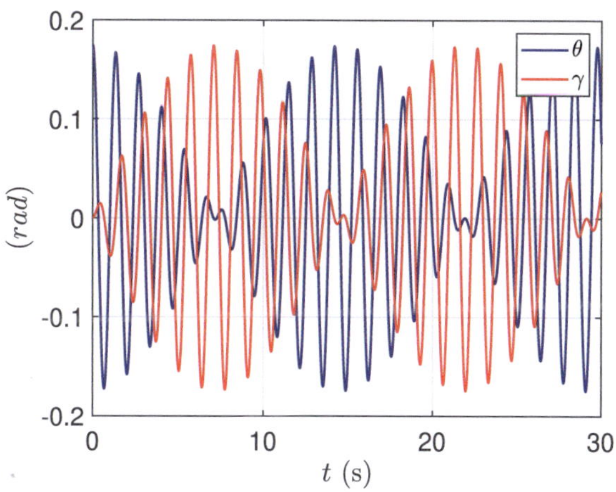

(b) Response for k = 2 N/m.

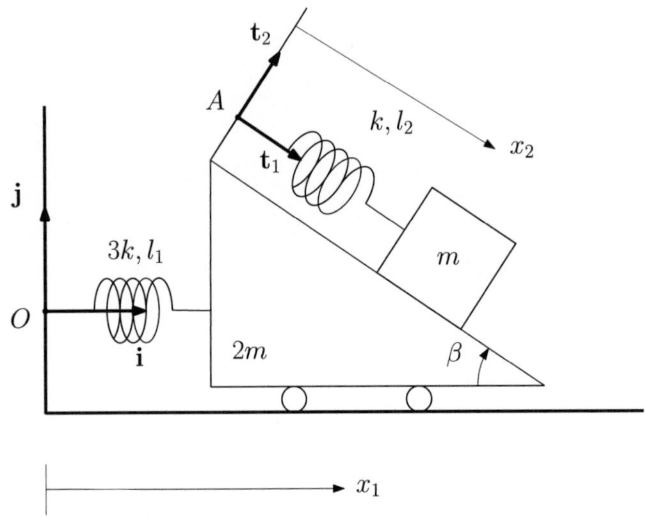

Fig. 5.12 A block with mass m slinging on the smooth face of a triangular cart

illustrated in Fig. 5.11b, where kinetic energy is transferred between the pendulums and the amplitudes fluctuate.

5.2.4.3 A Particle Moving on a Cart

In this example, consider the system shown in Fig. 5.12. It consists of a block with mass m sliding on the smooth inclined face of a cart with mass $2m$. This triangular-shaped cart is free to move on a smooth horizontal surface, and it is attached to a fixed point O by a linear spring with stiffness $3k$ and unstretched length l_1. The block is connected to the cart by another spring with stiffness k and unstretched length l_2. The objective is to analyze how the coupling between the cart and the block, ultimately influenced by the incline angle β, affects the natural frequencies of the system.

Kinematic Analysis

Using the coordinate system depicted in Fig. 5.12, the position vector of the triangular cart has the representation

$$\mathbf{r}_1 = x_1 \mathbf{i}. \tag{5.95}$$

The position of the block can be expressed as

$$\mathbf{r}_2 = \mathbf{r}_1 + \mathbf{r}_0 + x_2 \mathbf{t}_1, \tag{5.96}$$

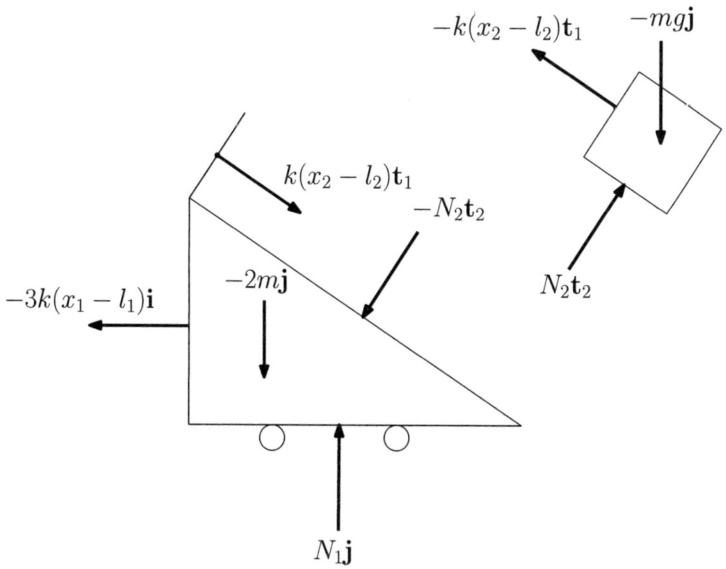

Fig. 5.13 FBDs of the block and the cart

where $\mathbf{r}_0$ is a constant vector. The orthonormal vectors $\mathbf{t}_1$ and $\mathbf{t}_2$ are fixed at A and are given by

$$\mathbf{t}_1 = \cos\beta\mathbf{i} - \sin\beta\mathbf{j}, \tag{5.97}$$

$$\mathbf{t}_2 = \sin\beta\mathbf{i} + \cos\beta\mathbf{j}. \tag{5.98}$$

The acceleration vectors of the cart and block are

$$\mathbf{a}_1 = \ddot{x}_1\mathbf{i}, \quad \mathbf{a}_2 = \ddot{x}_1\mathbf{i} + \ddot{x}_2\mathbf{t}_1, \tag{5.99}$$

respectively.

Equations of Motion

The FBDs of the particles are shown in Fig. 5.13. The normal force $N_2\mathbf{t}_2$ is orthogonal with respect to the inclined plane and acts on the block. This and the other normal force, $\mathbf{N}_1$, are unknowns that need to be eliminated so only equations involving the coordinates are obtained. To derive the equations of motion, start by applying Newton's second law to the block:

$$m\ddot{x}_1\mathbf{i} + m\ddot{x}_2\mathbf{t}_1 = -mg\mathbf{j} - k(x_2 - l_2)\mathbf{t}_1 + N_2\mathbf{t}_2. \tag{5.100}$$

Take the projection of this equation along $\mathbf{t}_1$ to reveal one of the equations of motion:

$$m\ddot{x}_1 \cos\beta + m\ddot{x}_2 = mg\sin\beta - k(x_2 - l_2). \tag{5.101}$$

Newton's second law applied to the cart yields

$$2m\ddot{x}_1\mathbf{i} = -2mg\mathbf{j} + k(x_2 - l_2)\mathbf{t}_1 - 3k(x_1 - l_1)\mathbf{i} + N_1\mathbf{j} - N_2\mathbf{t}_2. \tag{5.102}$$

The two unknown normal forces appear in Eq. (5.102). The normal force N_2 can be eliminated by adding this equation to Eq. (5.100). In the sequence, take the projection along $\mathbf{i}$ to eliminate N_1 and obtain the other equation of motion:

$$3m\ddot{x}_1 + m\ddot{x}_2 \cos\beta = -3k(x_1 - l_1). \tag{5.103}$$

The equilibrium position can be obtained by substituting the constant solutions $x_1 = x_1^e$ and $x_2 = x_2^e$ into Eqs. (5.101) and (5.103):

$$x_1^e = l_1, \quad x_2^e = l_2 + \frac{mg\sin\beta}{k}. \tag{5.104}$$

Define the new generalized coordinates $q_1 = x_1 - x_1^e$ and $q_2 = x_2 - x_2^e$ to rewrite the equations in the format

$$\begin{bmatrix} 3m & m\cos\beta \\ m\cos\beta & m \end{bmatrix} \begin{bmatrix} \ddot{q}_1 \\ \ddot{q}_2 \end{bmatrix} + \begin{bmatrix} 3k & 0 \\ 0 & k \end{bmatrix} \begin{bmatrix} q_1 \\ q_2 \end{bmatrix} = \begin{bmatrix} 0 \\ 0 \end{bmatrix}. \tag{5.105}$$

Natural Frequencies

It can be verified that the characteristic polynomial is

$$(\omega^2)^2 - \frac{6}{3 - (\cos\beta)^2}\frac{k}{m}(\omega^2) + \frac{3}{3 - (\cos\beta)^2}\frac{k^2}{m^2} = 0. \tag{5.106}$$

The two natural frequencies are given by

$$\omega_1 = \sqrt{\frac{3 - \sqrt{3}\cos\beta}{3 - (\cos\beta)^2}}\sqrt{\frac{k}{m}}, \quad \omega_2 = \sqrt{\frac{3 + \sqrt{3}\cos\beta}{3 - (\cos\beta)^2}}\sqrt{\frac{k}{m}}. \tag{5.107}$$

The dependence of the natural frequencies on the incline angle β is shown in Fig. 5.14. The cart and the block move along perpendicular directions when $\beta = 90°$. In this configuration, the coupling terms in the equations of motion vanish, causing the block and cart to behave like independent harmonic oscillators, each with natural frequency $\omega_n = \sqrt{k/m}$. The off-diagonal entries of the mass matrix increase as β decreases. Coupling between the two DOFs becomes more significant

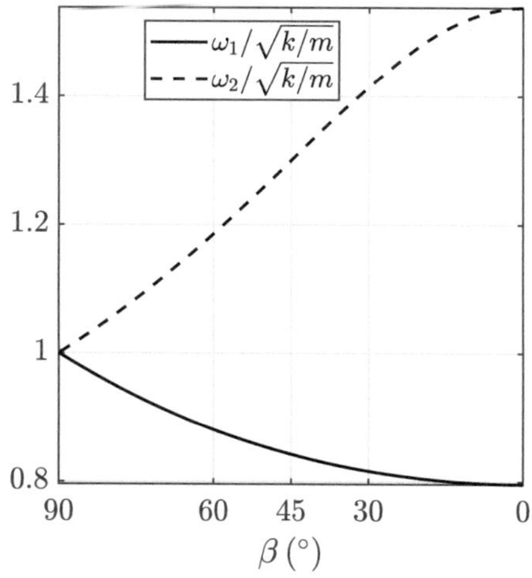

Fig. 5.14 Dependence of the natural frequencies on the inclined plane angle β

and causes the natural frequencies to diverge. The difference between the natural frequencies reaches its maximum when the block and cart move in the same direction ($\beta = 0$).

5.3 Free Vibration of Damped Systems

As discussed in Sect. 5.2, the free response of undamped MDOF systems consists of sustained oscillations resulting from the superposition of normal modes of vibrations. However, real systems are damped. Damping dissipates energy and causes the oscillations to gradually decay. After some time, the system eventually returns to its equilibrium position. To understand damped free vibration, one must solve the system of differential equations

$$\mathbf{M}\ddot{\mathbf{q}} + \mathbf{C}\dot{\mathbf{q}} + \mathbf{K}\mathbf{q} = \mathbf{0} \tag{5.108}$$

for a known set of ICs $\mathbf{q}(0) = \mathbf{q}_0$ and $\dot{\mathbf{q}}(0) = \dot{\mathbf{q}}_0$. Similar to the undamped case, it is useful to consider a simple example before presenting the general theory.

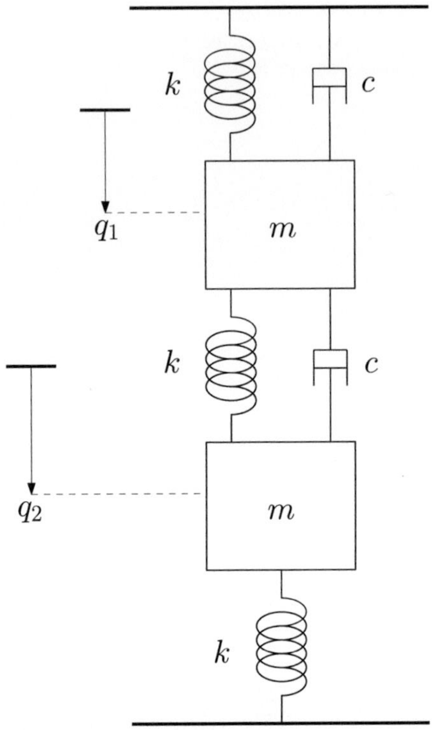

Fig. 5.15 A damped system with two particles

5.3.1 A Damped System with Two Particles

Consider the system shown in Fig. 5.15. The free vibration of the particles is governed by the differential equations

$$\begin{bmatrix} m & 0 \\ 0 & m \end{bmatrix} \begin{bmatrix} \ddot{q}_1 \\ \ddot{q}_2 \end{bmatrix} + \begin{bmatrix} 2c & -c \\ -c & c \end{bmatrix} \begin{bmatrix} \dot{q}_1 \\ \dot{q}_2 \end{bmatrix} + \begin{bmatrix} 2k & -k \\ -k & 2k \end{bmatrix} \begin{bmatrix} q_1 \\ q_2 \end{bmatrix} = \begin{bmatrix} 0 \\ 0 \end{bmatrix}, \tag{5.109}$$

where q_1 and q_2 are coordinates measured from the equilibrium position. Without loss of generality, assume that $m = 1\,\text{kg}$, $c = 2\,\text{Ns/m}$, and $k = 1\,\text{N/m}$ in the remainder of the example.

5.3.1.1 Solution Candidate

The purely oscillatory solution assumed in Sect. 5.2 cannot be applied here because damping may cause the particles to oscillate with decaying amplitudes or even exhibit non-oscillatory behavior. To move forward, it is useful to recall the physics of the damped free vibration of SDOF systems. As discussed in Chap. 2, the solution to the equation

$$m\ddot{q} + c\dot{q} + kq = 0 \tag{5.110}$$

takes the form

$$q = ve^{\lambda t}, \tag{5.111}$$

where $v \neq 0$ is a constant amplitude, and λ is a complex exponent. The real part of λ is negative and results in the exponential decay of the response, while the imaginary part corresponds to the damped frequency of oscillation. Interpreting Eq. (5.110) as a special case of Eq. (5.108) when $n = 1$, assume the solution of Eq. (5.109) to be

$$\mathbf{q} = \mathbf{v}e^{\lambda t}, \tag{5.112}$$

where

$$\mathbf{v} = \begin{bmatrix} v_1 \\ v_2 \end{bmatrix} \neq \mathbf{0} \tag{5.113}$$

is a vector of constant amplitudes. It will be shown that, similar to the SDOF case, the imaginary part of the complex exponent λ is responsible for the oscillatory behavior, and its negative real part is linked to the exponential decay of the response.

5.3.1.2 Computing the Solution Parameters

The values of $\mathbf{v}$ and λ that satisfy as solutions can be determined by substituting the assumed response (5.112) into the equations of motion:

$$\begin{bmatrix} \lambda^2 + 4\lambda + 2 & -2\lambda - 1 \\ -2\lambda - 1 & \lambda^2 + 2\lambda + 2 \end{bmatrix} \begin{bmatrix} v_1 \\ v_2 \end{bmatrix} = \begin{bmatrix} 0 \\ 0 \end{bmatrix}. \tag{5.114}$$

The trivial solution $\mathbf{v} = \mathbf{0}$ is avoided by imposing that the matrix $\lambda^2\mathbf{M} + \lambda\mathbf{C} + \mathbf{K}$ be singular:

$$\det\left(\begin{bmatrix} \lambda^2 + 4\lambda + 2 & -2\lambda - 1 \\ -2\lambda - 1 & \lambda^2 + 2\lambda + 2 \end{bmatrix}\right) = 0. \tag{5.115}$$

Expand the determinant to obtain the fourth-degree characteristic polynomial

$$\lambda^4 + 6\lambda^3 + 8\lambda^2 + 8\lambda + 3 = 0. \tag{5.116}$$

The four roots of this polynomial can be found numerically using MATLAB's code `Chapt5_code1` from the *Electronic Supplementary Material*. There are two real roots and a complex conjugate pair that are ordered arbitrarily as

$$\begin{aligned} &\lambda_1 = -0.4055 - 0.98i, \quad \lambda_2 = -4.6106, \\ &\lambda_3 = \bar{\lambda}_1, \quad \lambda_4 = -0.5784. \end{aligned} \tag{5.117}$$

The computation of the associated amplitude vectors follows a procedure similar to that used for determining the mode shapes. The entries v_{11} and v_{21} of the vector $\mathbf{v}_1$ are solutions to the system of linear equations

$$\begin{bmatrix} -0.4180 + 3.1254i & -0.1890 - 1.9601i \\ -0.1890 - 1.9601i & 0.3929 + 1.1653i \end{bmatrix} \begin{bmatrix} v_{11} \\ v_{21} \end{bmatrix} = \begin{bmatrix} 0 \\ 0 \end{bmatrix}. \tag{5.118}$$

This system is obtained by substituting λ_1 into Eq. (5.114). Since the matrix $\lambda_1^2\mathbf{M} + \lambda_1\mathbf{C} + \mathbf{K}$ is singular, there are infinitely many solutions for $\mathbf{v}_1$, all of which satisfy:

$$v_{11} = (0.6082 - 0.1418i)v_{21}. \tag{5.119}$$

The solutions are complex because λ_1 is complex. A normalization with $v_{21} = 1$ can be chosen:

$$\mathbf{v}_1 = \begin{bmatrix} 0.6082 - 0.1418i \\ 1 \end{bmatrix}. \tag{5.120}$$

The previous result can serve as a shortcut for computing $\mathbf{v}_3$. Simply take the complex conjugate of the equation

$$\left(\lambda_1^2\mathbf{M} + \lambda_1\mathbf{C} + \mathbf{K}\right)\mathbf{v}_1 = \mathbf{0} \tag{5.121}$$

to obtain

$$\left(\bar{\lambda}_1^2\mathbf{M} + \bar{\lambda}_1\mathbf{C} + \mathbf{K}\right)\bar{\mathbf{v}}_1 = \mathbf{0}. \tag{5.122}$$

Since $\lambda_3 = \bar{\lambda}_1$, the obvious conclusion is that

$$\mathbf{v}_3 = \bar{\mathbf{v}}_1 = \begin{bmatrix} 0.6082 + 0.1418i \\ 1 \end{bmatrix}. \tag{5.123}$$

In a similar fashion, the entries v_{12} and v_{22} of the vector $\mathbf{v}_2$ are solutions of

$$\begin{bmatrix} 4.8152 & 8.2212 \\ 8.2212 & 14.0364 \end{bmatrix} \begin{bmatrix} v_{12} \\ v_{22} \end{bmatrix} = \begin{bmatrix} 0 \\ 0 \end{bmatrix}. \tag{5.124}$$

A solution is

$$\mathbf{v}_2 = \begin{bmatrix} -1.7073 \\ 1 \end{bmatrix}. \tag{5.125}$$

It can be verified that the vector associated with λ_4 is

$$\mathbf{v}_4 = \begin{bmatrix} -7.5090 \\ 1 \end{bmatrix}. \tag{5.126}$$

5.3.1.3 System Response

The general response is the linear combination of all solutions found previously:

$$\mathbf{q}(t) = A_1\mathbf{v}_1 e^{\lambda_1 t} + A_2\mathbf{v}_2 e^{\lambda_2 t} + A_3\mathbf{v}_3 e^{\lambda_3 t} + A_4\mathbf{v}_4 e^{\lambda_4 t}, \tag{5.127}$$

where the constants A_1, A_2, A_3, and A_4 are determined from the IC. If $\mathbf{q}(0) = [1 \quad -1]^T$ and $\dot{\mathbf{q}}(0) = \mathbf{0}$, these constants are

$$\begin{bmatrix} A_1 \\ A_2 \\ A_3 \\ A_4 \end{bmatrix} = \begin{bmatrix} -0.4410 - 0.0148i \\ 0.1128 \\ -0.4410 + 0.0148i \\ -0.2308 \end{bmatrix}, \tag{5.128}$$

and the system response becomes

$$\mathbf{q}(t) = \begin{bmatrix} -0.2703 + 0.0535i \\ \\ -0.4410 - 0.0148i \end{bmatrix} e^{-0.4055t+i0.98t} + \begin{bmatrix} -0.2703 - 0.0535i \\ \\ -0.4410 + 0.0148i \end{bmatrix} e^{-0.4055t-i0.98t} + \begin{bmatrix} -0.1927 \\ \\ 0.1128 \end{bmatrix} e^{-4.6106t} + \begin{bmatrix} 1.7333 \\ \\ -0.2308 \end{bmatrix} e^{-0.5784t}. \tag{5.129}$$

The complex parts of the response form a complex conjugate pair. Combine them according to

$$\left\{ \begin{bmatrix} -0.2703 + 0.0535i \\ \\ -0.4410 - 0.0148i \end{bmatrix} e^{i0.98t} + \begin{bmatrix} -0.2703 - 0.0535i \\ \\ -0.4410 + 0.0148i \end{bmatrix} e^{-i0.98t} \right\} e^{-0.4055t} = Re \left\{ \begin{bmatrix} 0.5511e^{(0.98t-3.3371)i} \\ \\ 0.8825e^{(0.98t-3.1080)i} \end{bmatrix} \right\} e^{-0.4055t} \tag{5.130}$$

to express $\mathbf{q}(t)$ as

$$\mathbf{q}(t) = \begin{bmatrix} 0.5511\cos(0.98t - 3.3371) \\ \\ 0.8825\cos(0.98t - 3.1080) \end{bmatrix} e^{-0.4055t} + \begin{bmatrix} -0.1927 \\ \\ 0.1128 \end{bmatrix} e^{-4.6106t} + \begin{bmatrix} 1.7333 \\ \\ -0.2308 \end{bmatrix} e^{-0.5784t}. \tag{5.131}$$

The system vibrates as a superposition of two types of motion, often termed damped modes of vibration. In the damped mode

$$\mathbf{s}_1(t) = \begin{bmatrix} 0.5511\cos(0.98t - 3.3371) \\ \\ 0.8825\cos(0.98t - 3.1080) \end{bmatrix} e^{-0.4055t}, \tag{5.132}$$

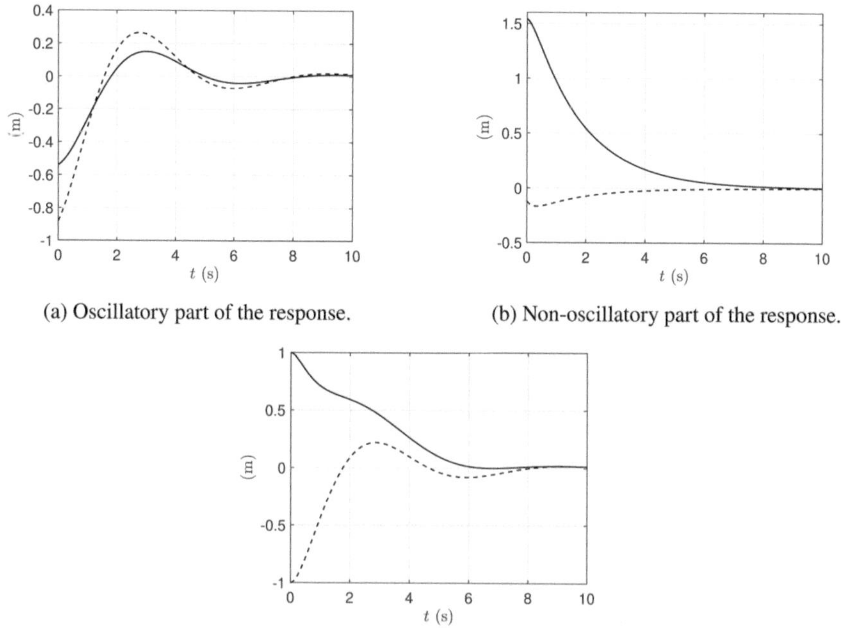

(a) Oscillatory part of the response.

(b) Non-oscillatory part of the response.

(c) System response is a combination of oscillatory and non-oscillatory behaviors.

Fig. 5.16 A plot of the solution from Eq. (5.131). The figures show the generalized coordinates q_1 (—) and q_2 (- -)

both particles perform harmonic vibration with identical damped frequency and with the same exponential decay. They do not pass through their equilibrium positions at the same instant because the phase angles are different. The other damped mode is characterized by non-oscillatory behavior:

$$\mathbf{s}_2(t) = \begin{bmatrix} -0.1927 \\ \\ 0.1128 \end{bmatrix} e^{-4.6106t} + \begin{bmatrix} 1.7333 \\ \\ -0.2308 \end{bmatrix} e^{-0.5784t}. \tag{5.133}$$

The complete response is a combination of the two damped modes, as illustrated in Fig. 5.16. This is a notable distinction from the behavior of SDOF systems. In those systems, the damped free response is either oscillatory with decaying amplitude or non-oscillatory.

5.3.1.4 Influence of Damping

It is generally of interest to determine the effect of damping on the overall response of the system, as it can significantly alter the dynamic behavior. This can be achieved by examining the plot shown in Fig. 5.17. The figure displays the root loci of the

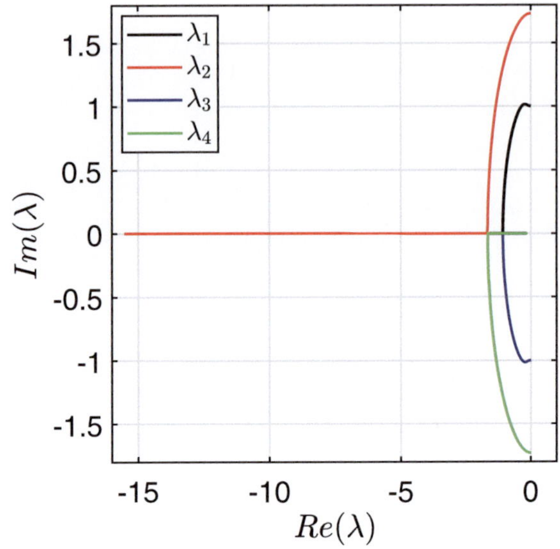

Fig. 5.17 Loci of roots of the characteristic polynomial for $c \in [0,\ 6]$. The plot is generated with $m = 1$ kg and $k = 1$ N/m

characteristic polynomial $\det(\lambda^2\mathbf{M}+\lambda\mathbf{C}+\mathbf{K}) = 0$ for various values of the viscous damping coefficient c, while m and k remain constant.

Then the system is undamped ($c = 0$), Eq. (5.114) reduces to

$$(\lambda^2\mathbf{M} + \mathbf{K})\mathbf{v} = \mathbf{0}. \tag{5.134}$$

This corresponds to the generalized eigenproblem from Eq. (5.59), with $\lambda^2 = -\omega^2$ and $\mathbf{v} = \mathbf{u}$. Consequently, all roots of the characteristic polynomial lie on the imaginary axis and are equal to

$$\lambda_1 = \bar{\lambda}_3 = i\omega_1, \quad \lambda_2 = \bar{\lambda}_4 = i\omega_2, \tag{5.135}$$

where the natural frequencies are $\omega_1 = 1$ rad/s and $\omega_2 = 1.7321$ rad/s. The response is purely oscillatory, with the system vibrating as a superposition of the normal modes determined by $\mathbf{v}_1 = \mathbf{v}_3 = \mathbf{u}_1 = [1 \quad 1]^T$ and $\mathbf{v}_2 = \mathbf{v}_4 = \mathbf{u}_2 = [1 \quad -1]^T$:

$$\mathbf{q}(t) = \begin{bmatrix} 1 \\ 1 \end{bmatrix} X_1 \cos(t - \phi_1) + \begin{bmatrix} 1 \\ -1 \end{bmatrix} X_2 \cos(1.7321t - \phi_2), \tag{5.136}$$

where X_1, X_2, ϕ_1, and ϕ_2 are determined from the ICs. As c is increased, the roots turn into two complex conjugate pairs with negative real parts. For instance, the eigenvalues are

$$\lambda_1 = \bar{\lambda}_3 = -0.2238 + 1.0149i, \quad \lambda_2 = \bar{\lambda}_4 = -1.2762 + 1.0718i, \tag{5.137}$$

when $c = 1$ Ns/m. The corresponding eigenvectors

$$\begin{aligned} \mathbf{v}_1 = \overline{\mathbf{v}}_3 &= \begin{bmatrix} -0.5270 - 0.2995i \\ -0.5419 - 0.5822i \end{bmatrix}, \\ \mathbf{v}_2 = \overline{\mathbf{v}}_4 &= \begin{bmatrix} 0.8801 - 0.0158i \\ -0.4385 + 0.1812i \end{bmatrix} \end{aligned} \tag{5.138}$$

can be used to express $\mathbf{q}(t)$ as

$$\begin{aligned} \mathbf{q}(t) = X_1 &\begin{bmatrix} 1.2124\cos(1.0149t - 2.6248 - \phi_1) \\ 1.5906\cos(1.0149t - 2.3204 - \phi_1) \end{bmatrix} e^{-0.2238t} \\ + X_2 &\begin{bmatrix} 1.7606\cos(1.0718t - 0.0179 - \phi_2) \\ 0.9489\cos(1.0718t + 2.7496 - \phi_2) \end{bmatrix} e^{-1.2762t}. \end{aligned} \tag{5.139}$$

The free vibration response is now a superposition of two damped modes. In each damped mode, the particles oscillate harmonically at the same damped frequency, and the amplitudes are modulated by the same exponential decay.

When the value of the viscous damping coefficient reaches $c \approx 1.285$, the roots λ_2 and λ_4 collapse on the real axis while the others still form a complex conjugate pair. This condition persists for any value of damping in $c \in (1, 285,\ 5.548)$. In this range, the system exhibits both oscillatory and non-oscillatory damped modes, an example of which consists of the response from Eq. (5.131). All roots are real, and the response becomes non-oscillatory for any value $c \geq 5.549$. For instance, all eigenvalues are equal to

$$\begin{aligned} \lambda_1 &= -1.2228, \quad \lambda_2 = -14.4609, \\ \lambda_3 &= -0.9348, \quad \lambda_4 = -0.1815. \end{aligned} \tag{5.140}$$

when $c = 5.6$. The response is purely non-oscillatory:

$$\begin{aligned} \mathbf{q}(t) = A_1 &\begin{bmatrix} -0.4974 \\ -0.8675 \end{bmatrix} e^{-1.2228t} + A_2 \begin{bmatrix} -0.8520 \\ 0.5236 \end{bmatrix} e^{-14.4609t} \\ + A_3 &\begin{bmatrix} 0.4870 \\ 0.8734 \end{bmatrix} e^{-0.9348t} + A_4 \begin{bmatrix} 0.9999 \\ -0.0161 \end{bmatrix} e^{-0.1815t}. \end{aligned} \tag{5.141}$$

5.3.2 *Damped Free Vibrations*

The previous example introduced several important aspects involved in the analysis of the free response of damped MDOF systems. Problems of this class require solving Eq. (5.108) using the steps delineated as follows.

5.3.2.1 The Quadratic Eigenvalue Problem

The general procedure to solve these equations is to postulate that the response is

$$\mathbf{q} = \mathbf{v}e^{\lambda t}, \tag{5.142}$$

where $\mathbf{v}$ is an n-dimensional constant vector, and λ is a scalar parameter. Substitute the assumed solution into Eq. (5.108) to obtain

$$(\lambda^2\mathbf{M} + \lambda\mathbf{C} + \mathbf{K})\mathbf{v} = \mathbf{0}. \tag{5.143}$$

This expression represents a nonlinear (quadratic) eigenproblem with eigenvalue λ and eigenvector $\mathbf{v}$. The eigenvalues are the roots of the characteristic polynomial

$$\det(\lambda^2\mathbf{M} + \lambda\mathbf{C} + \mathbf{K}) = 0. \tag{5.144}$$

The expansion of the determinant yields a polynomial of degree $2n$. According to the fundamental theorem of algebra, this polynomial has $2n$ eigenvalues, counting multiplicities of repeated roots. Some eigenvalues may be real and negative, while others can be complex with negative real parts. A complex eigenvalue must appear with its conjugate pair since the characteristic polynomial has real coefficients. The corresponding eigenvectors can be computed by solving the linear system of equations in (5.143), with each eigenvector determined up to a multiplicative constant. An eigenvector associated with a real eigenvalue can always be chosen to be real, and the eigenvectors corresponding to complex conjugate eigenvalues will themselves be complex conjugates.

5.3.2.2 System Response

The solution of Eq. (5.108) is given by

$$\mathbf{q}(t) = \sum_{j=1}^{2n} A_j\mathbf{v}_j e^{\lambda_j t}, \tag{5.145}$$

Table 5.2 Solution procedure for damped free vibration for the case of distinct eigenvalues

	$n = 1$	$n > 1$
Equations of motion	$m\ddot{q} + c\dot{q} + kq = 0$	$\mathbf{M}\ddot{\mathbf{q}} + \mathbf{C}\dot{\mathbf{q}} + \mathbf{K}\mathbf{q} = \mathbf{0}$
Assumed solution	$q = ve^{\lambda t}$	$\mathbf{q} = \mathbf{v}e^{\lambda t}$
Characteristic polynomial	$m\lambda^2 + c\lambda + k = 0$	$\det(\mathbf{M}\lambda^2 + \mathbf{C}\lambda + \mathbf{K}) = 0$
Eigenvalues	λ_1, λ_2	$\lambda_1, \lambda_2, \ldots, \lambda_{2n}$
Eigenvectors	$v_1 = v_2 = 1$	$(\mathbf{M}\lambda_j^2 + \mathbf{C}\lambda_j + \mathbf{K})\mathbf{v}_j = \mathbf{0}$
Free response	$q(t) = A_1 e^{\lambda_1 t} + A_2 e^{\lambda_2 t}$	$\mathbf{q}(t) = \sum_{j=1}^{2n} A_j \mathbf{v}_j e^{\lambda_j t}$

assuming that there are no repeated eigenvalues. The constants A_j $(j = 1, \ldots, 2n)$ are determined from the ICs. Real eigenvalues contribute to the non-oscillatory part of the response, and complex conjugate eigensolutions combine to form the oscillatory part. In the latter case, the imaginary part of the eigenvalues corresponds to a damped frequency of vibration, and the negative real part governs the exponential decay of the response. An overview of the solution procedure for damped free vibration is presented in Table 5.2 together with a comparison between SDOF and MDOF systems.

5.3.2.3 Computer Implementation

Eigenproblems, like the one in Eq. (5.143), are typically solved numerically in practical applications. The following MATLAB command solves the quadratic eigenvalue problem:.

```
% Solve quadratic eigenvalue problem:
[V,L] = polyeig(K,C,M);
```

One of the outputs of this function is the n-dimensional vector $\mathbf{L}$ that contains the eigenvalues:

$$\mathbf{L} = \begin{bmatrix} \lambda_1 \\ \lambda_2 \\ \vdots \\ \lambda_{2n} \end{bmatrix}. \tag{5.146}$$

The other output is a matrix $\mathbf{V}$ of size $n \times 2n$. The columns in this matrix are occupied by the corresponding eigenvectors:

$$\mathbf{V} = \begin{bmatrix} \mathbf{v}_1 & \mathbf{v}_2 & \dots & \mathbf{v}_{2n} \end{bmatrix}. \tag{5.147}$$

5.4 Forced Vibrations of MDOF Systems

Now that the free vibration has been covered, it is appropriate to discuss the response of MDOF systems in the presence of applied forces. Forced vibration analysis involves computing the solution of

$$\mathbf{M}\ddot{\mathbf{q}} + \mathbf{C}\dot{\mathbf{q}} + \mathbf{K}\mathbf{q} = \mathbf{f}(t), \tag{5.148}$$

for the ICs $\mathbf{q}(0) = \mathbf{q}_0$ and $\dot{\mathbf{q}}(0) = \dot{\mathbf{q}}_0$. The next section presents a simple example that introduces the physics of forced vibrations in an easy-to-visualize manner before the general case is addressed.

5.4.1 A System with Two Particles Subject to a Harmonic Force

Consider the system composed of two identical particles shown in Fig. 5.18. One of the particles is subjected to a harmonic force with amplitude F_0 and frequency ω. Using the generalized coordinates q_1 and q_2 measured from the static equilibrium position, the equations of motion of this system can be expressed as

$$\begin{aligned} &\begin{bmatrix} m & 0 \\ 0 & m \end{bmatrix} \begin{bmatrix} \ddot{q}_1 \\ \ddot{q}_2 \end{bmatrix} + \begin{bmatrix} 2c & -c \\ -c & 2c \end{bmatrix} \begin{bmatrix} \dot{q}_1 \\ \dot{q}_2 \end{bmatrix} \\ &+ \begin{bmatrix} 2k & -k \\ -k & 2k \end{bmatrix} \begin{bmatrix} q_1 \\ q_2 \end{bmatrix} = \begin{bmatrix} F_0 \\ 0 \end{bmatrix} e^{i\omega t}. \end{aligned} \tag{5.149}$$

The objective of this example is to find the response of the system.

5.4.1.1 Particular Solution

Recall from Chap. 3 that the steady-state response of an SDOF system excited by a harmonic force is harmonic with the same frequency as the excitation. Using the analogy where SDOF systems are special cases of MDOF systems when $n = 1$,

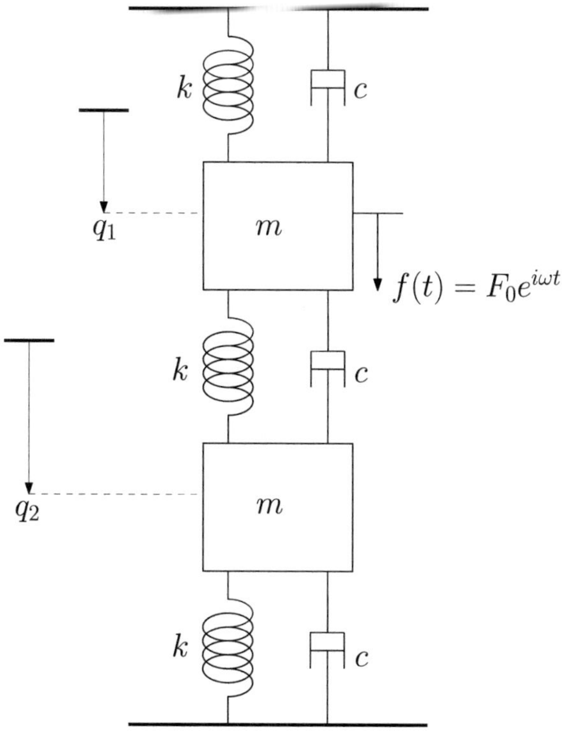

Fig. 5.18 A system with identical particles. The particles have mass m and are connected to linear springs with stiffness k and dampers with viscous damping coefficient c. One of the particles is acted upon by the harmonic force $F_0 e^{i\omega t}$

postulate that the particular solution of Eq. (5.149) is

$$\mathbf{q}_p(t) = \begin{bmatrix} q_{p_1} \\ q_{p_2} \end{bmatrix} = \begin{bmatrix} Q_1 \\ Q_2 \end{bmatrix} e^{i\omega t}, \tag{5.150}$$

where Q_1 and Q_2 are constant amplitudes of vibration that need to be determined. Differentiate this expression two times and substitute into the equations of motion to obtain

$$\begin{bmatrix} 2k - \omega^2 m + i2c\omega & -k - ic\omega \\ -k - ic\omega & 2k - \omega^2 m + i2c\omega \end{bmatrix} \begin{bmatrix} Q_1 \\ Q_2 \end{bmatrix} = \begin{bmatrix} F_0 \\ 0 \end{bmatrix}. \tag{5.151}$$

Compute the inverse to verify that

$$
\begin{aligned}
\mathbf{q}_p(t) = \frac{F_0}{2k} \Bigg\{ & \frac{1}{1 - \left(\dfrac{\omega}{\omega_1}\right)^2 + i2\zeta\left(\dfrac{\omega}{\omega_1}\right)} \begin{bmatrix} 1 \\ 1 \end{bmatrix} \\
& + \frac{1}{3 - \left(\dfrac{\omega}{\omega_1}\right)^2 + i6\zeta\left(\dfrac{\omega}{\omega_1}\right)} \begin{bmatrix} 1 \\ -1 \end{bmatrix} \Bigg\} e^{i\omega t},
\end{aligned}
\tag{5.152}
$$

where ω_1 is the first natural frequency of this system as computed in Sect. 5.2.1, and the damping ratio is defined to be $\zeta = c/(2m\omega_1)$. The functions q_{p_1} and q_{p_2} are complex because $\mathbf{C} \neq \mathbf{0}$. However, the physical responses must be real: they are either the real or imaginary parts of the complex solution depending on whether $f(t) = F_0 \cos(\omega t)$ or $f(t) = F_0 \sin(\omega t)$, respectively.

5.4.1.2 The Complete Solution

The response for forced vibrations is

$$
\mathbf{q}(t) = \mathbf{q}_h(t) + \mathbf{q}_p(t), \tag{5.153}
$$

where $\mathbf{q}_h$ is the general homogeneous solution that satisfies

$$
\mathbf{M}\ddot{\mathbf{q}}_h + \mathbf{C}\dot{\mathbf{q}}_h + \mathbf{K}\mathbf{q}_h = \mathbf{0}. \tag{5.154}
$$

The homogeneous solution contains two constants of integration, which are determined from the ICs. As illustrated in Fig. 5.19, any level of damping results in $\mathbf{q}_h \to \mathbf{0}$ as time progresses. Eventually, the system vibrates in a steady-state pattern $\mathbf{q} = \mathbf{q}_p$ sustained by the harmonic force. The steady-state vibrations of the particles are harmonic oscillations with the same frequency as the excitation.

5.4.1.3 The FRF Matrix

In an SDOF system, the steady-state amplitude of vibration Q and the force input F_0 are related through the FRF by $Q = H(\omega)F_0$. The FRF characterizes the frequency-dependent behavior of the system. A similar concept applies to MDOF systems. The system of equations in (5.151) can also be interpreted as a relationship between the

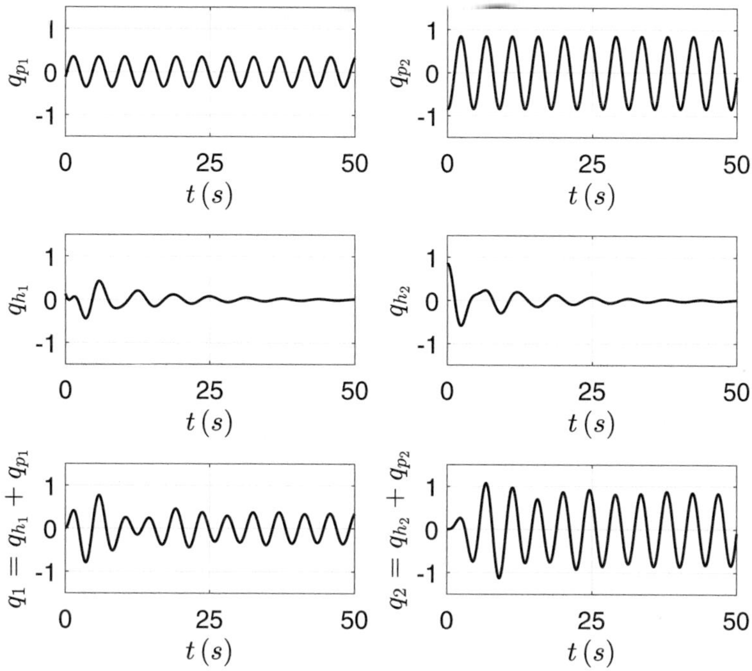

Fig. 5.19 Particular, homogeneous, and complete solutions of Eq. (5.149) with $m = 1$ kg, $k = 1$ N/m, and $\zeta = 0.075$. The plots were produced with $\mathbf{q}(0) = \dot{\mathbf{q}}(0) = \mathbf{0}$ and $f(t) = \cos\left(\sqrt{2}\omega_1 t\right)$

input vector $[F_0 \quad 0]^T$ and the output vector $[Q_1 \quad Q_2]^T$, achieved through the *FRF matrix*:

$$\begin{aligned}\mathbf{H}(\omega) &= \begin{bmatrix} H_{11}(\omega) & H_{12}(\omega) \\ H_{21}(\omega) & H_{22}(\omega) \end{bmatrix} \\ &= \begin{bmatrix} 2k - \omega^2 m + i2c\omega & -k - ic\omega \\ -k - ic\omega & 2k - \omega^2 m + i2c\omega \end{bmatrix}^{-1},\end{aligned} \tag{5.155}$$

where

$$H_{11} = H_{22} = \frac{1}{k} \frac{2 - \left(\dfrac{\omega}{\omega_1}\right)^2 + i4\zeta\left(\dfrac{\omega}{\omega_1}\right)}{\left[1 - \left(\dfrac{\omega}{\omega_1}\right)^2 + i2\zeta\left(\dfrac{\omega}{\omega_1}\right)\right] \times \left[3 - \left(\dfrac{\omega}{\omega_1}\right)^2 + i6\zeta\left(\dfrac{\omega}{\omega_1}\right)\right]}, \tag{5.156}$$

$$H_{21} = H_{12} = \frac{1}{k} \frac{1 + i2\zeta\left(\dfrac{\omega}{\omega_1}\right)}{\left[1 - \left(\dfrac{\omega}{\omega_1}\right)^2 + i2\zeta\left(\dfrac{\omega}{\omega_1}\right)\right] \times \left[3 - \left(\dfrac{\omega}{\omega_1}\right)^2 + i6\zeta\left(\dfrac{\omega}{\omega_1}\right)\right]}. \tag{5.157}$$

The steady-state response can then be written as

$$\mathbf{q}(t) = \begin{bmatrix} H_{11} F_0 e^{i\omega t} \\ \\ H_{21} F_0 e^{i\omega t} \end{bmatrix}. \tag{5.158}$$

The physical meaning of the entries in the FRF matrix follows directly from this expression. The entry $H_{11}(\omega)$ represents the FRF of the top particle when the harmonic force is applied to it and is therefore called the direct FRF. On the other hand, $H_{21}(\omega)$ represents the FRF of the bottom particle when the force is applied to the top particle and is referred to as the cross FRF. This interpretation can be similarly extended to the other entries.

5.4.1.4 Dependence of Steady-State Vibrations on the Excitation Frequency

To understand how the excitation frequency affects the steady-state response, express the entries in the FRF matrix in polar form according to

$$H_{11} = |H_{11}|e^{-i\phi_{11}}, \quad H_{21} = |H_{21}|e^{-i\phi_{21}}. \tag{5.159}$$

The steady-state solution becomes

$$\mathbf{q}(t) = \begin{bmatrix} |H_{11}| F_0 e^{i(\omega t - \phi_{11})} \\ \\ |H_{21}| F_0 e^{i(\omega t - \phi_{21})} \end{bmatrix}. \tag{5.160}$$

In steady state, the particles oscillate harmonically with the same frequency as the excitation. The absolute values of H_{11} and H_{21} magnify the amplitude of vibration, while the phase angles are responsible for a temporal delay with respect to the excitation, as shown in Fig. 5.20.

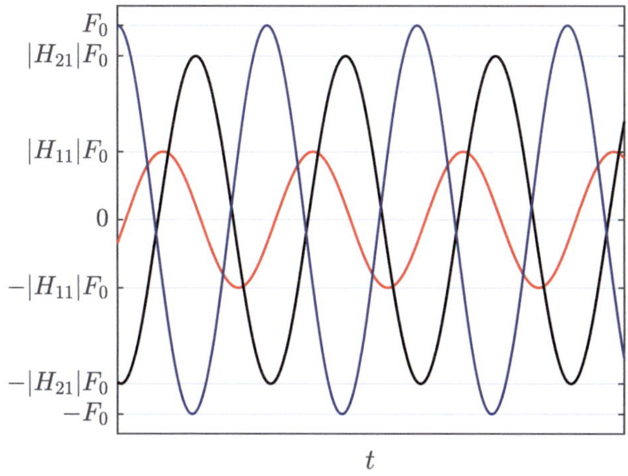

Fig. 5.20 Steady-state solution for q_1 (—) and q_2 (—) . The absolute values $|H_{11}|$ and $|H_{21}|$ magnify/filter the amplitude of vibration of the first and second particles, respectively. The phase angles ϕ_{11} and ϕ_{21} delay the response of each of them with respect to the excitation $f(t)$ (—)

The dependence of the steady-state behavior on the excitation frequency can be inferred from the frequency response curves from Figs. 5.21. These plots were generated with MATLAB code `Chapt5_code2`, provided in the *Electronic Supplementary Material*, considering the parameters $m = 1\,\text{kg}$, $k = 1\,\text{N/m}$, and $c = 0.2\,\text{Ns/m}$.

The first notable behavior from these figures is that $|H_{11}|$ and $|H_{21}|$ exhibit two peaks that grow to infinity when the undamped system ($\zeta = 0$) is excited at one of its natural frequencies: there are two resonance frequencies because this system has two DOFs. The effect of increasing damping is to reduce the resonance peaks and shift the frequencies at which they occur. If ζ is too high, one of the peaks disappears, and the particles vibrate as if they were an SDOF system.

The FRF H_{11} of an undamped system exhibits an interesting behavior when driven at $\omega = \sqrt{2}\omega_1$. At this frequency, known as antiresonance, $|H_{11}| = 0$. As damping increases, the amplitude at antiresonance also increases.

At both high ($\omega >> \omega_2$) and low ($\omega << \omega_1$) frequencies, the magnitude curves closely follow those of an undamped system. At low frequencies, the frequency response functions are approximately $H_{11} \approx 2/(3k)$ and $H_{21} \approx 1/(3k)$. The particles also vibrate in phase with the excitation. Conversely, both $H_{11}, H_{21} \to 0$ as $\omega \to \infty$. The particles exhibit low amplitudes at high frequencies. The top particle is out of phase with the excitation ($\phi_{11} \approx \pi$), while the lower particle remains in phase ($\phi_{21} \approx 0$).

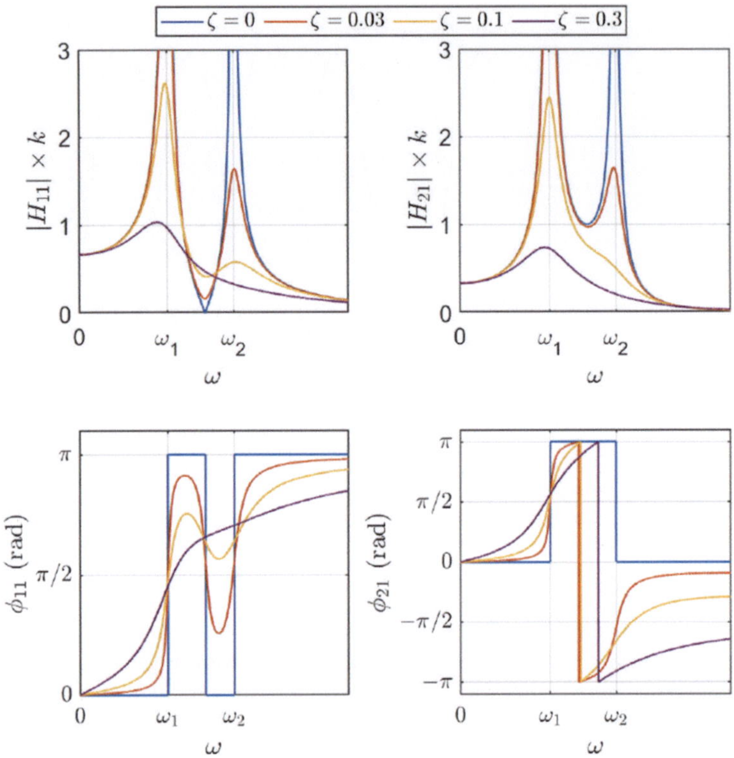

Fig. 5.21 Frequency response curves of H_{11} and H_{21} from Eqs. (5.156) and (5.157) for different values of ζ

Regarding the phase lag of the response, in the absence of damping, the particles are always either in phase or π rad out of phase with the excitation. This occurs because the frequency response functions are real when $\zeta = 0$. When damping is present, the particles experience phase lags within the range $[0, 2\pi)$. Note that the negative values for ϕ_{21} in the frequency response curve appear because MATLAB uses principal values in $[-\pi, \pi]$.

5.4.1.5 Profile of Vibration

The frequency response curves in Fig. 5.21 show that $|H_{11}|, |H_{21}| \to \infty$ when the undamped system is excited at one of its natural frequencies. Care is necessary to interpret the behavior of the system under these conditions. Without loss of generality, assume that $f(t) = F_0 \cos \omega t$. The complete solution of the associated undamped system ($\zeta = 0$) takes the form

$$\begin{bmatrix} q_1(t) \\ q_2(t) \end{bmatrix} = \begin{bmatrix} 1 \\ 1 \end{bmatrix} (a_1 \cos \omega_1 t + b_1 \sin \omega_1 t) + \begin{bmatrix} 1 \\ -1 \end{bmatrix} (a_2 \cos \omega_2 t + b_2 \sin \omega_2 t) + \left\{ \begin{bmatrix} 1 \\ 1 \end{bmatrix} \frac{F_0/2k}{1 - \left(\frac{\omega}{\omega_1}\right)^2} + \begin{bmatrix} 1 \\ -1 \end{bmatrix} \frac{F_0/2k}{3 - \left(\frac{\omega}{\omega_1}\right)^2} \right\} \cos \omega t. \tag{5.161}$$

A typical response predicted by this equation is shown in Fig. 5.22, which also illustrates the contributions of the homogeneous and particular parts of the solution. The complete solution is neither harmonic nor periodic because the homogeneous part does not decay over time.

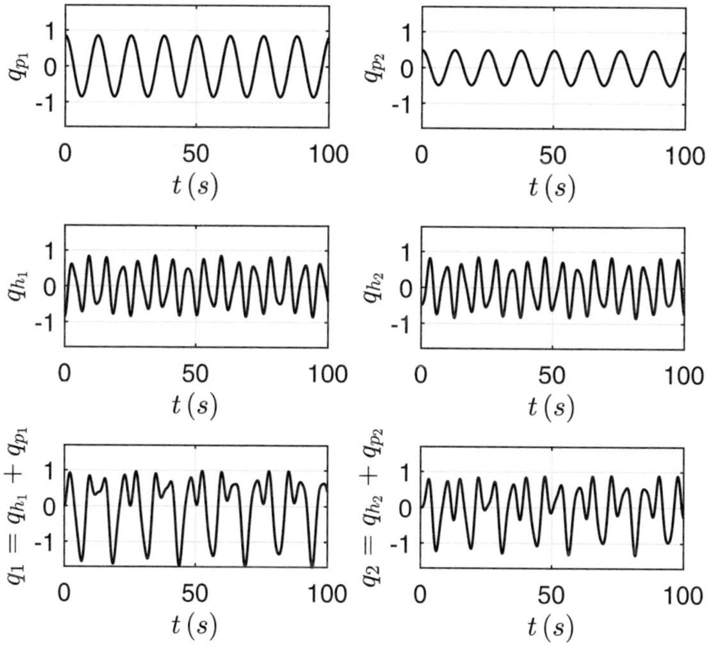

Fig. 5.22 Particular, homogeneous, and complete solutions of an undamped system with $k = 1$ N/m, $m = 1$ kg, $F_0 = 1$ N, $\omega = \omega_1/2$, and the ICs $\mathbf{q}_0 = \dot{\mathbf{q}}_0 = \mathbf{0}$

To understand how the system vibrates when excited at its first natural frequency, assume that $\omega \approx \omega_1$ but with $\omega \neq \omega_1$. The following condition follows from this assumption:

$$\left|\frac{1}{3-\left(\frac{\omega}{\omega_1}\right)^2}\right| \approx \frac{1}{2} << \left|\frac{1}{1-\left(\frac{\omega}{\omega_1}\right)^2}\right|. \tag{5.162}$$

Therefore, the complete solution from Eq. (5.161) can be approximated by

$$\begin{aligned} \begin{bmatrix} q_1(t) \\ q_2(t) \end{bmatrix} \approx & \begin{bmatrix} 1 \\ 1 \end{bmatrix} (a_1 \cos \omega_1 t + b_1 \sin \omega_1 t) \\ & + \begin{bmatrix} 1 \\ -1 \end{bmatrix} (a_2 \cos \omega_2 t + b_2 \sin \omega_2 t) \\ & + \begin{bmatrix} 1 \\ 1 \end{bmatrix} \frac{F_0/2m}{\omega_1^2 - \omega^2} \cos \omega t. \end{aligned} \tag{5.163}$$

For the sake of simplicity, take the ICs to be $\mathbf{q}_0 = \dot{\mathbf{q}}_0 = \mathbf{0}$. One can solve for the integration constants to obtain

$$\begin{bmatrix} q_1(t) \\ q_2(t) \end{bmatrix} = \frac{F_0/2m}{\omega_1^2 - \omega^2} (\cos \omega t - \cos \omega_1 t) \begin{bmatrix} 1 \\ 1 \end{bmatrix}. \tag{5.164}$$

Define $\Delta_1 = (\omega_1 - \omega)/2$ and rewrite the response as

$$\begin{bmatrix} q_1(t) \\ q_2(t) \end{bmatrix} = \frac{F_0/2m}{\omega_1 + \omega} \sin \left(\frac{\omega_1 + \omega}{2} t\right) \frac{\sin(\Delta_1 t)}{\Delta_1} \begin{bmatrix} 1 \\ 1 \end{bmatrix}. \tag{5.165}$$

As shown in Fig. 5.23a, both particles experience a fluctuation in amplitude known as beating. This is akin to the behavior of an undamped SDOF system excited close to its natural frequency. Interestingly, the motion of the particles follows the vibration profile of the first mode $\mathbf{u}_1 = [1 \quad 1]^T$ when the system is excited with $\omega \approx \omega_1$.

One can now take the limit $\omega \to \omega_1$ to obtain the response at resonance:

$$\begin{bmatrix} q_1(t) \\ q_2(t) \end{bmatrix} = \frac{F_0/2m}{2\omega_1} t \sin \omega_1 \begin{bmatrix} 1 \\ 1 \end{bmatrix}. \tag{5.166}$$

Fig. 5.23 Beating and resonance for the system. Plots are composed of curves for q_1 (—), q_2 (- -), and $f(t)$ (—). These plots were generated with the ICs $\mathbf{q}_0 = \dot{\mathbf{q}}_0 = \mathbf{0}$ together with $m = 1$ kg, $k = 1$ N/m, and $F_0 = 1$ N

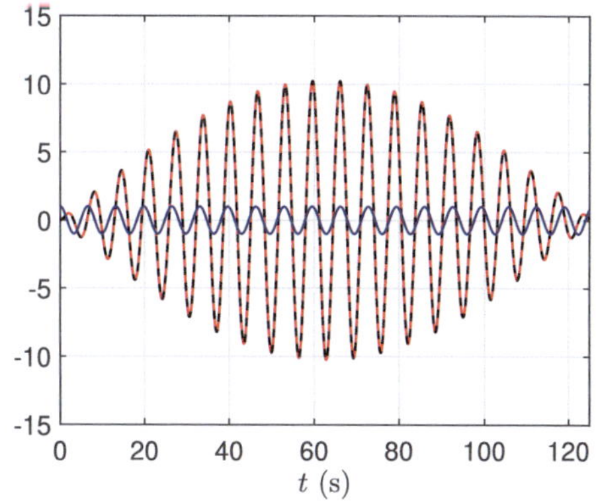

(a) Response for $\omega \approx \omega_1$.

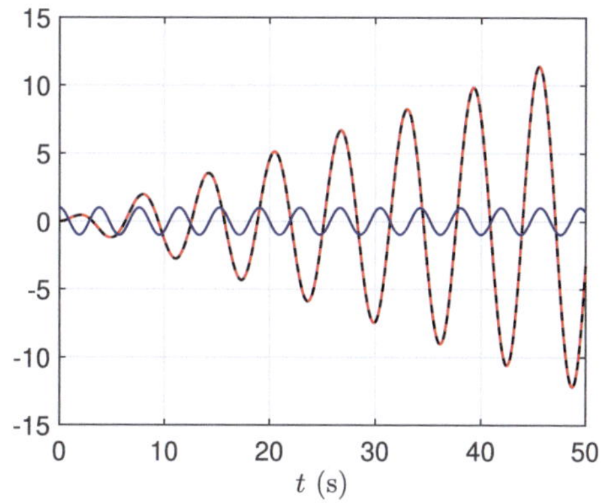

(b) Response for $\omega = \omega_1$.

A plot of this response is shown in Fig. 5.23b. The amplitude of oscillation of both particles grows linearly, as happened to an undamped SDOF system in resonance. The particles follow the vibration profile of the first mode: they move in phase with each other and maintain the same amplitude at all times. However, the phase lag between the particles and the excitation $f(t) = F_0 \cos \omega t$ is $\pi/2$.

The same steps can be applied to compute the response of the system when $\omega = \omega_2 = \sqrt{3}\omega_1$. It can be argued that

$$\left| \frac{1}{1 - \left(\dfrac{\omega}{\omega_1}\right)^2} \right| \approx \frac{1}{2} << \left| \frac{1}{3 - \left(\dfrac{\omega}{\omega_1}\right)^2} \right| \tag{5.167}$$

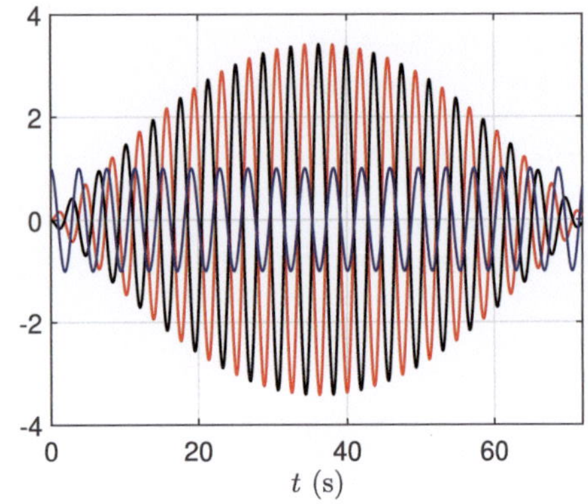

(a) Response for $\omega \approx \omega_2$.

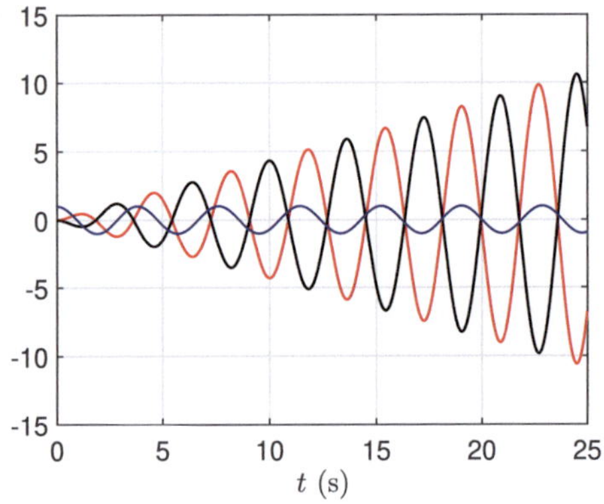

(b) Response for $\omega = \omega_2$.

Fig. 5.24 Beating and resonance for the system. Plots are composed of curves for q_1 (—), q_2 (—), and $f(t)$ (—). These plots were generated with the ICs $\mathbf{q}_0 = \dot{\mathbf{q}}_0 = \mathbf{0}$ together with $m = 1$ kg, $k = 1$ N/m, and $F_0 = 1$ N

when $\omega \approx \omega_2$, but with $\omega \neq \omega_2$. Therefore, the solution for $\mathbf{q}_0 = \dot{\mathbf{q}}_0 = \mathbf{0}$ can be approximated by

$$\begin{bmatrix} q_1(t) \\ \\ q_2(t) \end{bmatrix} \approx \frac{F_0/2m}{\omega_2 + \omega} \sin\left(\frac{\omega_2 + \omega}{2} t\right) \frac{\sin(\Delta_2 t)}{\Delta_2} \begin{bmatrix} 1 \\ \\ -1 \end{bmatrix}, \tag{5.168}$$

where $\Delta_2 = (\omega_2 - \omega)/2$. In this case, the system undergoes beating as illustrated in Fig. 5.24a. When the system is driven at $\omega = \omega_2$, it experiences resonance according to

$$\begin{bmatrix} q_1(t) \\ \\ q_2(t) \end{bmatrix} = \frac{F_0/2m}{2\omega_2} t \sin \omega_2 \begin{bmatrix} 1 \\ \\ -1 \end{bmatrix}. \tag{5.169}$$

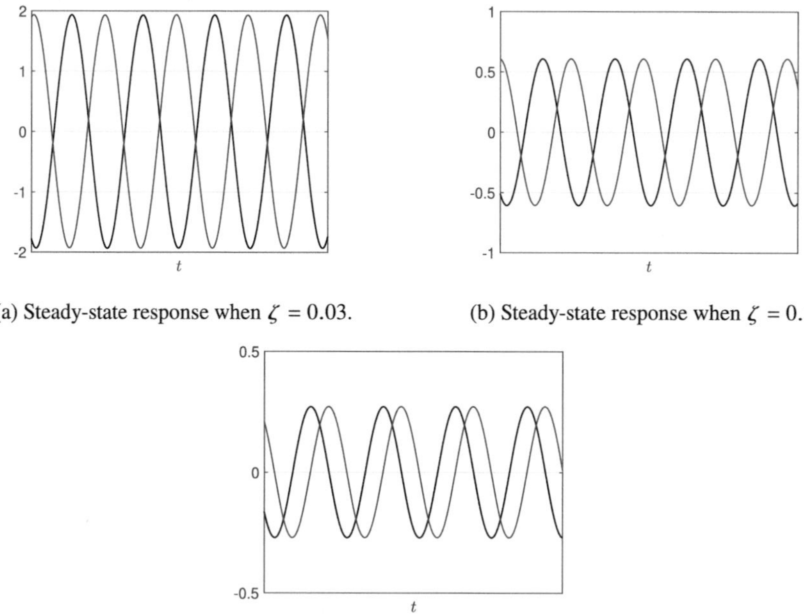

(a) Steady-state response when $\zeta = 0.03$.

(b) Steady-state response when $\zeta = 0.1$

(c) Steady-state response when $\zeta = 0.3$

Fig. 5.25 Effect of damping on the vibration profile when the system is excited with $\omega = \omega_2$ for $m = 1$ kg, $k = 1$ N/m, and $F_0 = 1$ N. The plots show the steady-state solutions q_1 (—) and q_2 (—)

This response is plotted in Fig. 5.24b. The amplitude of oscillation of the particles grows linearly, and they follow the vibration profile of the second mode, i.e., they vibrate out of phase with each other and have the same amplitude.

The approximations in Eqs. (5.162) and (5.167) do not hold if damping is present. If damping is small, the steady-state vibrations will follow closely the pattern of a mode. This pattern will break as the level of damping increases. These observations are summarized in Fig. 5.25.

5.4.2 Forced Vibrations of a System with Three Particles

It is now appropriate to apply the methodology presented in the previous section to systems with more DOFs. Consider the system with three particles shown in

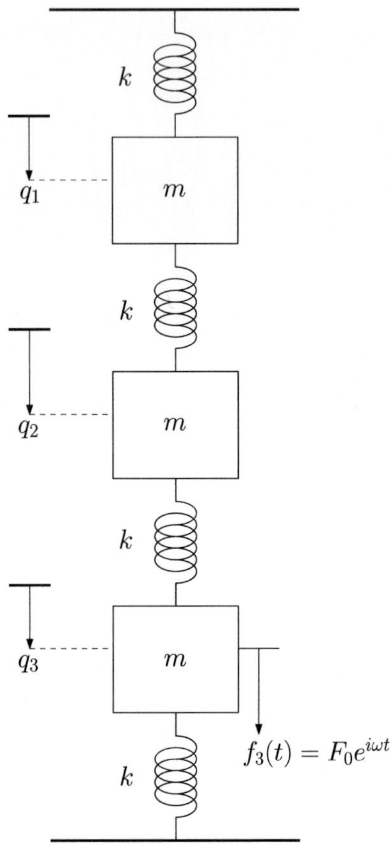

Fig. 5.26 A system with three particles excited by a harmonic force

Fig. 5.26. The vibrations of the particles are governed by the system of differential equations

$$\begin{bmatrix} m & 0 & 0 \\ 0 & m & 0 \\ 0 & 0 & m \end{bmatrix} \begin{bmatrix} \ddot{q}_1 \\ \ddot{q}_2 \\ \ddot{q}_3 \end{bmatrix} + \begin{bmatrix} 2k & -k & 0 \\ -k & 2k & -k \\ 0 & -k & 2k \end{bmatrix} \begin{bmatrix} q_1 \\ q_2 \\ q_3 \end{bmatrix} = \begin{bmatrix} 0 \\ 0 \\ F_0 \end{bmatrix} e^{i\omega t}. \tag{5.170}$$

The particular solution is given by

$$\begin{bmatrix} q_1(t) \\ q_2(t) \\ q_3(t) \end{bmatrix} = \mathbf{H}(\omega) \begin{bmatrix} 0 \\ 0 \\ F_0 \end{bmatrix} e^{i\omega t} = \begin{bmatrix} H_{13}F_0 \\ H_{23}F_0 \\ H_{33}F_0 \end{bmatrix} e^{i\omega t}, \tag{5.171}$$

It can be verified that the entries of interest from the FRF matrix $\mathbf{H}(\omega)$ are

$$\begin{aligned} H_{13}(\omega) &= \frac{1/(4k)}{\left[1-\left(\frac{\omega}{\omega_1}\right)^2\right]\left[1-\left(\frac{\omega}{\omega_2}\right)^2\right]\left[1-\left(\frac{\omega}{\omega_3}\right)^2\right]} \\ &= \frac{(2+\sqrt{2}/(8k)}{1-\left(\frac{\omega}{\omega_1}\right)^2} - \frac{1/(4k)}{1-\left(\frac{\omega}{\omega_2}\right)^2} + \frac{(2-\sqrt{2}/(8k)}{1-\left(\frac{\omega}{\omega_3}\right)^2}, \end{aligned} \tag{5.172}$$

$$\begin{aligned} H_{23}(\omega) &= \frac{1/(2k)}{\left[1-\left(\frac{\omega}{\omega_1}\right)^2\right]\left[1-\left(\frac{\omega}{\omega_3}\right)^2\right]} \\ &= \frac{\sqrt{2}(2+\sqrt{2})/(8k)}{1-\left(\frac{\omega}{\omega_1}\right)^2} - \frac{\sqrt{2}(2-\sqrt{2})/(8k)}{1-\left(\frac{\omega}{\omega_3}\right)^2}, \end{aligned} \tag{5.173}$$

and

$$\begin{aligned} H_{33}(\omega) &= \frac{\left[\frac{1}{2}-\left(\frac{\omega}{\omega_2}\right)^2\right]\left[\frac{3}{2}-\left(\frac{\omega}{\omega_2}\right)^2\right]\frac{1}{k}}{\left[1-\left(\frac{\omega}{\omega_1}\right)^2\right]\left[1-\left(\frac{\omega}{\omega_2}\right)^2\right]\left[1-\left(\frac{\omega}{\omega_3}\right)^2\right]} \\ &= \frac{(2+\sqrt{2})/(8k)}{1-\left(\frac{\omega}{\omega_1}\right)^2} + \frac{1/(4k)}{1-\left(\frac{\omega}{\omega_2}\right)^2} + \frac{(2-\sqrt{2})/(8k)}{1-\left(\frac{\omega}{\omega_3}\right)^2}. \end{aligned} \tag{5.174}$$

The frequency response curves of H_{13}, H_{23}, and H_{33} are shown in Fig. 5.27. The FRF H_{13} and H_{23} exhibit no antiresonances, but H_{33} shows them at $\omega = \sqrt{k/m}$ and $\omega = \sqrt{3k/m}$. The vibration amplitudes grow linearly when the system is excited at the resonant frequencies $\omega = \omega_1$ or $\omega = \omega_3$. The FRF H_{23} does not show a peak growing to infinity at $\omega = \omega_2$ because the intermediate particle acts as a node in

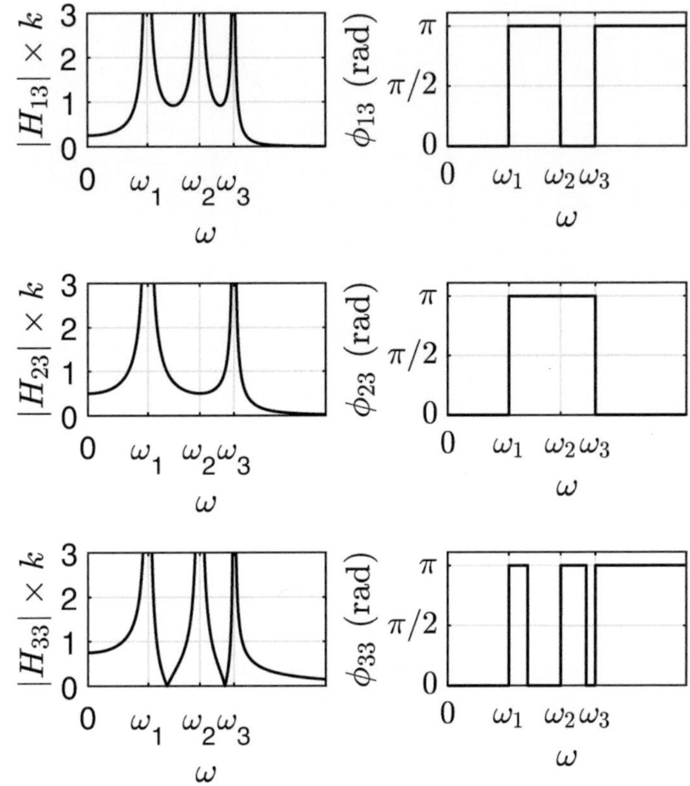

Fig. 5.27 Frequency response curves of H_{13}, H_{23}, and H_{33}

the second mode, leading to a smaller vibration amplitude compared to the other particles as shown in Fig. 5.28. Even though the intermediate particle is a node in the second mode, its vibration amplitude is not zero when the system is excited with $\omega = \omega_2$. This occurs because of the contributions from the first and third modes in this particle's response.

5.4.3 Forced Vibrations and the Frequency Response Function Matrix

The previous examples illustrated how to compute the response of two and three-DOFs systems under the action of a harmonic external force. For a general case,

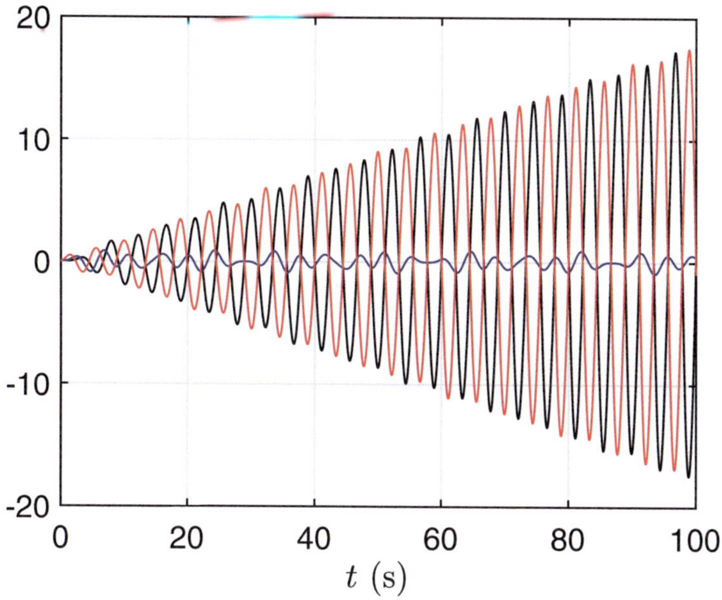

Fig. 5.28 System in resonance with $\omega = \omega_2$. The intermediate particle has smaller amplitude of vibration because it is a node for the second mode. Plots are composed of curves for q_1 (—), q_2 (—), and q_3 (—). The ICs $\mathbf{q}_0 = \dot{\mathbf{q}}_0 = \mathbf{0}$ were considered

consider an n-DOFs system governed by Eq. (5.148) and suppose it is excited by a harmonic excitation

$$\mathbf{f}(t) = \mathbf{F}_0 e^{i\omega t}, \tag{5.175}$$

where $\mathbf{F}_0$ is a constant $n \times 1$ vector. The steady-state solution is

$$\mathbf{q}(t) = \mathbf{Q} e^{i\omega t}, \tag{5.176}$$

where $\mathbf{Q}$ is the $n \times 1$ vector of amplitudes. The steady-state response to harmonic excitation is also harmonic and has the same frequency as the excitation.

Substituting the assumed solution (5.176) into the equations of motion (5.148) allows the computation of amplitude vector $\mathbf{Q}$ according to

$$\mathbf{Q} = \mathbf{H}(\omega)\mathbf{F}_0, \tag{5.177}$$

where $\mathbf{H}(\omega) = (\mathbf{K} - \omega^2\mathbf{M} + i\omega\mathbf{C})^{-1}$ is the FRF matrix of order $n \times n$. These results are summarized in Table 5.3, demonstrating that the solution process for forced vibrations in systems with multiple degrees of freedom closely resembles that of SDOF systems.

Table 5.3 Solution procedure for forced vibration

	$n = 1$	$n > 1$
Equations of motion	$m\ddot{q} + c\dot{q} + kq = F_0 e^{i\omega t}$	$\mathbf{M}\ddot{\mathbf{q}} + \mathbf{C}\dot{\mathbf{q}} + \mathbf{K}\mathbf{q} = \mathbf{F}_0 e^{i\omega t}$
FRF	$H = 1/(k - m\omega^2 + i\omega c)$	$\mathbf{H} = (\mathbf{K} - \omega^2\mathbf{M} + i\omega\mathbf{C})^{-1}$
Steady-state solution	$q_p(t) = H(\omega)F_0 e^{i\omega t}$	$\mathbf{q}_p(t) = \mathbf{H}(\omega)\mathbf{F}_0 e^{i\omega t}$
Complete solution	$q(t) = q_h(t) + q_p(t)$	$\mathbf{q}(t) = \mathbf{q}_h(t) + \mathbf{q}_p(t)$

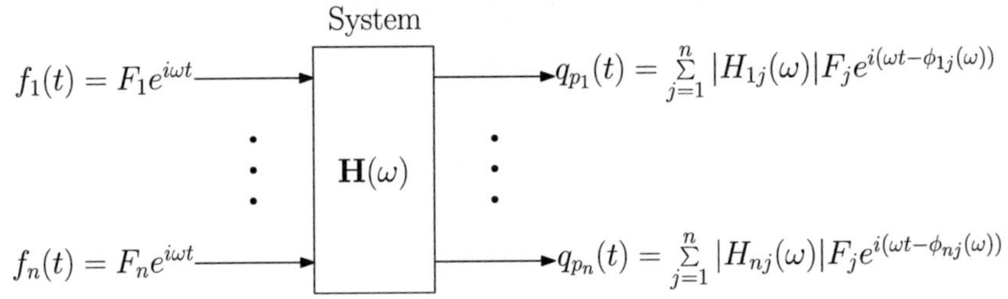

Fig. 5.29 The forced vibrations of an MDOF system is characterized by the FRF matrix. It connects the force inputs to all output DOFs

The FRF matrix fully characterizes MDOF systems, providing information on how the response is magnified and delayed relative to the excitation. It allows an MDOF mechanical system to be interpreted as a multi-input, multi-output system. When a single input, say $f_1(t)$, is applied to input channel 1 (with other elements of $f(t)$ set to zero), a response is generated in all output channels, as illustrated in Fig. 5.29.

The FRF matrix is symmetric due to the symmetry of **M**, **C**, and **K**. Therefore, only $n(n+1)/2$ entries are needed out of a total of n^2. An entry H_{ij} of **H** can be interpreted as the FRF of the i-th DOF due to a harmonic force applied at the j-th DOF. The diagonal entries in **H** are thus called direct FRF, and the off-diagonal ones are termed cross FRF. Frequency response curves can be plotted for each entry.

In an undamped system, the absolute value plot typically shows n peaks with infinite height at each natural frequency (resonance). The system oscillates according to the vibration profile of the mode corresponding to the excited natural frequency. If a DOF is a node for a particular mode, the FRF for that DOF will not have a peak at the natural frequency associated with that mode. Some entries in **H** may exhibit antiresonance, occurring at certain frequencies where the absolute value of the entry in **H** is zero.

Damping reduces the height of the peaks at resonance and increases the amplitude at antiresonances. A lightly damped system retains resonant peaks at the natural frequencies, and its steady-state vibration closely follows the corresponding mode shape profile. If damping is too high, some resonant peaks may disappear.

5.4.4 *Illustrative Example: The Dynamic Vibration Absorber*

Unchecked vibrations can lead to material fatigue, failure, excessive wear, and even catastrophic collapse. For instance, bridges and high-rise buildings subjected to wind or seismic forces can experience harmful oscillations that jeopardize structural integrity. Similarly, in precision manufacturing and aerospace, vibrations can diminish the accuracy and performance of sensitive equipment.

Vibration control is essential in both mechanical equipment and large infrastructure. Engineers employ various strategies, categorized into passive and active methods. Active control uses sensors and actuators to dynamically counteract vibrations in real time, while passive control operates without external energy, relying on the intrinsic dynamic characteristics of the system. A widely used passive technique is the dynamic vibration absorber (DVA), which reduces vibrations at specific frequencies by adding a tuned mechanical appendage that absorbs unwanted vibrations. These absorbers are widely used in applications such as tall buildings to combat wind and seismic forces, vehicle suspension systems to enhance ride comfort, and rotating machinery to minimize vibrations at critical speeds. This example illustrates the operation of a DVA and explores key design considerations.

5.4.4.1 Modeling the Primary and Secondary Systems

Consider a primary mass-spring system governed by

$$m_1\ddot{q}_1 + k_1 q_1 = f(t) = F_0 e^{i\omega t}. \tag{5.178}$$

With the intent of reducing the steady-state vibrations caused by the harmonic force $f(t)$, a secondary absorber system that consists of a mass m_2 and a spring k_2 is attached to the primary system as illustrated in Fig. 5.30.

It can be shown that the equations of motion of the composite system are

$$\begin{bmatrix} m_1 & 0 \\ 0 & m_2 \end{bmatrix} \begin{bmatrix} \ddot{q}_1 \\ \ddot{q}_2 \end{bmatrix} + \begin{bmatrix} k_1 + k_2 & -k_2 \\ -k_2 & k_2 \end{bmatrix} \begin{bmatrix} q_1 \\ q_2 \end{bmatrix} = \begin{bmatrix} F_0 \\ 0 \end{bmatrix} e^{i\omega t}, \tag{5.179}$$

where q_1 and q_2 are generalized coordinates measured from the static equilibrium position.

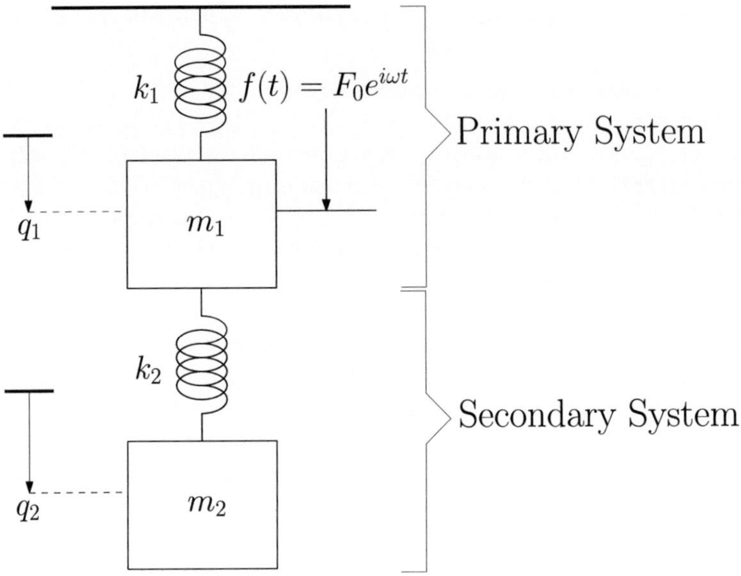

Fig. 5.30 Primary system attached to a secondary absorber system

5.4.4.2 The Steady-State Solution

The steady-state solution is

$$\mathbf{q} = \begin{bmatrix} Q_1 \\ Q_2 \end{bmatrix} e^{i\omega t} = \mathbf{Q}e^{i\omega t}. \tag{5.180}$$

Substitute into the equations of motion to verify that

$$\begin{aligned} \mathbf{Q} &= \begin{bmatrix} k_1 + k_2 - m_1\omega^2 & -k_2 \\ -k_2 & k_2 - m_2\omega^2 \end{bmatrix}^{-1} \begin{bmatrix} F_0 \\ 0 \end{bmatrix} \\ &= \frac{F_0}{(k_1 + k_2 - m_1\omega^2)(k_2 - m_2\omega^2) - k_2^2} \begin{bmatrix} k_2 - m_2\omega^2 \\ k_2 \end{bmatrix}. \end{aligned} \tag{5.181}$$

The steady-state amplitudes are commonly expressed as follows:

$$\frac{Q_1}{\delta_e} = \frac{1 - \frac{1}{r^2}\left(\frac{\omega}{\omega_n}\right)^2}{\left[1 + \mu r^2 - \left(\frac{\omega}{\omega_n}\right)^2\right]\left[1 - \frac{1}{r^2}\left(\frac{\omega}{\omega_n}\right)^2\right] - \mu r^2}, \tag{5.182}$$

$$\frac{Q_2}{\delta_e} = \frac{1}{\left[1 + \mu r^2 - \left(\frac{\omega}{\omega_n}\right)^2\right]\left[1 - \frac{1}{r^2}\left(\frac{\omega}{\omega_n}\right)^2\right] - \mu r^2}, \tag{5.183}$$

where $\omega_n = \sqrt{\frac{k_1}{m_1}}$ is the natural frequency of the primary system and $\omega_a = \sqrt{\frac{k_2}{m_2}}$ is the natural frequency of the secondary absorber system. The other variables are nondimensional parameters defined by

- Static deflection of the primary system due to F_0: $\delta_e = \frac{F_0}{k_1}$;
- Mass ratio: $\mu = \frac{m_2}{m_1}$;
- Frequency ratio: $r = \frac{\omega_a}{\omega_n}$.

5.4.4.3 Tuning of Absorber

If the absorber is *tuned* to the excitation frequency ω, i.e., k_2 and m_2 are chosen so that

$$\omega_a = \omega, \tag{5.184}$$

then the steady-state amplitude of the primary system reduces to $Q_1 = 0$. Why would this happen? Note that the steady-state amplitude of the absorber is

$$Q_2 = -\frac{F_0}{k_2} \tag{5.185}$$

when $\omega = \omega_a$. The applied harmonic force and the force exerted by the absorber on the primary system combine to generate

$$F_0 e^{i\omega t} + k_2 q_2 = 0. \tag{5.186}$$

This shows that the absorber system exerts a force equal and opposite to the impressed force and neutralizes it, thus reducing Q_1 to zero. A direct consequence from Eq. (5.185) is that the effectiveness of the absorber depends on the allowable clearance $|Q_2|_{max}$.

5.4.4.4 Tuning to Primary Natural Frequency

To reduce the amplitude of the primary system at its resonant frequency ω_n, the absorber should be tuned to match

$$\omega_a = \omega_n. \tag{5.187}$$

The frequency response curves of Q_1 and Q_2 are shown in Fig. 5.31 when the absorber is tuned to the primary system natural frequency. Previously, the primary system exhibited a resonant peak at ω_n. However, the introduction of the DVA reduced its amplitude to $Q_1 = 0$ at $\omega = \omega_n$. The energy from the primary system is transferred to vibrate the absorber instead.

The addition of the absorber to the primary system increases the number of DOFs. The composite system now has two new frequencies, Ω_1 and Ω_2, at which resonance can occur. These frequencies are the roots of the characteristic polynomial

$$0 = \left(\frac{\Omega}{\omega_n}\right)^4 - (r^2 + \mu r^2 + 1)\left(\frac{\Omega}{\omega_n}\right)^2 + r^2. \tag{5.188}$$

It can be checked that

$$\frac{\Omega_{1,2}}{\omega_n} = \sqrt{\frac{r^2 + \mu r^2 + 1 \mp \sqrt{(r^2 + \mu r^2 + 1)^2 - 4r^2}}{2}}. \tag{5.189}$$

The dependence of $\Omega_{1,2}$ on the mass and frequency ratios is shown in Fig. 5.32. The separation between frequencies increases as μ increases. If the mass ratio is too low, the interval (Ω_1, Ω_2) might be too narrow for the DVA to operate adequately in the presence of small fluctuations in the excitation frequency ω.

5.5 Modal Analysis

As previously discussed, the equations governing small vibrations in mechanical systems form a system of second-order, linear ODEs. These equations are typically coupled, i.e., the j-th component equation involves not only the generalized coordinate q_j and its derivatives but also other coordinates and their derivatives. Coupling introduces both mathematical and physical challenges in problem analysis, as seen in the treatment of free and forced vibrations from the previous sections. Engineers address this challenge using a technique called modal analysis, which involves a

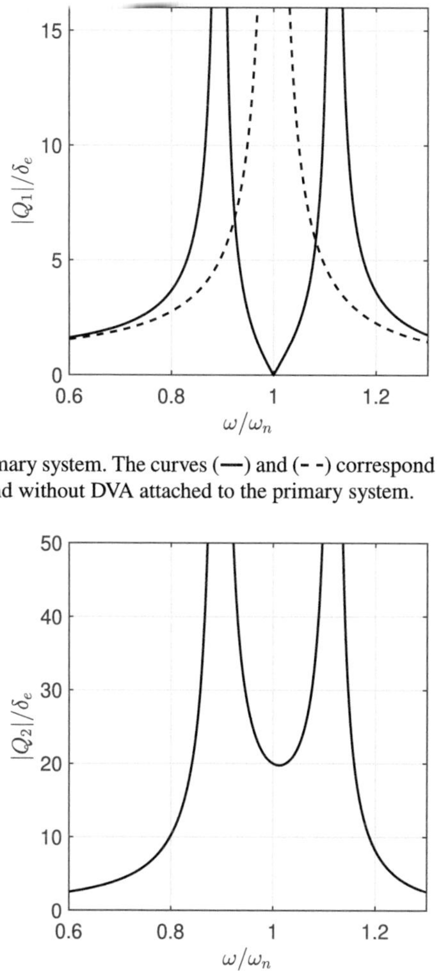

(a) Primary system. The curves (—) and (- -) correspond to the cases with and without DVA attached to the primary system.

(b) Secondary absorber system.

Fig. 5.31 Frequency response curves of steady-state amplitude of vibration

change of coordinates that decouples the equations of motion, transforming an MDOF system into simpler, independent subsystems, each with a single degree of freedom. Decoupling the equations through modal analysis simplifies response calculations and provides additional advantages. For example, it allows engineers to identify the modes that significantly impact the response, enabling them to ignore less influential modes and thus reduce the computational cost of obtaining the response. This section provides a detailed overview of the theory of modal analysis. Firstly, an example is provided to introduce the main concepts. The general theory

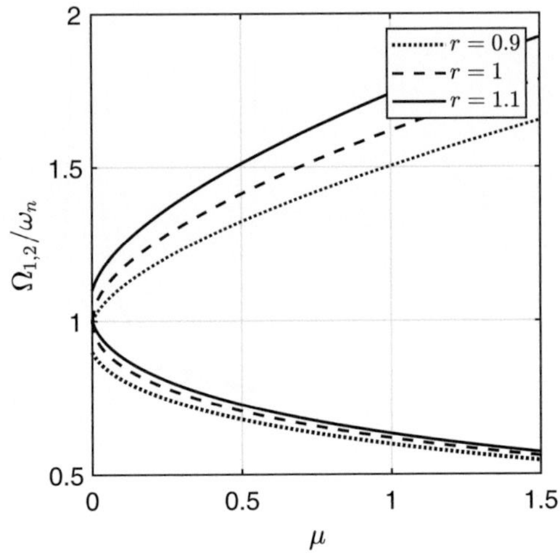

Fig. 5.32 Dependence of the composite system natural frequencies on the mass and frequency ratios

is presented in the sequence, and two additional examples are supplied to further clarify the applicability of modal analsys.

5.5.1 Decoupling the Equations of a System with Two Particles

Consider the two-DOF system shown from Fig. 5.30. For the sake of simplicity, take $m_1 = 2m$, $m_2 = m$, $k_1 = 2k$, and $k_2 = k$. The equations of motion in (5.179) reduce to

$$\begin{bmatrix} 2m & 0 \\ 0 & m \end{bmatrix} \begin{bmatrix} \ddot{q}_1 \\ \ddot{q}_2 \end{bmatrix} + \begin{bmatrix} 3k & -k \\ -k & k \end{bmatrix} \begin{bmatrix} q_1 \\ q_2 \end{bmatrix} = \begin{bmatrix} F_0 \\ 0 \end{bmatrix} e^{i\omega t}. \tag{5.190}$$

The objective of this example is to explore how the properties of the eigensolutions can transform this coupled system into two independent SDOF systems.

5.5.1.1 Orthogonality of Mode Shapes

The characteristic polynomial is

$$2k^2 - 5km(\omega^2) + 2m^2(\omega^2)^2 = 0. \tag{5.191}$$

The two natural frequencies can be obtained from the roots of this polynomial:

$$\omega_1 = \frac{\sqrt{2}}{2}\sqrt{\frac{k}{m}}, \quad \omega_2 = \sqrt{2}\sqrt{\frac{k}{m}}. \tag{5.192}$$

Use the natural frequencies to compute the associated mode shapes as

$$\mathbf{u}_1 = \begin{bmatrix} 1 \\ 2 \end{bmatrix}, \quad \mathbf{u}_2 = \begin{bmatrix} 1 \\ -1 \end{bmatrix}. \tag{5.193}$$

A simple check reveals that the mode shapes are not orthogonal in the usual sense:

$$\mathbf{u}_1^T \mathbf{u}_2 = -1 \neq 0. \tag{5.194}$$

However, they are orthogonal with respect to a different inner product. Define an inner product weighted by the mass matrix according with[4]

$$\mathbf{x}^T \mathbf{M} \mathbf{y}, \tag{5.195}$$

for any two-dimensional vectors $\mathbf{x}$ and $\mathbf{y}$. With respect to this modified inner product, the mode shapes satisfy

$$\mathbf{u}_1^T \mathbf{M} \mathbf{u}_2 = 0. \tag{5.196}$$

A similar result is obtained if one uses an inner product weighted by the stiffness matrix:

$$\mathbf{u}_1^T \mathbf{K} \mathbf{u}_2 = 0. \tag{5.197}$$

The conclusion is that, even though the mode shapes are not orthogonal in the usual sense, they are orthogonal with respect to the mass and stiffness matrices.

5.5.1.2 Mass Normalization of Mode Shapes

Insert the first mode shape into the two vector slots from the inner product weighted by the mass matrix to check that

$$\mathbf{u}_1^T \mathbf{M} \mathbf{u}_1 = 6m. \tag{5.198}$$

[4] This operation is an inner product because it satisfies the four axioms of inner products listed in Appendix B.

The mode shape $\mathbf{u}_1$ is said to be mass-normalized if

$$\mathbf{u}_1^T \mathbf{M} \mathbf{u}_1 = 1. \tag{5.199}$$

The scalar constant α that yields a mass-normalized mode shape $\alpha \mathbf{u}_1$ is given by

$$\alpha = \frac{1}{\sqrt{6m}}. \tag{5.200}$$

The choice of normalization scheme is arbitrary, but mass normalization is commonly assumed in vibration analysis. The mass-normalized modes can be organized in the modal matrix:

$$\mathbf{U} = \begin{bmatrix} \dfrac{1}{\sqrt{6m}} & \dfrac{1}{\sqrt{3m}} \\ \dfrac{2}{\sqrt{6m}} & -\dfrac{1}{\sqrt{3m}} \end{bmatrix}. \tag{5.201}$$

5.5.1.3 Orthogonality Relationships and Decoupling

The mass-normalized modal matrix can be used to express the orthogonality relationships as

$$\mathbf{U}^T \mathbf{M} \mathbf{U} = \begin{bmatrix} 1 & 0 \\ 0 & 1 \end{bmatrix}, \quad \mathbf{U}^T \mathbf{K} \mathbf{U} = \boldsymbol{\Omega}, \tag{5.202}$$

where $\boldsymbol{\Omega}$ is the spectral matrix:

$$\boldsymbol{\Omega} = \begin{bmatrix} k/2m & 0 \\ 0 & 2k/m \end{bmatrix}. \tag{5.203}$$

The orthogonality relationships, as applied in Eq. (5.202), are used to diagonalize both the mass and stiffness matrices. With this in mind, define a coordinate transformation by

$$\begin{bmatrix} q_1 \\ q_2 \end{bmatrix} = \mathbf{U} \begin{bmatrix} p_1 \\ p_2 \end{bmatrix}. \tag{5.204}$$

This transformation can be interpreted as expressing the system response as a linear combination of the mode shape vectors with time-dependent coefficients:

$$\begin{bmatrix} q_1 \\ q_2 \end{bmatrix} = p_1(t)\mathbf{u}_1 + p_2(t)\mathbf{u}_2. \tag{5.205}$$

Thus, it connects the physical coordinates q_1 and q_2 to the so-called *modal coordinates* p_1 and p_2. Substitute the coordinate transformation into the equations of motion and premultiply the result by $\mathbf{U}^T$. The orthogonality relationships in Eq. (5.202) imply that

$$\begin{bmatrix} \ddot{p}_1 \\ \ddot{p}_2 \end{bmatrix} + \begin{bmatrix} k/(2m) & 0 \\ 0 & 2k/m \end{bmatrix} \begin{bmatrix} p_1 \\ p_2 \end{bmatrix} = \mathbf{g}(t), \tag{5.206}$$

where the vector of applied forces for the modal coordinates is

$$\mathbf{g}(t) = \mathbf{U}^T \mathbf{f}(t) = \begin{bmatrix} F_0/\sqrt{6m} \\ F_0/\sqrt{3m} \end{bmatrix} e^{i\omega t}. \tag{5.207}$$

The modal transformation (5.204) decoupled the system into two independent SDOF subsystems, both of which have unit mass because the mode shapes are mass-normalized. The subsystem described by the modal coordinate p_1 has "stiffness" $k/(2m)$ and is acted upon by the harmonic force $g_1(t)$, while the other one has "stiffness" $2k/m$ and is under the force $g_2(t)$. This process is illustrated in Fig. 5.33.

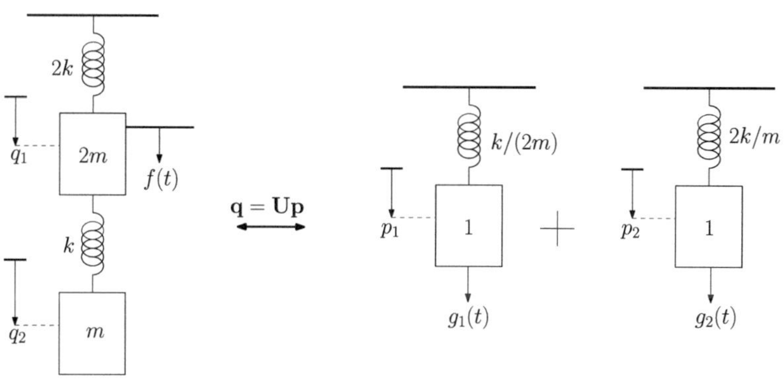

Fig. 5.33 Modal analysis transforms the coupled MDOF system into two independent SDOF systems

The two differential equations in Eq. (5.206) can be solved independently using the techniques from previous chapters. The general solutions are

$$p_1(t) = a_1 \cos \omega_1 t + b_1 \sin \omega_1 t + \frac{F_0/\sqrt{6m}}{\omega_1^2 - \omega^2} e^{i\omega t}, \tag{5.208}$$

$$p_2(t) = a_2 \cos \omega_2 t + b_2 \sin \omega_2 t + \frac{F_0/\sqrt{3m}}{\omega_2^2 - \omega^2} e^{i\omega t}. \tag{5.209}$$

The integration constants are determined from the ICs for the modal coordinates. These ICs can be obtained from the inverse modal transformation

$$\begin{bmatrix} p_1 \\ p_2 \end{bmatrix} = \mathbf{U}^T \mathbf{M} \begin{bmatrix} q_1 \\ q_2 \end{bmatrix}. \tag{5.210}$$

If $\mathbf{q}(0) = [1 \quad -1]^T$ and $\dot{\mathbf{q}}(0) = \mathbf{0}$, it can be verified that $\mathbf{p}(0) = [0 \quad \sqrt{3m}]^T$ and $\dot{\mathbf{p}}(0) = \mathbf{0}$. Assuming that $f(t) = F_0 \cos \omega t$, the solutions become

$$p_1(t) = \frac{F_0/\sqrt{6m}}{\omega_1^2 - \omega^2} (\cos \omega t - \cos \omega_1 t), \tag{5.211}$$

$$p_2(t) = \frac{F_0/\sqrt{3m}}{\omega_2^2 - \omega^2} \left\{ \left[\frac{3m(\omega_2^2 - \omega^2)}{F_0} - 1 \right] \cos \omega_2 t + \cos \omega t \right\}. \tag{5.212}$$

Once $p_1(t)$ and $p_2(t)$ are known, the physical coordinates can be recovered with the modal transformation:

$$\begin{aligned} \begin{bmatrix} q_1(t) \\ q_2(t) \end{bmatrix} &= \begin{bmatrix} 1 \\ 2 \end{bmatrix} \frac{F_0/6m}{\omega_1^2 - \omega^2} (\cos \omega t - \cos \omega_1 t) \\ &+ \begin{bmatrix} 1 \\ -1 \end{bmatrix} \left\{ \left(1 - \frac{F_0/3m}{\omega_2^2 - \omega^2} \right) \cos \omega_2 t + \frac{F_0/3m}{\omega_2^2 - \omega^2} \cos \omega t \right\}. \end{aligned} \tag{5.213}$$

The responses of the system in physical and modal coordinates are shown in Fig. 5.34.

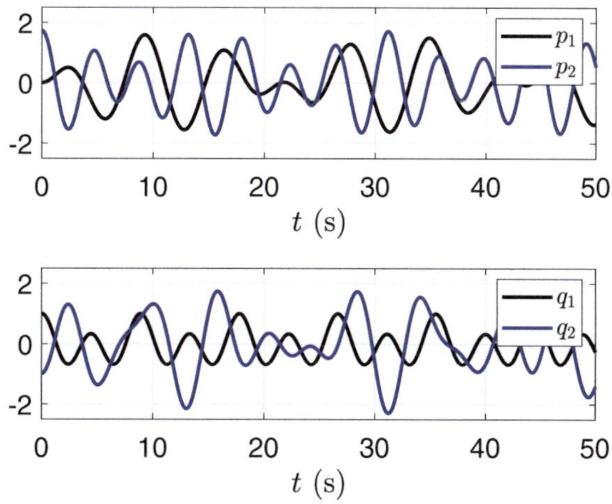

Fig. 5.34 Comparison between the solutions for the modal coordinates, from Eqs. (5.211) and (5.212), and the solutions in physical coordinates from Eq. (5.213). Plots generated with $m = 1$ kg, $k = 1$ N/m, $F_0 = 1$ N, and $\omega = 1$ rad/s

5.5.2 *Decoupling by Modal Analysis*

In the previous example, the eigensolutions were used to decouple the equations of motion. This section details how this procedure can be generalized to decouple the equations of motion of any n-DOFs system governed by

$$\mathbf{M}\ddot{\mathbf{q}} + \mathbf{K}\mathbf{q} = \mathbf{f}(t). \tag{5.214}$$

5.5.2.1 Orthogonality Relationships

For simplicity, assume that system (5.214) possesses n distinct natural frequencies, ordered as $\omega_1 < \omega_2 < \ldots < \omega_n$. The associated mode shape vectors are generally not orthogonal in the usual sense. However, they are always orthogonal with respect to the mass matrix. To demonstrate this, consider two distinct eigensolutions

$$\mathbf{K}\mathbf{u}_j = \omega_j^2\mathbf{M}\mathbf{u}_j, \tag{5.215}$$

$$\mathbf{K}\mathbf{u}_l = \omega_l^2\mathbf{M}\mathbf{u}_l. \tag{5.216}$$

Premultiply Eqs. (5.215) and (5.216) by $\mathbf{u}_l^T$ and $\mathbf{u}_j^T$, respectively:

$$\mathbf{u}_l^T \mathbf{K}\mathbf{u}_j = \omega_j^2 \mathbf{u}_l^T \mathbf{M}\mathbf{u}_j, \tag{5.217}$$

$$\mathbf{u}_j^T \mathbf{K}\mathbf{u}_l = \omega_l^2 \mathbf{u}_j^T \mathbf{M}\mathbf{u}_l. \tag{5.218}$$

To move forward, note that the terms $\mathbf{u}_j^T \mathbf{K}\mathbf{u}_l$ and $\mathbf{u}_j^T \mathbf{M}\mathbf{u}_j$ are scalars since each has dimension of 1×1. Therefore, they are unaffected by matrix transposition and must satisfy

$$\mathbf{u}_l^T \mathbf{K}\mathbf{u}_j = \mathbf{u}_j^T \mathbf{K}\mathbf{u}_l, \tag{5.219}$$

$$\mathbf{u}_l^T \mathbf{M}\mathbf{u}_j = \mathbf{u}_j^T \mathbf{M}\mathbf{u}_l. \tag{5.220}$$

Use this expressions to rewrite Eq. (5.218) as

$$\mathbf{u}_l^T \mathbf{K}\mathbf{u}_j = \omega_j^2 \mathbf{u}_l^T \mathbf{M}\mathbf{u}_j. \tag{5.221}$$

Taking the difference between Eqs. (5.221) and (5.217) shows that

$$0 = (\omega_l^2 - \omega_j^2)\mathbf{u}_l^T \mathbf{M}\mathbf{u}_j. \tag{5.222}$$

Since the frequency difference $\omega_l^2 - \omega_j^2 \neq 0$ for distinct eigensolutions ($l \neq j$), this equation implies that the mode shapes are orthogonal with respect to the mass matrix: $\mathbf{u}_l^T \mathbf{M}\mathbf{u}_j = 0$. If the mode shapes are mass-normalized, then the orthogonality relationship can be cast as

$$\mathbf{u}_l^T \mathbf{M}\mathbf{u}_j = \delta_{lj}, \tag{5.223}$$

where $\delta_{lj} = 0$ if $l \neq j$ or $\delta_{lj} = 1$ if $l = j$. Substitute this result into Eq. (5.217) to conclude that the mode shapes are also orthogonal with respect to the stiffness matrix:

$$\mathbf{u}_l^T \mathbf{K}\mathbf{u}_j = \omega_l^2 \delta_{lj}. \tag{5.224}$$

The orthogonality relationships from Eqs. (5.223) and (5.224) can be used to derive a transformation that simultaneously diagonalizes $\mathbf{M}$ and $\mathbf{K}$. Use block matrix multiplication to easily verify that

$$\mathbf{U}^T\mathbf{M}\mathbf{U} = \begin{bmatrix} \mathbf{u}_1^T\mathbf{M}\mathbf{u}_1 & \mathbf{u}_1^T\mathbf{M}\mathbf{u}_2 & \dots & \mathbf{u}_1^T\mathbf{M}\mathbf{u}_n \\ \mathbf{u}_2^T\mathbf{M}\mathbf{u}_1 & \mathbf{u}_2^T\mathbf{M}\mathbf{u}_2 & \dots & \mathbf{u}_2^T\mathbf{M}\mathbf{u}_n \\ \vdots & \vdots & \ddots & \vdots \\ \mathbf{u}_n^T\mathbf{M}\mathbf{u}_1 & \mathbf{u}_n^T\mathbf{M}\mathbf{u}_2 & \dots & \mathbf{u}_n^T\mathbf{M}\mathbf{u}_n \end{bmatrix} = \begin{bmatrix} 1 & 0 & \dots & 0 \\ 0 & 1 & \dots & 0 \\ \vdots & \vdots & \ddots & \vdots \\ 0 & 0 & \dots & 1 \end{bmatrix} = \mathbf{I}. \tag{5.225}$$

The same steps can be used to show that

$$\mathbf{U}^T\mathbf{K}\mathbf{U} = \begin{bmatrix} \omega_1^2 & 0 & \dots & 0 \\ 0 & \omega_2^2 & \dots & 0 \\ \vdots & \vdots & \ddots & \vdots \\ 0 & 0 & \dots & \omega_n^2 \end{bmatrix} = \mathbf{\Omega}. \tag{5.226}$$

5.5.2.2 Modal Transformation

Define a coordinate transformation from the physical coordinates $\mathbf{q}$ to the *principal or modal coordinates* $\mathbf{p}$ by

$$\mathbf{q}(t) = \sum_{j=1}^{n} p_j(t)\mathbf{u}_j = \begin{bmatrix} \mathbf{u}_1 & \mathbf{u}_2 & \cdots & \mathbf{u}_n \end{bmatrix} \begin{bmatrix} p_1 \\ p_2 \\ \vdots \\ p_n \end{bmatrix} = \mathbf{U}\mathbf{p}(t). \tag{5.227}$$

The aforementioned expression is also referred to as modal expansion, mode superposition, or eigenvector expansion of the system response $\mathbf{q}(t)$. Substitute this transformation into the equations of motion (5.214) and use the orthogonality relationships to obtain the decoupled system of equations

$$\ddot{\mathbf{p}} + \boldsymbol{\Omega}\mathbf{p} = \mathbf{g}(t). \tag{5.228}$$

The initial values are transformed in accordance with $\mathbf{p}(0) = \mathbf{U}^T\mathbf{M}\mathbf{q}(0)$ and $\dot{\mathbf{p}}(0) = \mathbf{U}^T\mathbf{M}\dot{\mathbf{q}}(0)$. The excitation $\mathbf{f}(t)$ is transformed into

$$\mathbf{g}(t) = \mathbf{U}^T\mathbf{f}(t) = \begin{bmatrix} \mathbf{u}_1^T\mathbf{f}(t) \\ \mathbf{u}_2^T\mathbf{f}(t) \\ \vdots \\ \mathbf{u}_n^T\mathbf{f}(t) \end{bmatrix}. \tag{5.229}$$

The modal equation associated with p_j is

$$\ddot{p}_j + \omega_j^2 p_j = \mathbf{u}_j^T\mathbf{f}(t), \quad j = 1,\ 2,\ \ldots,\ n. \tag{5.230}$$

Each decoupled modal equation can be solved independently, and solution of the original equation is readily recovered from $\mathbf{q} = \mathbf{U}\mathbf{p}$.

5.5.3 Modal Analysis Applied to Damped Systems

In the previous section, modal analysis was used to demonstrate that an undamped system can always be decoupled through modal analysis. Applying the modal transformation from Eq. (5.227) to a damped system governed by

$$\mathbf{M}\ddot{\mathbf{q}} + \mathbf{C}\dot{\mathbf{q}} + \mathbf{K}\mathbf{q} = \mathbf{f}(t) \tag{5.231}$$

results in the system of differential equations

$$\ddot{\mathbf{p}} + \mathbf{D}\dot{\mathbf{p}} + \boldsymbol{\Omega}\mathbf{p} = \mathbf{g}(t). \tag{5.232}$$

These equations are completely decoupled if and only if the modal damping matrix $\mathbf{D} = \mathbf{U}^T\mathbf{C}\mathbf{U}$ is diagonal. Unfortunately, not every damped system can be decoupled through modal analysis. For illustration purposes, suppose the coefficient matrices are

$$\mathbf{M} = \mathbf{I}, \quad \mathbf{C} = \frac{1}{10}\begin{bmatrix} 7 & -1 \\ -1 & 2 \end{bmatrix}, \quad \mathbf{K} = \begin{bmatrix} 2 & -1 \\ -1 & 2 \end{bmatrix}. \tag{5.233}$$

Compute the mass-normalized mode shape vectors to construct the modal transformation

$$\mathbf{q} = \frac{1}{\sqrt{2}}\begin{bmatrix} 1 & 1 \\ 1 & -1 \end{bmatrix}\mathbf{p}. \tag{5.234}$$

Modal analysis transforms this system into

$$\ddot{\mathbf{p}} + \frac{1}{20}\begin{bmatrix} 7 & 5 \\ 5 & 11 \end{bmatrix}\dot{\mathbf{p}} + \begin{bmatrix} 1 & 0 \\ 0 & 3 \end{bmatrix}\mathbf{p} = \mathbf{U}^T\mathbf{f}(t). \tag{5.235}$$

Both **M** and **K** are diagonalized, but the modal damping matrix is not diagonal.

In general, damped systems cannot be decoupled through modal analysis. This limitation arises because modal analysis involves the simultaneous diagonalization of three matrices, which is impossible to achieve without imposing severe mathematical restriction on **C**. The subclass of damped system that can be decoupled by modal analysis is termed *classically damped systems*. For these systems, **D** is diagonal and, therefore, must commute with $\boldsymbol{\Omega}$:

$$\mathbf{D}\boldsymbol{\Omega} = \boldsymbol{\Omega}\mathbf{D}. \tag{5.236}$$

Substitute $\mathbf{D} = \mathbf{U}^T\mathbf{C}\mathbf{U}$ and $\mathbf{U}^{-1}\mathbf{U} = \mathbf{I}$ into Eq. (5.236) to obtain

$$\mathbf{C}\mathbf{U}\mathbf{U}^T\mathbf{U}^{-T}\mathbf{\Omega}\mathbf{U}^{-1} = \mathbf{U}^{-T}\mathbf{\Omega}\mathbf{U}^{-1}\mathbf{U}\mathbf{U}^T\mathbf{C}. \tag{5.237}$$

This expression can be simplified by substituting $\mathbf{M}^{-1} = \mathbf{U}\mathbf{U}^T$ and $\mathbf{K} = \mathbf{U}^{-T}\mathbf{\Omega}\mathbf{U}^{-1}$. It can then be concluded that a damped system is decoupled through modal analysis if and only if

$$\mathbf{C}\mathbf{M}^{-1}\mathbf{K} = \mathbf{K}\mathbf{M}^{-1}\mathbf{C}. \tag{5.238}$$

This condition encompasses the well-known *proportionally damped systems*, where the damping matrix is given by

$$\mathbf{C} = \eta_0\mathbf{M} + \eta_1\mathbf{K} \tag{5.239}$$

for some scalar constants η_0 and η_1. A system with proportional damping satisfies the restriction (5.238) and, for this reason, can always be decoupled by modal analysis. The modal damping matrix is

$$\mathbf{D} = \eta_0\mathbf{I} + \eta_1\mathbf{\Omega}. \tag{5.240}$$

A flowchart for decoupling a mechanical system using modal analysis is presented in Fig. 5.35.

5.5.4 Illustrative Examples

This section provides two additional illustrative examples to clarify the use of modal analysis.

5.5.4.1 Vibrations of Railway Carriages

Consider the model of two railway carriages shown in Fig. 5.36. This is a two-DOF model for a railway convoy represented as two particles that move on a smooth horizontal surface. Each particle has mass m, and they are connected by a linear spring with stiffness k and unstretched length l_0. The objective of this example is to derive the modal equations of motion and use them to analyze the dynamics of the system.

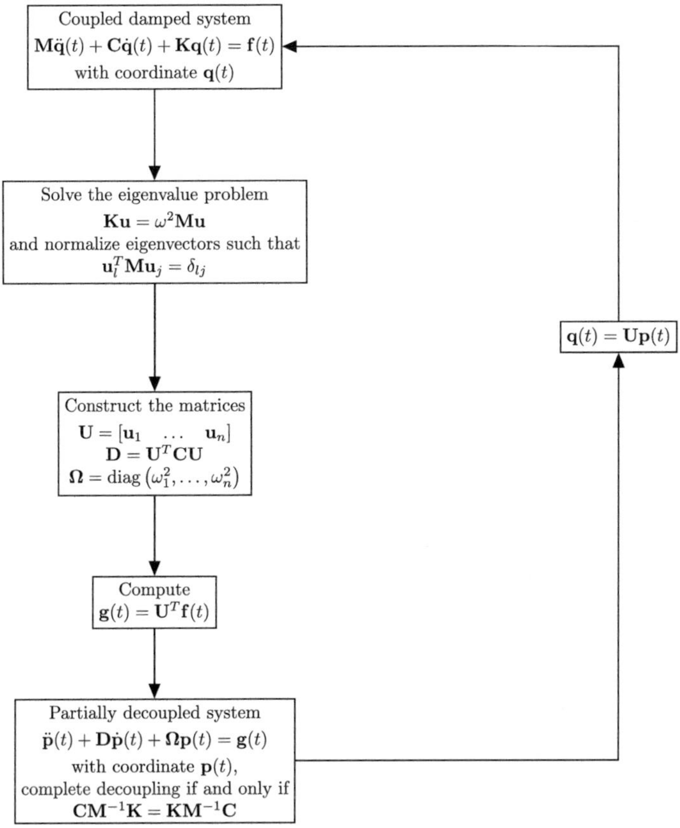

Fig. 5.35 Algorithm for decoupling by modal analysis

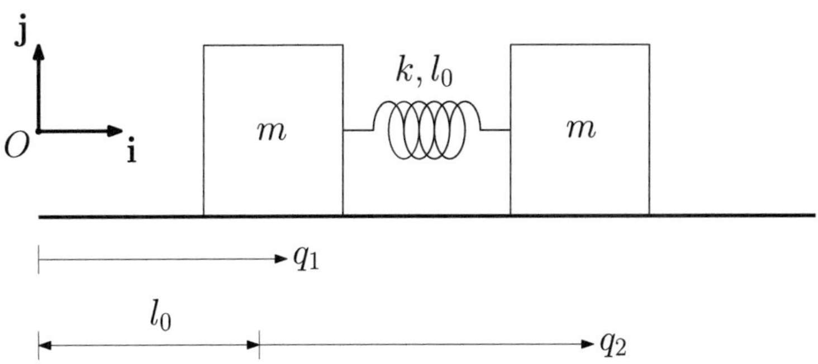

Fig. 5.36 Simplified model of a railway convoy composed of two carriages

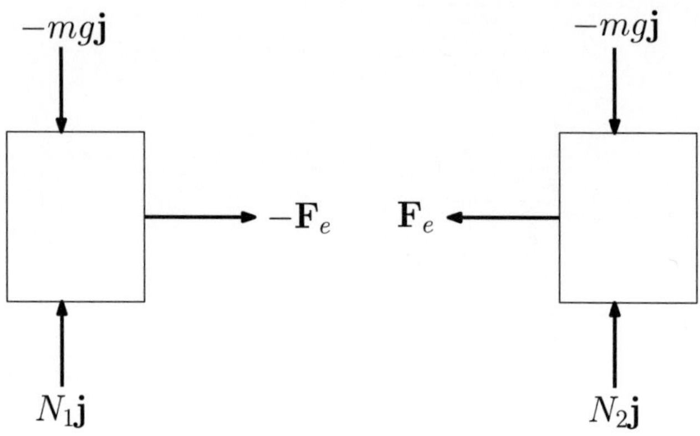

Fig. 5.37 FBDs of the railway carriages

Equations of Motion
Using the generalized coordinates shown in Fig. 5.36, the position vectors for the particle on the left and on the right are

$$\mathbf{r}_1 = q_1\mathbf{i}, \quad \mathbf{r}_2 = (l_0 + q_2)\mathbf{i}, \tag{5.241}$$

respectively. The acceleration vectors are given by

$$\mathbf{a}_1 = \ddot{q}_1\mathbf{i}, \quad \mathbf{a}_2 = \ddot{q}_2\mathbf{i}. \tag{5.242}$$

The FBDs of the particles are shown in Fig. 5.37. The force due to the spring is

$$\mathbf{F}_e = -k(q_2 - q_1)\mathbf{i}. \tag{5.243}$$

Apply Newton's second law to the particles to show that the equations of motion are

$$\begin{bmatrix} m & 0 \\ 0 & m \end{bmatrix} \begin{bmatrix} \ddot{q}_1 \\ \ddot{q}_2 \end{bmatrix} + \begin{bmatrix} k & -k \\ -k & k \end{bmatrix} \begin{bmatrix} q_1 \\ q_2 \end{bmatrix} = \begin{bmatrix} 0 \\ 0 \end{bmatrix}. \tag{5.244}$$

Natural Frequencies and Mode Shapes
The characteristic polynomial is

$$0 = (\omega^2)^2 - 2\frac{k}{m}(\omega^2). \tag{5.245}$$

Compute the roots to find the two natural frequencies:

$$\omega_1 = 0, \quad \omega_2 = \sqrt{2k/m}. \tag{5.246}$$

The associated mass-normalized mode shapes are

$$\mathbf{u}_1 = \frac{1}{\sqrt{2m}}\begin{bmatrix}1\\1\end{bmatrix}, \quad \mathbf{u}_2 = \frac{1}{\sqrt{2m}}\begin{bmatrix}1\\-1\end{bmatrix}. \tag{5.247}$$

Obtaining the Response from Modal Coordinates
The equations of motion in modal coordinates are

$$\ddot{p}_1 = 0, \tag{5.248}$$

$$\ddot{p}_2 + \frac{2k}{m}p_2 = 0. \tag{5.249}$$

The modal equation for p_1 has no zeroth-order term because $\omega_1 = 0$. The solutions of these equations are

$$p_1(t) = a_1 t + b_1, \tag{5.250}$$

$$p_2(t) = a_2 \cos\omega_2 t + b_2 \sin\omega_2 t. \tag{5.251}$$

The constants a_1, b_1, a_2, and b_2 are determined from the ICs $\mathbf{p}(0)$ and $\dot{\mathbf{p}}(0)$. The modal transformation can be used to retrieve the response in the physical coordinates:

$$\mathbf{q}(t) = (a_1 t + b_1)\mathbf{u}_1 + (a_2 \cos\omega_2 t + b_2 \sin\omega_2 t)\mathbf{u}_2. \tag{5.252}$$

This equation facilitates a proper interpretation of the first mode, which is shown in Fig. 5.38a. This mode represents motions where both particles move in the same direction at equal speeds. The system behaves like a rigid body without changing the length of the spring, resulting in no variation in elastic potential energy. This is known as a *rigid body mode*, and it occurs because there are no restraints or supports

(a) First mode. (b) Second mode.

Fig. 5.38 Modes of the carriages

that allow the system to oscillate around an equilibrium position for arbitrary ICs. Specific sets of ICs can excite the second mode, illustrated in Fig. 5.38b, where the particles oscillate with equal amplitude but in phase opposition.

5.5.4.2 A Classically Damped System

Consider a damped system in free vibration with coefficient matrices given by

$$\mathbf{M} = \mathbf{I}, \quad \mathbf{C} = \begin{bmatrix} 3 & -1 \\ -1 & 3 \end{bmatrix}, \quad \mathbf{K} = \frac{1}{2}\begin{bmatrix} 5 & -3 \\ -3 & 5 \end{bmatrix}, \tag{5.253}$$

and ICs $\mathbf{q}(0) = \mathbf{0}$ and $\dot{\mathbf{q}}(0) = [1 \quad 0]^T$.

This system can be decoupled by modal analysis because

$$\mathbf{C}\mathbf{M}^{-1}\mathbf{K} = \mathbf{K}\mathbf{M}^{-1}\mathbf{C} = \begin{bmatrix} 9 & -7 \\ -7 & 9 \end{bmatrix}. \tag{5.254}$$

The mass-normalized modal matrix

$$\mathbf{U} = \frac{1}{\sqrt{2}}\begin{bmatrix} 1 & 1 \\ 1 & -1 \end{bmatrix} \tag{5.255}$$

can be used to decouple the system into

$$\ddot{\mathbf{p}} + \begin{bmatrix} 2 & 0 \\ 0 & 4 \end{bmatrix}\dot{\mathbf{p}} + \begin{bmatrix} 1 & 0 \\ 0 & 4 \end{bmatrix}\mathbf{p} = \mathbf{0}. \tag{5.256}$$

The ICs for the modal coordinates are $\mathbf{p}(0) = \mathbf{0}$ and

$$\dot{\mathbf{p}}(0) = \frac{1}{\sqrt{2}}\begin{bmatrix} 1 \\ 1 \end{bmatrix}. \tag{5.257}$$

The modal damping ratios are $\zeta_1 = \zeta_2 = 1$, so both modal DOFs are critically damped:

$$p_1(t) = \frac{1}{\sqrt{2}} t e^{-t}, \tag{5.258}$$

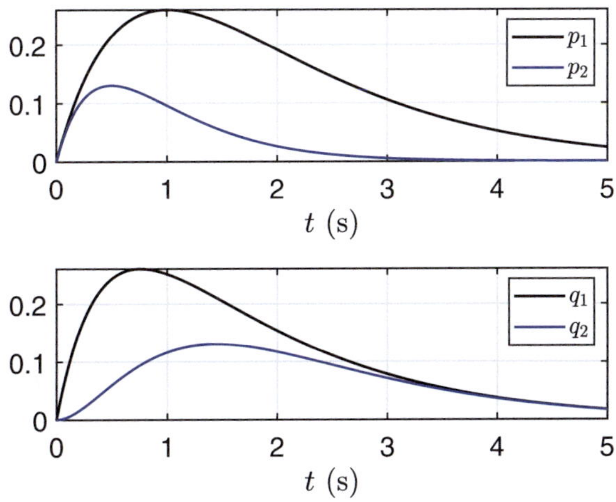

Fig. 5.39 Solution for the physical and modal coordinates

$$p_1(t) = \frac{1}{\sqrt{2}} t e^{-2t}. \tag{5.259}$$

The response in physical coordinates is

$$\mathbf{q}(t) = \frac{1}{2}\begin{bmatrix} 1 \\ 1 \end{bmatrix} t e^{-t} + \frac{1}{2}\begin{bmatrix} 1 \\ -1 \end{bmatrix} t e^{-2t}. \tag{5.260}$$

The responses in physical and modal coordinates are shown in Fig. 5.39.

5.6 Summary

The free vibration of an undamped n-DOFs system is governed by the equations

$$\mathbf{M}\ddot{\mathbf{q}} + \mathbf{K}\mathbf{q} = \mathbf{0}.$$

The response is a superposition of the normal modes

$$\mathbf{q}(t) = \sum_{j=1}^{n} X_j \mathbf{s}_j(t - \tau_j) = \sum_{j=1}^{n} A_j \cos(\omega_j t - \phi_j)\, \mathbf{u}_j,$$

where the $2n$ constants A_j and ϕ_j $(j = 1, \ldots, n)$ are determined from the ICs. A natural frequency ω_j and its corresponding mode shape $\mathbf{u}_j$ are solutions of the generalized eigenproblem

$$\mathbf{K}\mathbf{u}_j = \omega_j^2 \mathbf{M}\mathbf{u}_j.$$

Consequently, a normal mode $\mathbf{s}_j(t)$ represents synchronous motion where all the DOFs oscillate harmonically with the same natural frequency. The amplitude ratio and phase lag between any two system components are determined from the corresponding entries in the mode shape.

When damping is present, the equations of motion for free vibration become

$$\mathbf{M}\ddot{\mathbf{q}} + \mathbf{C}\dot{\mathbf{q}} + \mathbf{K}\mathbf{q} = \mathbf{0}.$$

The general solution is given by

$$\mathbf{q}(t) = \sum_{j=1}^{2n} c_j \mathbf{v}_j e^{\lambda_j t},$$

for distinct values of λ_j $(j = 1, 2, \ldots, 2n)$. The constants c_j are determined from the ICs, while the amplitude vectors $\mathbf{v}_j$ and the scalar parameters λ_j are solutions of the quadratic eigenproblem

$$(\lambda^2 \mathbf{M} + \lambda \mathbf{C} + \mathbf{K})\mathbf{v} = \mathbf{0}.$$

The eigenvalues can be real or complex. The real part is associated with the exponential decay of the response, while the imaginary part represents the damped frequency of oscillation. Eigenvectors associated with real eigenvalues can always be chosen to be real, and eigenvectors associated with complex conjugate eigenvalues are also complex conjugates. The solution is generally composed of oscillatory and non-oscillatory parts.

There is particular interest in the response due to harmonic excitation. In this case, it is necessary to determine the solution of

$$\mathbf{M}\ddot{\mathbf{q}} + \mathbf{C}\dot{\mathbf{q}} + \mathbf{K}\mathbf{q} = \mathbf{F}_0 e^{i\omega t},$$

where $\mathbf{F}_0$ is a vector of amplitudes with dimension $n \times 1$. The steady-state response is given by

$$\mathbf{q}(t) = \mathbf{H}(\omega)\mathbf{F}_0 e^{i\omega t}, \tag{5.261}$$

where

$$\mathbf{H}(\omega) = [\mathbf{M}(i\omega)^2 + \mathbf{C}(i\omega) + \mathbf{K}]^{-1}$$

is the $n \times n$ FRF matrix. This matrix is symmetric, and its entries are complex when $\mathbf{C} \neq \mathbf{0}$. An entry $H_{lj}(i\omega)$ has, at most, n peaks, which occur at the system's natural frequencies if damping is small.

Modal analysis employs the modal transformation $\mathbf{q} = \mathbf{U}\mathbf{p}$ to convert the equations

$$\mathbf{M}\ddot{\mathbf{q}} + \mathbf{C}\dot{\mathbf{q}} + \mathbf{K}\mathbf{q} = \mathbf{f}(t)$$

into the equations of motion in modal coordinates:

$$\ddot{\mathbf{p}} + \mathbf{D}\dot{\mathbf{p}} + \boldsymbol{\Omega}\mathbf{p} = \mathbf{g}(t),$$

where $\mathbf{g}(t) = \mathbf{U}^T\mathbf{f}(t)$, $\mathbf{D} = \mathbf{U}^T\mathbf{C}\mathbf{U}$ and

$$\boldsymbol{\Omega} = \begin{bmatrix} \omega_1^2 & & & \\ & \omega_2^2 & & \\ & & \ddots & \\ & & & \omega_n^2 \end{bmatrix}. \tag{5.262}$$

An undamped system is always decoupled by modal analysis, but a damped one is decoupled if and only if $\mathbf{D}$ is diagonal. A necessary and sufficient condition for modal analysis to decouple a damped system is

$$\mathbf{C}\mathbf{M}^{-1}\mathbf{K} = \mathbf{K}\mathbf{M}^{-1}\mathbf{C}.$$

This condition is satisfied by the subclass known as proportionally damped systems, where

$$\mathbf{C} = \eta_0\mathbf{M} + \eta_1\mathbf{K}$$

for constants η_0 and η_1.

5.7 Problems

5.1 Find the natural frequencies and mode shapes of the system from Problem 4.3.

5.2 Find the natural frequencies and mode shapes of the system from Problem 4.4.

5.3 Find the natural frequencies and mode shapes of the system from Problem 4.5.

5.4 Find the natural frequencies and mode shapes of the system from Problem 4.6.

5.5 Find the natural frequencies and mode shapes of the system from Problem 4.7.

5.6 Find the natural frequencies and mode shapes of the system from Problem 4.10.

5.7 A two-DOF system in free vibration is governed by:

$$\begin{bmatrix} 2 & 0 \\ 0 & 1 \end{bmatrix} \begin{bmatrix} \ddot{q}_1 \\ \ddot{q}_2 \end{bmatrix} + \begin{bmatrix} 2 & -2 \\ -2 & 2 \end{bmatrix} \begin{bmatrix} \dot{q}_1 \\ \dot{q}_2 \end{bmatrix} + \begin{bmatrix} 3 & -1 \\ -1 & 1 \end{bmatrix} \begin{bmatrix} q_1 \\ q_2 \end{bmatrix} = \begin{bmatrix} 0 \\ 0 \end{bmatrix}. \tag{5.263}$$

Suppose the ICs are

$$\begin{bmatrix} q_1(0) \\ q_2(0) \end{bmatrix} = \begin{bmatrix} 0 \\ 0 \end{bmatrix}, \quad \begin{bmatrix} \dot{q}_1(0) \\ \dot{q}_2(0) \end{bmatrix} = \begin{bmatrix} -1/2 \\ 2 \end{bmatrix}. \tag{5.264}$$

Compute $q_1(t)$ and $q_2(t)$.

5.8 Suppose the system from Problem 4.9 has

$$\begin{aligned} m_1 &= 3\,\text{kg}, \quad m_2 = 5\,\text{kg}, \quad r = 0.3\,\text{m}, \\ a &= r/2, \quad k_1 = k_3 = 3500\,\text{N/m}, \quad k_2 = 4500\,\text{N/m}. \end{aligned} \tag{5.265}$$

If it is excited by a harmonic excitation with

$$\omega = 8\,\text{Hz}, \quad F_0 = 9\,\text{N}, \tag{5.266}$$

determine the steady-state amplitudes of vibration of the cart and the cylinder.

5.9 The rigid body shown in Fig. 5.40 is free to rotate around a fixed axis passing at point O. This body has mass M, moment of inertia I_{zz}^{O}, and is excited by a moment $M(t)\mathbf{k}$. The position of its mass center C has the representation

$$\mathbf{x}_C = l\mathbf{e}_1. \tag{5.267}$$

An absorber is attached to the body to suppress the torsional vibrations. The absorber has mass m and is connected to two springs with stiffness k and unstretched length l_0. The position vector of the particle is

$$\mathbf{r} = \mathbf{x}_C + a\mathbf{e}_1 + x(t)\mathbf{e}_2. \tag{5.268}$$

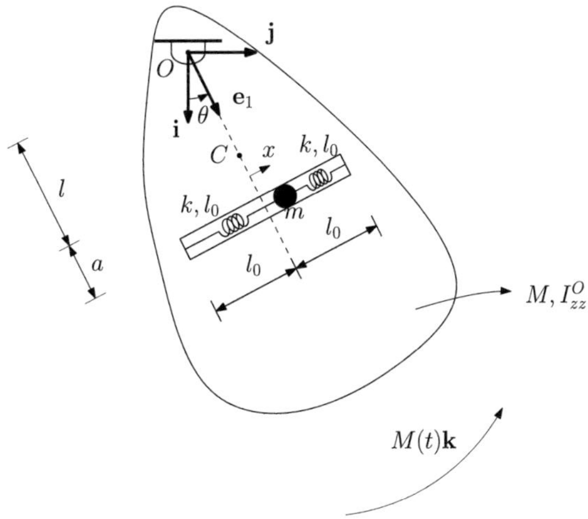

Fig. 5.40 A model for a torsional vibration absorber. Gravitational forces act along the **i** direction

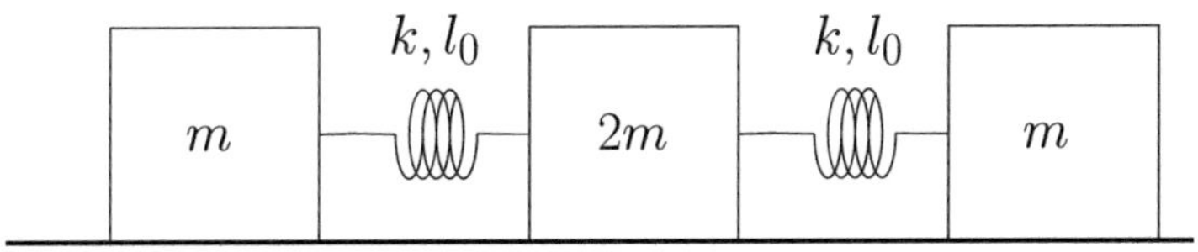

Fig. 5.41 An unrestrained system with three particles

The springs are unstretched when $x = 0$. Assuming that $M(t) = M_0 e^{i\omega t}$, show that the steady-state vibrations of the rigid body are suppressed if the absorber is tuned to $k/m = \omega^2/2$.

5.10 Consider the three-DOF system shown in Fig. 5.41. Find the modal equations of motion and use the solutions of those equations to obtain the response in physical coordinates.

Appendix A
Complex Numbers

This appendix offers a brief review of complex numbers and basic algebraic operations.

A.1 Vector Notation and Basic Algebra

A complex number $z \in \mathbb{C}$ is an ordered pair $z = (a, b)$ of two real numbers $a, b \in \mathbb{R}$. It is typically expressed in "vector" or rectangular form

$$z = a + ib, \tag{A.1}$$

where a is referred to as the real part of z, denoted as $Re(z)$, and b is the imaginary part of z, denoted as $Im(z)$. The symbol i represents the imaginary unit $i = \sqrt{-1}$. Because complex numbers are essentially ordered pairs of two real numbers, they can be visualized as points or vectors in the two-dimensional plane called complex plane or Argand plane, illustrated in Fig. A.1, where the horizontal axis represents the real part and the vertical axis represents the imaginary part of a complex number.

Complex numbers follow some algebraic operations defined according to standard vector rules. For instance, consider two complex numbers $z_1 = a_1 + ib_1$ and $z_2 = a_2 + ib_2$. These two complex numbers are equal, $z_1 = z_2$, if and only if $a_1 = a_2$ and $b_1 = b_2$. Similarly, complex number addition is defined as:

$$z_1 + z_2 = (a_1 + a_2) + i(b_1 + b_2). \tag{A.2}$$

In contrast to vectors in $\mathbb{R}^2$, the multiplication of z_1 and z_2 can be achieved using standard distributive property and taking into account that $i^2 = -1$:

$$z_1 z_2 = (a_1 a_2 - b_1 b_2) + i(a_1 b_2 + b_1 a_2). \tag{A.3}$$

R. Gonçalves Salsa Junior, *Fundamentals of the Theory of Mechanical Vibrations*,
https://doi.org/10.1007/978-3-031-83995-5

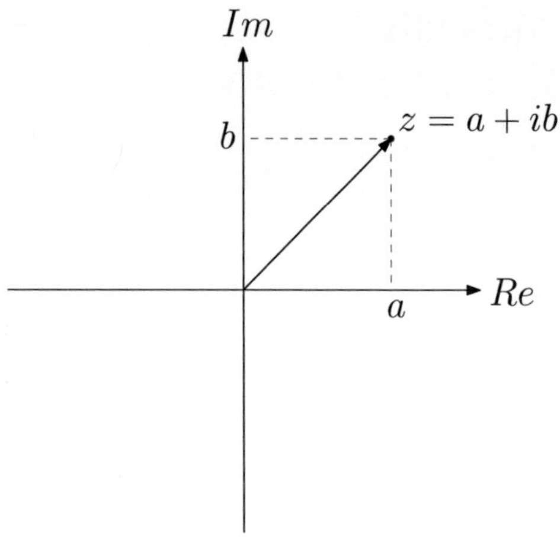

Fig. A.1 A complex number $z = a + ib$ represented on the complex plane

Division of complex numbers is well-defined by letting the inverse of $z = a+ib \neq 0$ be

$$z^{-1} = \frac{1}{z} = \frac{a - ib}{a^2 + b^2}. \tag{A.4}$$

A.2 Complex Conjugation

The complex conjugate of $z = a + ib$ is defined to be $\overline{z} = a - ib$. Figure A.2 illustrates that, geometrically, taking the complex conjugate corresponds to a reflection about the real axis. Using complex conjugation, the real and imaginary parts of z can be expressed as

$$a = \frac{z + \overline{z}}{2}, \quad ib = \frac{z - \overline{z}}{2}. \tag{A.5}$$

The following properties also hold:

$$\overline{z_1 + z_2} = \overline{z}_1 + \overline{z}_2, \tag{A.6}$$

$$\overline{z_1 z_2} = \overline{z}_1 \overline{z}_2, \tag{A.7}$$

$$\overline{\left(\frac{z_1}{z_2}\right)} = \frac{\overline{z}_1}{\overline{z}_2}, \quad \text{for } z_2 \neq 0. \tag{A.8}$$

Fig. A.2 Geometrical interpretation of complex conjugation

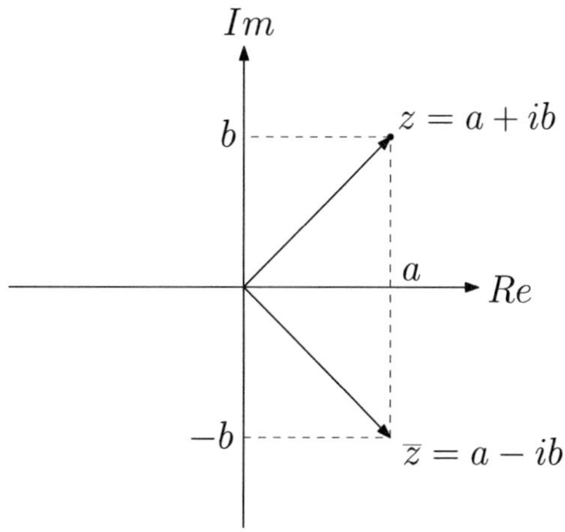

The magnitude (or modulus) of $z = a + ib$ represents the length of the "vector" z in the complex plane and can be computed according to

$$|z| = \sqrt{a^2 + b^2} = \sqrt{z\overline{z}}. \tag{A.9}$$

Consequently, the inverse of z can be expressed as

$$z^{-1} = \frac{\overline{z}}{|z|^2}. \tag{A.10}$$

A.3 Polar Form

In relation to the angle θ depicted in Fig. A.3, a complex number z can be expressed in its polar form by

$$z = |z|(\cos\theta + i \sin\theta). \tag{A.11}$$

Applying Euler's identity

$$e^{i\theta} = \cos\theta + i \sin\theta, \tag{A.12}$$

the polar form becomes

$$z = |z|e^{i\theta}. \tag{A.13}$$

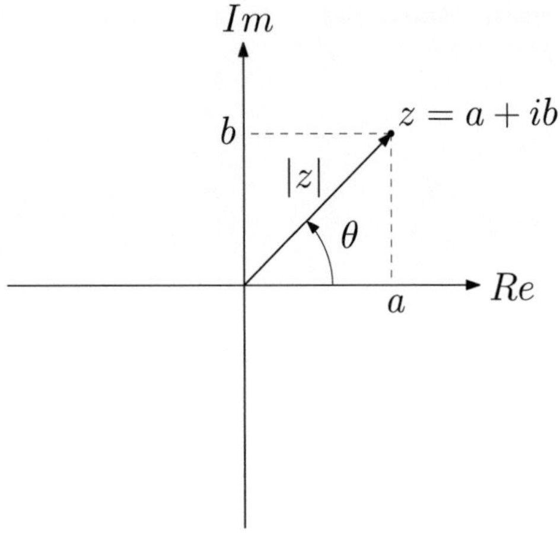

Fig. A.3 Polar coordinates of a complex number

The angle θ is also referred to as the argument of z and can be computed with

$$\theta = \arctan \frac{b}{a}. \tag{A.14}$$

The polar form of complex numbers is particularly advantageous due to the algebraic properties of exponential functions. For example, the product of two complex numbers $z_1 = |z_1|e^{i\theta_1}$ and $z_2 = |z_2|e^{i\theta_2}$ is given by

$$z_1 z_2 = |z_1||z_2|e^{i(\theta_1+\theta_2)}, \tag{A.15}$$

while division is

$$\frac{z_1}{z_2} = \frac{|z_1|}{|z_2|}e^{i(\theta_1-\theta_2)} \tag{A.16}$$

provided that $z_2 \neq 0$. In addition, the complex conjugate of z is simply

$$\overline{z} = |z|e^{-i\theta}. \tag{A.17}$$

The only drawback with the polar form is that the determination of θ for a given z is not unique. A practical solution is to adopt a convention that relies on principal values. In this convention, θ is chosen within $0 \leq \theta < 2\pi$ or any other interval of length 2π.

Appendix B
Linear Spaces and Vectors

This appendix offers a succinct overview of vector properties and linear spaces.

B.1 Linear Spaces

A *linear space* (or vector space) is a set $V = \{\mathbf{a},\ \mathbf{b},\ \mathbf{c},\ \ldots\}$ whose elements are called *vectors*. A linear space V is endowed with two operations, being vector addition and scalar multiplication, which are defined by the following axioms for any vectors $\mathbf{a},\ \mathbf{b},\ \mathbf{c} \in V$ and scalars $\alpha,\ \beta \in \mathbb{R}$:

(a) $(\alpha\mathbf{a} + \beta\mathbf{b}) \in V$.
(b) Commutativity of vector addition:

$$\mathbf{a} + \mathbf{b} = \mathbf{b} + \mathbf{a}. \tag{B.1}$$

(c) Associativity of vector addition:

$$\mathbf{a} + (\mathbf{b} + \mathbf{c}) = (\mathbf{a} + \mathbf{b}) + \mathbf{c}. \tag{B.2}$$

(d) Existence of a zero element in V: there exists a vector $\mathbf{0}$ such that

$$\mathbf{0} + \mathbf{a} = \mathbf{a}. \tag{B.3}$$

(e) Existence of additive inverses in V: for each $\mathbf{a} \in V$, there exists a unique $\mathbf{b} \in V$ such that

$$\mathbf{a} + \mathbf{b} = \mathbf{0}. \tag{B.4}$$

The vector $\mathbf{b}$ is commonly written as $\mathbf{b} = -\mathbf{a}$.

R. Gonçalves Salsa Junior, *Fundamentals of the Theory of Mechanical Vibrations*,
https://doi.org/10.1007/978-3-031-83995-5

(f) Commutativity of scalar multiplication:

$$\alpha(\beta\mathbf{a}) = \beta(\alpha\mathbf{a}). \tag{B.5}$$

(g) Distributive property of scalar multiplication:

$$(\alpha + \beta)\mathbf{a} = \alpha\mathbf{a} + \beta\mathbf{a}. \tag{B.6}$$

(h) Distributive property of scalar multiplication over vector addition:

$$\alpha(\mathbf{a} + \mathbf{b}) = \alpha\mathbf{a} + \alpha\mathbf{b}. \tag{B.7}$$

(i) $1\mathbf{a} = \mathbf{a}$.

B.2 Linear Independence of Vectors

A *linear combination* of vectors $\mathbf{v}_1, \mathbf{v}_2, \ldots, \mathbf{v}_m$ of V is another vector $\mathbf{v} \in V$ given by

$$\mathbf{v} = \alpha_1\mathbf{v}_1 + \alpha_2\mathbf{v}_2 + \ldots + \alpha_m\mathbf{v}_m, \tag{B.8}$$

where the scalars $\alpha_1, \alpha_2, \ldots, \alpha_m \in \mathbb{R}$ are called the *coefficients* of this linear combination. A list of vectors $\mathbf{v}_1, \mathbf{v}_2, \ldots, \mathbf{v}_m \in V$ is said to be *linearly independent* if and only if the linear combination capable of representing the zero vector

$$\mathbf{0} = \alpha_1\mathbf{v}_1 + \alpha_2\mathbf{v}_2 + \ldots + \alpha_m\mathbf{v}_m \tag{B.9}$$

is possible only when all the coefficients are equal to zero:

$$\alpha_1 = \alpha_2 = \ldots = \alpha_m = 0. \tag{B.10}$$

If Eq. (B.9) is satisfied with at least one of the coefficients $\alpha_1, \alpha_2, \ldots, \alpha_m$ different than zero, then the vectors are *linearly dependent*. This implies that a list of vectors is linearly dependent if and only if one of these vectors is a linear combination of the remaining vectors. A list that includes the zero vector $\mathbf{0}$ is necessarily linearly dependent.

B.2.1 *Basis of a Finite Dimensional Linear Space*

A linear space V is finite-dimensional when there exists a positive integer n such that all linearly independent lists of vectors in V contain, at most, n vectors. In this case, the dimension of V is denoted as n, and it is written as V^n.

A *basis* for V^n is defined as any set of n linearly independent vectors. Suppose the list of vectors $\mathbf{v}_1, \mathbf{v}_2, \ldots, \mathbf{v}_n$ constitutes a basis for V^n. To simplify notation, this basis will be referred to as $\{\mathbf{v}_j\}$. Given any vector $\mathbf{v}$ in V^n, there exist unique scalars $\alpha_1, \alpha_2, \ldots, \alpha_n \in \mathbb{R}$ such that

$$\mathbf{v} = \sum_{j=1}^{n} \alpha_j \mathbf{v}_j. \tag{B.11}$$

The scalars $\alpha_1, \alpha_2, \ldots, \alpha_n$ are the coefficients of $\mathbf{v}$ in the basis $\{\mathbf{v}_j\}$.

B.3 Euclidean Spaces

A *Euclidean space* of dimension n, denoted as $\mathcal{E}^n$, is a linear space of dimension n equipped with the *inner product* operation:

$$f(\mathbf{u}, \mathbf{v}) = \mathbf{u} \cdot \mathbf{v} \in \mathbb{R}, \tag{B.12}$$

which must satisfy, for any vectors $\mathbf{u}, \mathbf{v}, \mathbf{w} \in \mathcal{E}^n$,

(i) $\mathbf{u} \cdot \mathbf{u} \geq 0$, and $\mathbf{u} \cdot \mathbf{u} = 0$ if and only if $\mathbf{u} = \mathbf{0}$.
(ii) $\mathbf{u} \cdot \mathbf{v} = \mathbf{v} \cdot \mathbf{u}$.
(iii) $\mathbf{u} \cdot (\mathbf{v} + \mathbf{w}) = \mathbf{u} \cdot \mathbf{v} + \mathbf{u} \cdot \mathbf{w}$.
(iv) $(\alpha\mathbf{u}) \cdot \mathbf{v} = \alpha(\mathbf{u} \cdot \mathbf{v}) = \mathbf{u} \cdot (\alpha\mathbf{v})$, for all $\alpha \in \mathbb{R}$.

The (Euclidean) norm of a vector $\mathbf{v} \in \mathcal{E}^n$ is related to the inner product through

$$\|\mathbf{v}\| = \sqrt{\mathbf{v} \cdot \mathbf{v}}. \tag{B.13}$$

Property (i) for the inner product implies that $\|\mathbf{v}\| \geq 0$ while $\|\mathbf{v}\| = 0$ if and only if $\mathbf{v} = \mathbf{0}$.

The angle $\theta \in [0, \pi]$ between any two vectors $\mathbf{u}$ and $\mathbf{v}$ satisfies

$$\mathbf{u} \cdot \mathbf{v} = \|\mathbf{u}\| \|\mathbf{v}\| \cos\theta. \tag{B.14}$$

Vectors $\mathbf{u}$ and $\mathbf{v}$ are orthogonal if and only if $\theta = \pi/2$, that is, when $\mathbf{u} \cdot \mathbf{v} = 0$. Similarly, these vectors are parallel when $\theta = 0$.

B.3.1 Orthonormal Basis

A list of vectors $\{\mathbf{e}_1, \ldots, \mathbf{e}_m\}$ is orthonormal when each vector in the list is a unit vector and orthogonal to the other vectors in that list:

$$\mathbf{e}_l \cdot \mathbf{e}_j = \delta_{lj} = \delta_{jl} = \begin{cases} 1, & \text{if } l = j \\ 0, & \text{if } l \neq j \end{cases}, \tag{B.15}$$

where δ_{lj} is known as the *Kronecker delta*. It is important to note that any orthonormal list $\mathbf{e}_1, \ldots, \mathbf{e}_m \in \mathcal{E}^n$, $m \leq n$, is linearly independent. The representation of a vector $\mathbf{v}$ in the orthonormal basis $\{\mathbf{e}_j\}$ is

$$\mathbf{v} = \sum_{j=1}^{n} (\mathbf{v} \cdot \mathbf{e}_j)\mathbf{e}_j. \tag{B.16}$$

This representation simplifies operations such as norm and inner product and is therefore widely used. The inner product between $\mathbf{v}$ and $\mathbf{u}$ is

$$\mathbf{v} \cdot \mathbf{u} = \sum_{j=1}^{n} v_j u_j, \tag{B.17}$$

where v_j, u_j $(j = 1, \ldots, n)$ are the components of $\mathbf{v}$ and $\mathbf{u}$ in an orthonormal basis $\{\mathbf{e}_j\}$, respectively. If the components of these vectors are organized as $n \times 1$ column matrices

$$\mathbf{v} = \begin{bmatrix} v_1 \\ v_2 \\ \vdots \\ v_n \end{bmatrix}, \quad \mathbf{u} = \begin{bmatrix} u_1 \\ u_2 \\ \vdots \\ u_n \end{bmatrix}, \tag{B.18}$$

then the inner product (B.17) can be computed with the matrix product

$$\mathbf{v}^T \mathbf{u} = \sum_{j=1}^{n} v_j u_j. \tag{B.19}$$

The norm of a vector $\mathbf{v}$ in this basis is

$$\|\mathbf{v}\| = \sqrt{v_1^2 + v_2^2 + \ldots + v_n^2}. \tag{B.20}$$

B.4 Cross (or Vector) Product

The *cross product* is an operation

$$\mathbf{f}(\mathbf{u}, \mathbf{v}) = \mathbf{u} \times \mathbf{v} : \mathcal{E}^3 \longmapsto \mathcal{E}^3 \tag{B.21}$$

characterized by:

(1) For any vectors $\mathbf{u},\ \mathbf{v} \in \mathcal{E}^3$,

$$\mathbf{u} \times \mathbf{v} = -\mathbf{v} \times \mathbf{u}. \tag{B.22}$$

(2) For any $\mathbf{u},\ \mathbf{v},\ \mathbf{w} \in \mathcal{E}^3$ and $\alpha,\ \beta \in \mathbb{R}$,

$$(\alpha\mathbf{u} + \beta\mathbf{v}) \times \mathbf{w} = \alpha(\mathbf{u} \times \mathbf{w}) + \beta(\mathbf{v} \times \mathbf{w}). \tag{B.23}$$

(3) The angle θ between any two vectors $\mathbf{u},\ \mathbf{v} \in \mathcal{E}^3$ satisfies

$$\|\mathbf{u} \times \mathbf{v}\| = \|\mathbf{u}\|\,\|\mathbf{v}\| \sin\theta. \tag{B.24}$$

The operation $\mathbf{u} \cdot (\mathbf{v} \times \mathbf{w})$ is called *vector triple product* and it has the following shifting property:

$$\mathbf{u} \cdot (\mathbf{v} \times \mathbf{w}) = \mathbf{w} \cdot (\mathbf{u} \times \mathbf{v}) = \mathbf{v} \cdot (\mathbf{w} \times \mathbf{u}). \tag{B.25}$$

In addition, the triple product is linear in each of its arguments, and it changes sign whenever two of its entries are permuted. It can be shown that vectors $\mathbf{u}$, $\mathbf{v}$, and $\mathbf{w}$ are linearly independent if and only if $\mathbf{u} \cdot (\mathbf{v} \times \mathbf{w}) \neq 0$. A basis $\{\mathbf{v}_1,\ \mathbf{v}_2,\ \mathbf{v}_3\}$ is called right-handed when

$$(\mathbf{v}_1 \times \mathbf{v}_2) \cdot \mathbf{v}_3 > 0 \tag{B.26}$$

or left-handed when

$$(\mathbf{v}_1 \times \mathbf{v}_2) \cdot \mathbf{v}_3 < 0. \tag{B.27}$$

Consequently, an orthonormal basis $\mathbf{e}_1,\ \mathbf{e}_2,\ \mathbf{e}_3$ of $\mathcal{E}^3$ with

$$\mathbf{e}_1 \times \mathbf{e}_2 = \mathbf{e}_3 \tag{B.28}$$

is right-handed. If vectors **a**, **b**, and **c** can be represented by

$$\mathbf{a} = \sum_{j=1}^{3} a_i \mathbf{e}_j, \quad \mathbf{b} = \sum_{j=1}^{3} b_j \mathbf{e}_j, \quad \mathbf{c} = \sum_{j=1}^{3} c_j \mathbf{e}_j, \tag{B.29}$$

in an orthonormal right-handed basis, the triple product can be computed using the determinant of the matrix formed by the coefficients of these vectors:

$$(\mathbf{a} \times \mathbf{b}) \cdot \mathbf{c} = \begin{vmatrix} a_1 & b_1 & c_1 \\ a_2 & b_2 & c_2 \\ a_3 & b_3 & c_3 \end{vmatrix}. \tag{B.30}$$

B.5 Differentiating Vectors

Consider a vector $\mathbf{u} \in \mathcal{E}^n$ and assume that this vector depends on a real parameter $t \in \mathbb{R}$. The derivative of $\mathbf{u}$ with respect to t is defined as

$$\dot{\mathbf{u}} = \frac{d}{dt}\mathbf{u} = \lim_{\Delta t \to 0} \frac{\mathbf{u}(t + \Delta t) - \mathbf{u}(t)}{\Delta t}. \tag{B.31}$$

If a fixed orthonormal basis $\{\mathbf{e}_j\}$ is used to represent this vector, then $\dot{\mathbf{u}}$ becomes

$$\dot{\mathbf{u}} = \sum_{j=1}^{n} \lim_{\Delta t \to 0} \left[\frac{u_j(t + \Delta t) - u_j(t)}{\Delta t} \right] \mathbf{e}_j = \sum_{j=1}^{n} \dot{u}_j \mathbf{e}_j. \tag{B.32}$$

This implies that

$$\frac{d}{dt}(\alpha \mathbf{u}) = \dot{\alpha}\mathbf{u} + \alpha\dot{\mathbf{u}}, \tag{B.33}$$

$$\frac{d}{dt}(\mathbf{u} \cdot \mathbf{v}) = \dot{\mathbf{u}} \cdot \mathbf{v} + \mathbf{u} \cdot \dot{\mathbf{v}}, \tag{B.34}$$

$$\frac{d}{dt}(\mathbf{u} \times \mathbf{v}) = \dot{\mathbf{u}} \times \mathbf{v} + \mathbf{u} \times \dot{\mathbf{v}}, \tag{B.35}$$

for any vector functions **u**, **v** and scalar function α.

Appendix C
Matrices

This appendix provides a list of essential matrix operations.

C.1 Definition

A matrix $\mathbf{A}$ is a rectangular array of numbers, symbols, or expressions arranged in rows and columns. Each entry within a matrix is identified by its row and column position, often denoted as A_{lj}, where l represents the row number and j represents the column number. An $n \times m$ matrix $\mathbf{A}$ has nm entries A_{lj} that are arranged according with

$$\mathbf{A} = \begin{bmatrix} A_{11} & A_{12} & \cdots & A_{1m} \\ A_{21} & A_{22} & \cdots & A_{2m} \\ \vdots & \vdots & \ddots & \vdots \\ A_{n1} & A_{n2} & \cdots & A_{nm} \end{bmatrix}. \tag{C.1}$$

Matrices exhibit diverse forms, each possessing unique characteristics and applications. In the context of vibrations, particular emphasis is placed on square matrices, where the number of rows equals the number of columns, i.e., $m = n$. Unless explicitly stated otherwise, it is assumed that matrices are square.

The main diagonal of an $n \times n$ matrix $\mathbf{A}$ consists of the entries where the row number matches the column number: A_{jj} for $j = 1, \ldots, n$. Diagonal matrices have nonzero entries only along the main diagonal. Triangular matrices, encompassing both upper triangular and lower triangular matrices, contain zeros either above or

R. Gonçalves Salsa Junior, *Fundamentals of the Theory of Mechanical Vibrations*,
https://doi.org/10.1007/978-3-031-83995-5

below the main diagonal, respectively. The identity matrix $\mathbf{I}$ has entries $I_{lj} = \delta_{lj}$ $(l, j = 1, \ldots, n)$, essentially forming a diagonal matrix with all diagonal entries set to 1. Conversely, the zero matrix $\mathbf{0}$ has all entries equal to zero.

C.2 Basic Matrix Algebra

This section outlines the fundamental properties of matrix algebra.

C.2.1 *Matrix and Linear Spaces*

Consider $n \times n$ matrices $\mathbf{A}$, $\mathbf{B}$ with entries A_{lj} and B_{lj} $(l, j = 1, 2, \ldots, n)$, respectively. These matrices are equal, $\mathbf{A} = \mathbf{B}$, if and only if $A_{lj} = B_{lj}$. Moreover, the addition of these matrices gives $\mathbf{C} = \mathbf{A} + \mathbf{B}$, where the entries of $\mathbf{C}$ are $C_{lj} = A_{lj} + B_{lj}$. Matrix addition also satisfies:

(A) Commutativity of matrix addition: $\mathbf{A} + \mathbf{B} = \mathbf{B} + \mathbf{A}$.
(B) Associativity of matrix addition: $\mathbf{A} + (\mathbf{B} + \mathbf{C}) = (\mathbf{A} + \mathbf{B}) + \mathbf{C}$.
(C) Addition with zero matrix: $\mathbf{A} + \mathbf{0} = \mathbf{A}$.
(D) Existence of addition inverse: $\mathbf{A} - \mathbf{A} = \mathbf{0}$.

If a scalar $\alpha \in \mathbb{R}$ is given, the elements of the matrix $\alpha\mathbf{A}$ are defined as αA_{lj}. Consequently, the matrix-scalar multiplication meet the following properties for any matrices $\mathbf{A}$, $\mathbf{B}$ and scalars $\alpha, \beta \in \mathbb{R}$:

(E) Commutativity of scalar multiplication: $\alpha(\beta\mathbf{A}) = \alpha\beta(\mathbf{A}) = \beta(\alpha\mathbf{A})$.
(F) Distributive property of scalar multiplication: $(\alpha + \beta)\mathbf{A} = \alpha\mathbf{A} + \beta\mathbf{A}$.
(G) Distributive property of scalar multiplication over matrix addition: $\alpha(\mathbf{A}+\mathbf{B}) = \alpha\mathbf{A} + \alpha\mathbf{B}$.
(H) $1\mathbf{A} = \mathbf{A}$.

Properties (A)–(H) are analogous to the axioms described in Appendix B.1 for defining a vector space. Therefore, the set of all $n \times n$ matrices forms a linear space.

C.2.2 *Matrix Multiplication*

An $n \times n$ matrix $\mathbf{A}$ maps a vector $\mathbf{v} \in \mathcal{E}^n$ to another vector $\mathbf{u} = \mathbf{A}\mathbf{v}$ in $\mathcal{E}^n$ according to

$$u_l = \sum_{j=1}^{n} A_{lj} v_j. \quad (l = 1, 2, \ldots, n) \tag{C.2}$$

This operation is known as matrix-vector multiplication, and $\mathbf{A}$ is a linear mapping:

$$\mathbf{A}(\alpha\mathbf{v} + \beta\mathbf{u}) = \alpha\mathbf{A}\mathbf{v} + \beta\mathbf{A}\mathbf{u} \tag{C.3}$$

for any vectors $\mathbf{v}$, $\mathbf{u}$ and scalars α, β. Additionally, the identity matrix satisfies:

$$\mathbf{I}\mathbf{v} = \mathbf{v} \tag{C.4}$$

for any vector $\mathbf{v}$, while the zero matrix maps any vector $\mathbf{v}$ to the zero vector:

$$\mathbf{0}\mathbf{v} = \mathbf{0}. \tag{C.5}$$

The product of two $n \times n$ matrices $\mathbf{A}$ and $\mathbf{B}$ results in the $n \times n$ matrix:

$$\mathbf{C} = \mathbf{A}\mathbf{B} \tag{C.6}$$

with entries defined as

$$C_{lj} = \sum_{k=1}^{n} A_{lk} B_{kj}, \tag{C.7}$$

for $l, j = 1, \ldots, n$. Generally, the product of $\mathbf{A}$ and $\mathbf{B}$ is not commutative, i.e., $\mathbf{AB} \neq \mathbf{BA}$. However, every matrix commutes with the identity matrix $\mathbf{AI} = \mathbf{IA} = \mathbf{A}$ or the zero matrix $\mathbf{A0} = \mathbf{0A} = \mathbf{0}$. Finally, it can be verified that

$$\begin{aligned} &\mathbf{A}(\mathbf{B} + \mathbf{C}) = \mathbf{AB} + \mathbf{AC}, \quad (\mathbf{A} + \mathbf{B})\mathbf{C} = \mathbf{AC} + \mathbf{BC}, \\ &\mathbf{A}(\alpha\mathbf{B}) = \alpha(\mathbf{AB}) = (\alpha\mathbf{A})\mathbf{B}, \end{aligned} \tag{C.8}$$

for any scalar α and $n \times n$ matrices $\mathbf{A}$, $\mathbf{B}$, and $\mathbf{C}$.

C.3 Operations with Matrices

This section centers on a variety of operations concerning matrices.

C.3.1 Elementary Row/Column Operations

Elementary matrix operations encompass the following actions:

- Interchanging any two rows/columns of $\mathbf{A}$.
- Multiplying any row/column of $\mathbf{A}$ by a nonzero scalar.
- Adding any scalar multiple of a row/column of $\mathbf{A}$ to another row/column of $\mathbf{A}$.

These operations are commonly employed to transform **A** into its row reduced echelon form. This is a unique representation in which:

- Any row containing a nonzero entry precedes any row in which all the entries are zero.
- The first nonzero entry in each row is the only nonzero entry in its column (this entry is called pivot).
- The initial entry in each row is 1, and it is situated in a column to the right of the first nonzero entry in the preceding row.

The rank of a matrix, i.e., the maximum number of linearly independent rows or columns, corresponds to the number of nonzero rows in the row reduced echelon form.

C.3.2 Transpose of a Matrix

The transpose of an $n \times n$ matrix **A** is another $n \times n$ matrix **B**, denoted $\mathbf{B} = \mathbf{A}^T$, with entries given by

$$B_{lj} = A_{jl}. \quad (l, j = 1, \ldots, n) \tag{C.9}$$

This definition implies that $(\mathbf{A}^{\mathrm{T}})^{\mathrm{T}} = \mathbf{A}$. It can also be verified that

$$(\mathbf{AB})^{\mathrm{T}} = \mathbf{B}^{\mathrm{T}}\mathbf{A}^{\mathrm{T}} \tag{C.10}$$

and

$$(\alpha\mathbf{A} + \beta\mathbf{B})^{\mathrm{T}} = \alpha\mathbf{A}^{\mathrm{T}} + \beta\mathbf{B}^{\mathrm{T}}. \tag{C.11}$$

The concept of transposition leads to the categorization of matrices into symmetric and skew symmetric. A matrix **A** is symmetric when $\mathbf{A}^{\mathrm{T}} = \mathbf{A}$, i.e., $A_{lj} = A_{jl}$. Conversely, a matrix **A** is skew symmetric when $\mathbf{A}^{\mathrm{T}} = -\mathbf{A}$. The entries of a skew-symmetric matrix satisfy $A_{lj} = -A_{jl}$, and consequently, $A_{jj} = 0$ $(j = 1, \ldots, n)$.

Every matrix **A** can be uniquely expressed as the sum of symmetric and skew-symmetric matrices:

$$\mathbf{A} = \mathbf{S} + \mathbf{W}, \tag{C.12}$$

where $\mathbf{S}^{\mathrm{T}} = \mathbf{S}$ represents the symmetric part, and $\mathbf{W}^{\mathrm{T}} = -\mathbf{W}$ denotes the skew-symmetric part of **A**. Given **A**, these two parts can be computed with

$$\mathbf{S} = \frac{\mathbf{A} + \mathbf{A}^T}{2}, \quad \mathbf{W} = \frac{\mathbf{A} - \mathbf{A}^T}{2}. \tag{C.13}$$

C.3.3 Inverse of a Matrix

An $n \times n$ matrix $\mathbf{A}$ is invertible when the linear system of equations

$$\mathbf{Au} = \mathbf{v} \tag{C.14}$$

admits a unique solution $\mathbf{u} \in \mathcal{E}^n$ for a given $\mathbf{v} \in \mathcal{E}^n$. Otherwise, this matrix is termed *singular*. If $\mathbf{A}$ is invertible, the solution of the linear system is written as

$$\mathbf{u} = \mathbf{A}^{-1}\mathbf{v}, \tag{C.15}$$

where $\mathbf{A}^{-1}$ is the (unique) inverse of $\mathbf{A}$. The solution $\mathbf{u}$ of the linear system may be computed using elementary operations to transform the augmented $n \times (n + 1)$ matrix $(\mathbf{A}, \mathbf{v})$ into its row reduced echelon form. Alternatively, the inverse of $\mathbf{A}$ can be obtained by transforming the augmented $n \times 2n$ matrix $(\mathbf{A}, \mathbf{I})$ into $(\mathbf{I}, \mathbf{A}^{-1})$ by elementary operations.

There are several properties involving the inverse of matrices. For instance,

$$\mathbf{AA}^{-1} = \mathbf{I} = \mathbf{A}^{-1}\mathbf{A}. \tag{C.16}$$

and

$$(\mathbf{A}^{\mathrm{T}})^{-1} = (\mathbf{A}^{-1})^{\mathrm{T}}. \tag{C.17}$$

Property (C.17) implies that $(\mathbf{A}^{\mathrm{T}})^{-1}$ is written simply as $\mathbf{A}^{-\mathrm{T}}$. Finally, if another matrix $\mathbf{B}$ is also invertible, it follows that

$$(\mathbf{AB})^{-1} = \mathbf{B}^{-1}\mathbf{A}^{-1}. \tag{C.18}$$

C.3.4 Determinant of a Matrix

The determinant of a 2×2 matrix

$$\mathbf{A} = \begin{bmatrix} A_{11} & A_{12} \\ A_{21} & A_{22} \end{bmatrix} \tag{C.19}$$

is defined by

$$\det(\mathbf{A}) = A_{11}A_{22} - A_{12}A_{21}. \tag{C.20}$$

This result can be employed recursively to compute the determinant of an $n \times n$ matrix $\mathbf{A}$ using the cofactor expansion:

$$\det(\mathbf{A}) = \sum_{l=1}^{n} A_{lj}\mathrm{cof}_{lj}(\mathbf{A}) = \sum_{j=1}^{n} A_{lj}\mathrm{cof}_{lj}(\mathbf{A}), \tag{C.21}$$

where $\mathrm{cof}_{lj}(\mathbf{A}) = (-1)^{l+j}\det(\mathcal{M}_{lj}(\mathbf{A}))$ is referred to as the cofactor of $\mathbf{A}$. It is computed using the minor $\mathcal{M}_{lj}(\mathbf{A})$, which is the $(n-1) \times (n-1)$ submatrix of $\mathbf{A}$ obtained by "removing" the $l-$th row and $j-$th column from $\mathbf{A}$.

A key theoretical result regards the determinant of invertible matrices: a matrix $\mathbf{A}$ is invertible if and only if $\det(\mathbf{A}) \neq 0$. Consequently,

$$\det(\mathbf{A}^{-1}) = \frac{1}{\det(\mathbf{A})} \tag{C.22}$$

for any invertible matrix $\mathbf{A}$. It may also be checked that $\det(\mathbf{I}) = 1$, $\det(\mathbf{A}^T) = \det(\mathbf{A})$, and

$$\det(\mathbf{AB}) = \det(\mathbf{A})\det(\mathbf{B}). \tag{C.23}$$

C.4 Eigenvalues and Eigenvectors of a Matrix

A scalar λ is an eigenvalue of a matrix $\mathbf{A}$ if there exists a nonzero vector $\mathbf{u}$ such that

$$\mathbf{Au} = \lambda\mathbf{u}, \tag{C.24}$$

where $\mathbf{u}$ is known as eigenvector of $\mathbf{A}$ associated with λ. Equation (C.24) is equivalent to the expression

$$(\mathbf{A} - \lambda\mathbf{I})\mathbf{u} = \mathbf{0}. \tag{C.25}$$

Therefore, the matrix $\mathbf{A} - \lambda\mathbf{I}$ must be singular:

$$\det(\mathbf{A} - \lambda\mathbf{I}) = 0. \tag{C.26}$$

This results in a polynomial of degree n in λ, called the characteristic polynomial. This polynomial has real coefficients; thus, its roots are either all real, all complex, or mixed, with complex eigenvalues appearing in complex conjugate pairs. If $\mathbf{A}$ is symmetric, then all of its eigenvalues are real. Furthermore, if $\mathbf{A}$ is symmetric and also satisfies the condition

$$\mathbf{v}^T\mathbf{Av} > 0 \tag{C.27}$$

for any vector $\mathbf{v} \neq 0$, then $\mathbf{A}$ is said to be positive-definite, and it possesses only positive eigenvalues.

Once an eigenvalue λ is determined, the associated eigenvector can be found, up to a scalar multiplier, by solving the system of n equations in (C.25):

$$\sum_{j=1}^{n}(A_{lj} - \lambda\delta_{lj})u_j = 0, \tag{C.28}$$

for $l = 1, \ldots, n$. Typically, the eigenvalues are distinct: $\lambda_1 \neq \lambda_2 \neq \cdots \neq \lambda_n$. In such instances, all eigenvectors $\mathbf{u}_1, \ldots, \mathbf{u}_n$ are linearly independent. A real eigenvalue can always be associated with a real eigenvector. However, eigenvectors corresponding to complex eigenvalues may also be complex. In this way, if

$$\mathbf{u} = \begin{bmatrix} u_1 \\ \vdots \\ u_n \end{bmatrix} \tag{C.29}$$

is an eigenvector associated with λ, then the complex conjugate eigenvalue $\overline{\lambda}$ is associated with the complex conjugate eigenvector

$$\overline{\mathbf{u}} = \begin{bmatrix} \overline{u}_1 \\ \vdots \\ \overline{u}_n \end{bmatrix}. \tag{C.30}$$

Appendix D
Review of Newton–Euler Dynamics

In this appendix, Newton–Euler dynamics of particles, systems of particles, and rigid bodies is concisely reviewed.

D.1 Dynamics of a Single Particle

A particle is an idealized object that is assumed to have mass but occupies zero volume.

D.1.1 Dynamic Analysis Using Cartesian Coordinates

Consider a particle P with mass m moving in an inertial space with origin at O. Using a Cartesian coordinate system, the position vector of this particle has the general representation

$$\mathbf{r}(t) = x(t)\mathbf{i} + y(t)\mathbf{j} + z(t)\mathbf{k}, \tag{D.1}$$

where $\{\mathbf{i},\ \mathbf{j},\ \mathbf{k}\}$ are constant unit vectors. Differentiate $\mathbf{r}$ with respect to time to obtain the instantaneous velocity vector

$$\mathbf{v} = \dot{x}\mathbf{i} + \dot{y}\mathbf{j} + \dot{z}\mathbf{k}. \tag{D.2}$$

Differentiating $\mathbf{v}$ yields the instantaneous acceleration vector of P:

$$\mathbf{a} = \ddot{x}\mathbf{i} + \ddot{y}\mathbf{j} + \ddot{z}\mathbf{k}. \tag{D.3}$$

R. Gonçalves Salsa Junior, *Fundamentals of the Theory of Mechanical Vibrations*,
https://doi.org/10.1007/978-3-031-83995-5

If the particle is subject to the resultant force $\mathbf{F}$, application of Newton's second law

$$\mathbf{F} = m\mathbf{a} \tag{D.4}$$

gives the three differential equations

$$m\ddot{x} = \mathbf{F} \cdot \mathbf{i} = F_x, \tag{D.5}$$

$$m\ddot{y} = \mathbf{F} \cdot \mathbf{j} = F_y, \tag{D.6}$$

$$m\ddot{z} = \mathbf{F} \cdot \mathbf{k} = F_z, \tag{D.7}$$

that may be integrated to obtain $x(t)$, $y(t)$ e $z(t)$.

D.1.2 Dynamic Analysis Using Polar Coordinates

According to Fig. D.1, the Cartesian and polar coordinates are related by

$$x = r\cos\theta, \tag{D.8}$$

$$y = r\sin\theta. \tag{D.9}$$

The position vector of a particle on planar motion can then be described by

$$\mathbf{r} = x\mathbf{i} + y\mathbf{j} = r\mathbf{e}_r, \tag{D.10}$$

where the basis vectors associated with the polar coordinates are

$$\mathbf{e}_r = \cos\theta\mathbf{i} + \sin\theta\mathbf{j}, \tag{D.11}$$

$$\mathbf{e}_\theta = -\sin\theta\mathbf{i} + \cos\theta\mathbf{j}. \tag{D.12}$$

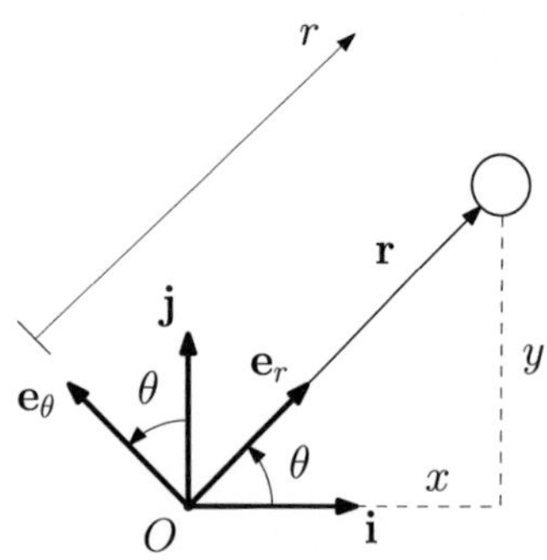

Fig. D.1 Using the polar coordinate system to describe the motion of a particle with position vector **r**

It can be checked that the derivative of these vectors with respect to θ are

$$\frac{d\mathbf{e}_r}{d\theta} = \mathbf{e}_\theta, \qquad \frac{d\mathbf{e}_\theta}{d\theta} = -\mathbf{e}_r. \tag{D.13}$$

In this manner, the velocity and acceleration vectors can be expressed as

$$\mathbf{v} = \dot{r}\mathbf{e}_r + r\dot{\theta}\mathbf{e}_\theta, \tag{D.14}$$

$$\mathbf{a} = (\ddot{r} - r\dot{\theta}^2)\mathbf{e}_r + (r\ddot{\theta} + 2\dot{r}\dot{\theta})\mathbf{e}_\theta, \tag{D.15}$$

respectively. Newton's second law results in differential equations of the form

$$m(\ddot{r} - r\dot{\theta}^2) = \mathbf{F} \cdot \mathbf{e}_r = F_r, \tag{D.16}$$

$$m(r\ddot{\theta} + 2\dot{r}\dot{\theta}) = \mathbf{F} \cdot \mathbf{e}_\theta = F_\theta. \tag{D.17}$$

D.1.3 Energy Relations for a Particle

The power theorem states that the motion of a particle satisfies

$$\dot{T} = \mathbf{F} \cdot \mathbf{v}, \tag{D.18}$$

where

$$T = \frac{1}{2} m\mathbf{v} \cdot \mathbf{v} \tag{D.19}$$

is its kinetic energy. Separating $\mathbf{F} = \mathbf{F}_{con} + \mathbf{F}_{ncon}$ in conservative and non-conservative parts, respectively, and recalling that

$$\mathbf{F}_{con} \cdot \mathbf{v} = -\dot{U} \tag{D.20}$$

where $U = U(\mathbf{r})$ is the potential energy of the particle, one obtains

$$\dot{E} = \mathbf{F}_{ncon} \cdot \mathbf{v}. \tag{D.21}$$

The total mechanical energy is $E = T + U$. If the only forces that do work are conservative,

$$\dot{E} = 0. \tag{D.22}$$

Therefore, E remains constant during the motion:

$$E = T + V = T_0 + V_0 = E_0, \tag{D.23}$$

and the constant value E_0 can be determined from the ICs.

D.2 Dynamics of a System of Particles

Now, a system composed of many particles, such as the one shown in Fig. D.2, is considered.

D.2.1 Balance of Linear and Angular Momenta

Let m_j be the mass of P_j and $\mathbf{r}_j$ be the position vector of P_j at time t, relative to a fixed origin O. The center of mass C of this system is located by the position vector

$$\mathbf{r}_C = \frac{1}{m}\sum_{j=1}^{N} m_j \mathbf{r}_j, \tag{D.24}$$

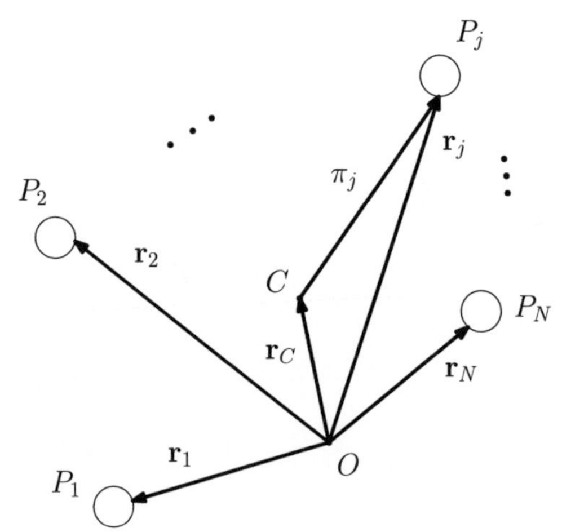

Fig. D.2 A system composed of particles P_j $(j = 1, 2, \ldots, N)$ subject to external and internal forces. The position of the center of mass C is $\mathbf{r}_C$, and the position of particle P_j relative to C is $\boldsymbol{\pi}_j = \mathbf{r}_j - \mathbf{r}_C$

where $m = \sum_{j=1}^{N} m_j$ is the mass of the system. The linear momentum $\mathbf{G}$ and the angular momentum $\mathbf{H}^O$ about O for the system are given by

$$\mathbf{G} = \sum_{j=1}^{N} m_j \mathbf{v}_j = m\dot{\mathbf{r}}_C = m\mathbf{v}_C, \tag{D.25}$$

$$\mathbf{H}^O = \sum_{j=1}^{N} \mathbf{r}_j \times m_j \mathbf{v}_j, \tag{D.26}$$

respectively. Let $\mathbf{F}$ be the resultant of the external forces that act on the system and $\mathbf{M}^O$ be the resultant torque due to the external forces and couples about O. Euler's laws state that

$$\mathbf{F} = \dot{\mathbf{G}} = m\ddot{\mathbf{r}}_C = m\mathbf{a}_C, \tag{D.27}$$

$$\mathbf{M}^O = \dot{\mathbf{H}}^O. \tag{D.28}$$

Euler's first law (D.27) is also called balance of linear momentum while (D.28) is the balance of angular momentum. It follows from Euler's two laws that the balance of angular momentum may also be expressed with respect to the mass center C:

$$\mathbf{M} = \dot{\mathbf{H}}, \tag{D.29}$$

where

$$\mathbf{H} = \sum_{j=1}^{N} \boldsymbol{\pi}_j \times m_j \dot{\boldsymbol{\pi}}_j \tag{D.30}$$

is the angular momentum of the system about C, and $\mathbf{M}$ is the resultant external torque about the center of mass.

D.2.2 Energy Relations for a System of Particles

The kinetic energy of the system is

$$T = \frac{1}{2} \sum_{j=1}^{N} m_j \mathbf{v}_j \cdot \mathbf{v}_j. \tag{D.31}$$

Alternatively, Koenig's decomposition theorem states that it can be expressed as

$$T = \frac{1}{2} m \mathbf{v}_C \cdot \mathbf{v}_C + \frac{1}{2} \sum_{j=1}^{N} m_j \dot{\boldsymbol{\pi}}_j \cdot \dot{\boldsymbol{\pi}}_j. \tag{D.32}$$

It follows from the power theorem for a particle that the kinetic energy of the system satisfies

$$\dot{T} = \sum_{j=1}^{N} \mathbf{F}_j \cdot \mathbf{v}_j, \tag{D.33}$$

where $\mathbf{F}_j$ is the resultant force acting on P_j. It is convenient to separate $\mathbf{F}_j = \mathbf{F}_{j,con} + \mathbf{F}_{j,ncon}$ in conservative and non-conservative parts, respectively. Letting $U(\mathbf{r}_1, \mathbf{r}_2, \ldots, \mathbf{r}_N)$ be the potential energy of the system, it follows that

$$\sum_{j=1}^{N} \mathbf{F}_{j,con} \cdot \mathbf{v}_j = -\dot{U} \tag{D.34}$$

and, therefore, the total mechanical energy of the system $E = T + U$ satisfies

$$\dot{E} = \sum_{j=1}^{N} \mathbf{F}_{j,ncon} \cdot \mathbf{v}_j. \tag{D.35}$$

D.3 Dynamics of a Rigid Body

A rigid body is a collection of particles where the distance between any two of them remains constant. For the purpose of this discussion, only rigid bodies undergoing planar motions are considered. In this particular type of motion, dynamic analysis is based on the "slice" of the body that lies on the plane of motion, as illustrated in Fig. D.3.

D.3.1 A Rigid Body Rotating Around a Fixed Axis

Referring to Fig. D.4, consider a rigid body that is rotating about an axle lying along a fixed axis passing through O. Let the unit vector $\mathbf{k}$ point along this axis, and $\{\mathbf{i}, \mathbf{j}, \mathbf{k}\}$ be a fixed orthonormal basis. Define a right-handed orthonormal basis

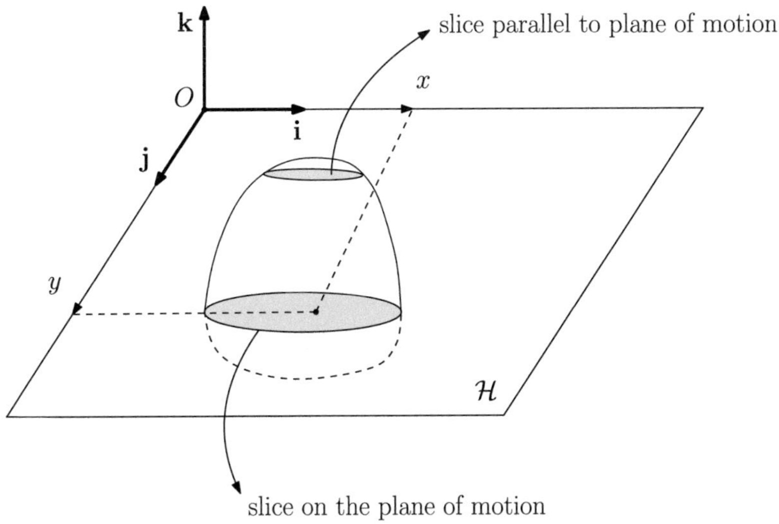

Fig. D.3 A rigid body undergoing planar motion. Slices of the body that are initially parallel to the plane of motion $\mathcal{H}$ remain parallel to it

Fig. D.4 Rigid body rotating about a fixed axis passing through O

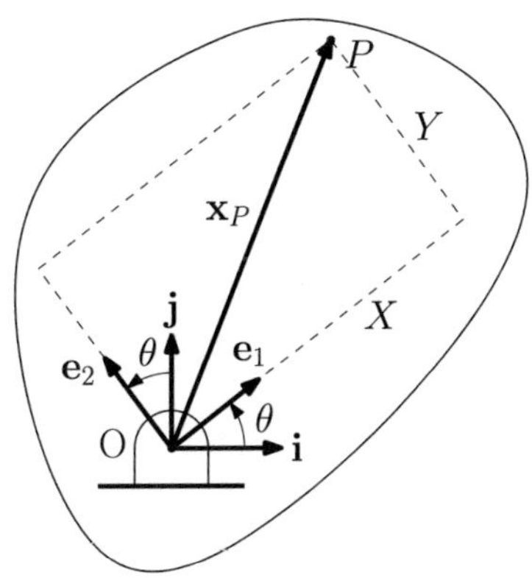

$\{\mathbf{e}_1, \mathbf{e}_2, \mathbf{e}_3\}$, which is fixed on the body, by

$$\begin{aligned} \mathbf{e}_1 &= \cos\theta\mathbf{i} + \sin\theta\mathbf{j}, \\ \mathbf{e}_2 &= -\sin\theta\mathbf{i} + \cos\theta\mathbf{j}, \\ \mathbf{e}_3 &= \mathbf{k}. \end{aligned} \tag{D.36}$$

The vectors $\mathbf{e}_1$, $\mathbf{e}_2$, $\mathbf{e}_3$ are called corotational basis vectors. The angular velocity and angular acceleration vectors of the body are

$$\boldsymbol{\omega} = \dot{\theta}\mathbf{k}, \quad \boldsymbol{\alpha} = \dot{\boldsymbol{\omega}} = \ddot{\theta}\mathbf{k}, \tag{D.37}$$

respectively. Let $\mathbf{x}_P = X\mathbf{e}_1 + Y\mathbf{e}_2 + Z\mathbf{k}$ be the position vector of any particle P of the body. Then X, Y, and Z are constants, and it can be shown that the velocity and acceleration fields are given in Eulerian form as

$$\mathbf{v} = \boldsymbol{\omega} \times \mathbf{x}_P, \quad \mathbf{a} = \dot{\boldsymbol{\omega}} \times \mathbf{x}_P + \boldsymbol{\omega} \times (\boldsymbol{\omega} \times \mathbf{x}_P), \tag{D.38}$$

respectively.

If the body has mass m, its linear momentum $\mathbf{G}$ is given by

$$\mathbf{G} = m\dot{\mathbf{x}}_C = m\mathbf{v}_C, \tag{D.39}$$

where $\mathbf{x}_C$ is the position of the center of mass of the body. The angular momentum $\mathbf{H}^O$ about O can be expressed as

$$\mathbf{H}^O = \left(-I_{xz}^O\,\mathbf{e}_1 - I_{yz}^O\,\mathbf{e}_2 + I_{zz}^O\,\mathbf{e}_3\right)\dot{\theta}, \tag{D.40}$$

where I_{zz}^O is the moment of inertia, and I_{xz}^O, I_{yz}^O are the products of inertia with respect to the axis $\{\mathbf{e}_1, \mathbf{e}_2, \mathbf{e}_3\}$ passing through O. These components of the inertia tensor can be computed using the following volume integrals:

$$I_{zz}^O = \int_V \rho(X^2 + Y^2)dV, \quad I_{xz}^O = \int_V \rho XZdV, \quad I_{yz}^O = \int_V \rho YZdV, \tag{D.41}$$

where V is the volume of the body and ρ is its mass per unit volume.

Let $\mathbf{F}$ be the resultant external force on the body and $\mathbf{M}^O$ be the resultant external torque about O. According to Euler's laws,

$$\mathbf{F} = \dot{\mathbf{G}} = m\ddot{\mathbf{x}}_C = m\mathbf{a}_C = m\left[\dot{\boldsymbol{\omega}} \times \mathbf{x}_C + \boldsymbol{\omega} \times (\boldsymbol{\omega} \times \mathbf{x}_C)\right], \quad \mathbf{M}^O = \dot{\mathbf{H}}^O. \tag{D.42}$$

Note that the balance of angular momentum takes the form

$$\mathbf{M}^O = \left(-I_{xz}^O\ddot{\theta} + I_{yz}^O\dot{\theta}^2\right)\mathbf{e}_1 + \left(-I_{yz}^O\ddot{\theta} - I_{xz}^O\dot{\theta}^2\right)\mathbf{e}_2 + I_{zz}^O\ddot{\theta}\mathbf{e}_3. \tag{D.43}$$

The components of $\mathbf{M}^O$ along $\mathbf{e}_1$ and $\mathbf{e}_2$ are constraint moments that maintain the planar motion about the fixed axis. If $\mathbf{e}_3$ is a principal axis of inertia, the products of inertia are $I_{xz}^O = I_{yz}^O = 0$, and no in-plane moment is necessary to maintain the motion: $\mathbf{M}^O \cdot \mathbf{e}_1 = \mathbf{M}^O \cdot \mathbf{e}_2 = 0$.

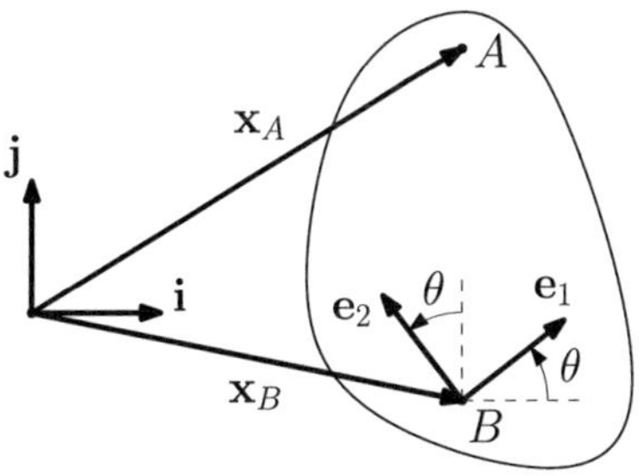

Fig. D.5 A rigid body in general plane motion. The unit orthonormal vectors $\{\mathbf{e}_1, \mathbf{e}_2, \mathbf{e}_3 = \mathbf{k}\}$ form the corotational basis

D.3.2 A Rigid Body Undergoing General Plane Motion

Figure D.5 depicts a rigid body undergoing general plane motion. Consider an arbitrary point B on the body, characterized by the position vector $\mathbf{x}_B$ in the current configuration. Now, for any other point A on the body, represented by its position vector $\mathbf{x}_A$, it can be demonstrated that

$$\mathbf{v}_A = \mathbf{v}_B + \boldsymbol{\omega} \times (\mathbf{x}_A - \mathbf{x}_B), \tag{D.44}$$

where $\boldsymbol{\omega} = \dot{\theta}\mathbf{k}$. Consequently, the acceleration field satisfies

$$\mathbf{a}_A = \mathbf{a}_B + \dot{\boldsymbol{\omega}} \times (\mathbf{x}_A - \mathbf{x}_B) + \boldsymbol{\omega} \times [\boldsymbol{\omega} \times (\mathbf{x}_A - \mathbf{x}_B)] . \tag{D.45}$$

For general plane motions of a rigid body, Euler's laws are

$$\mathbf{F} = \dot{\mathbf{G}} = m\dot{\mathbf{v}}_C = m\mathbf{a}_C, \tag{D.46}$$

and

$$\mathbf{M} = \dot{\mathbf{H}}, \tag{D.47}$$

where $\mathbf{H}$ is the angular momentum about the mass center, and $\mathbf{M}$ is the resultant external torque with respect to the center of mass. Recall $\mathbf{H}$ takes the form

$$\mathbf{H} = \left(-I_{xz}\mathbf{e}_1 - I_{yz}\mathbf{e}_2 + I_{zz}\mathbf{e}_3\right)\dot{\theta}, \tag{D.48}$$

where I_{zz} is the moment of inertia, and I_{xz}, I_{yz} are the products of inertia with respect to the axis $\{\mathbf{e}_1, \mathbf{e}_2, \mathbf{e}_3\}$ passing through the center of mass. If $\boldsymbol{\pi} = X\mathbf{e}_1 + Y\mathbf{e}_2 + Z\mathbf{k}$ is the position of a point of the body with respect to the mass center, the aforementioned components of the inertia tensor can be computed using the volume integrals

$$I_{zz} = \int_V \rho(X^2 + Y^2)dV, \quad I_{xz} = \int_V \rho XZdV, \quad I_{yz} = \int_V \rho YZdV. \tag{D.49}$$

D.3.3 Translational Motions of a Rigid Body

For a rigid body in translational motion, the velocity and acceleration fields are given by

$$\mathbf{v} = \mathbf{v}_C, \quad \mathbf{a} = \mathbf{a}_C, \tag{D.50}$$

respectively. In addition, $\boldsymbol{\omega} = \mathbf{0}$ and

$$\mathbf{H} = \mathbf{0}. \tag{D.51}$$

Therefore, Euler's laws require that

$$\mathbf{F} = \dot{\mathbf{G}} = m\mathbf{a}_C, \quad \mathbf{M} = \dot{\mathbf{H}} = \mathbf{0}. \tag{D.52}$$

D.3.4 Energy Relations for a Rigid Body

According to König's decomposition, the kinetic energy of a rigid body in planar motion can be expressed as

$$T = \frac{1}{2}m\mathbf{v}_C \cdot \mathbf{v}_C + \frac{1}{2}I_{zz}\dot{\theta}^2. \tag{D.53}$$

If the body is acted upon by a couple $\mathbf{M}_p$ and forces $\mathbf{F}_i$ concentrated on points with position $\mathbf{x}_i$ $(i = 1, \ldots, N)$, then the power theorem states that

$$\dot{T} = \sum_{j=1}^{N} \mathbf{F}_j \cdot \mathbf{v}_j + \mathbf{M}_p \cdot \boldsymbol{\omega}. \tag{D.54}$$

Again, write $\mathbf{F}_j = \mathbf{F}_{j,con} + \mathbf{F}_{j,ncon}$ in conservative and non-conservative parts, respectively. The power of the conservative forces is

$$\sum_{j=1}^{N} \mathbf{F}_{j,con} \cdot \mathbf{v}_j = -\dot{U} \tag{D.55}$$

where $U(\mathbf{x}_1, \mathbf{x}_2, \ldots, \mathbf{x}_N)$ is the potential energy of the body. Therefore, the total mechanical energy satisfies

$$\dot{E} = \sum_{j=1}^{N} \mathbf{F}_{j,ncon} \cdot \mathbf{v}_j + \mathbf{M}_p \cdot \boldsymbol{\omega}. \tag{D.56}$$

When the body is rotating about a fixed axis passing through O along the direction $\mathbf{e}_3 = \mathbf{k}$, the kinetic energy can be expressed by

$$T = \frac{1}{2} I_{zz}^{O} \dot{\theta}^2, \tag{D.57}$$

and the power theorem becomes

$$\dot{T} = \mathbf{M}^{O} \cdot \boldsymbol{\omega}. \tag{D.58}$$

If the body is translating, the kinetic energy is

$$T = \frac{1}{2} m \mathbf{v}_C \cdot \mathbf{v}_C, \tag{D.59}$$

and the power theorem

$$\dot{T} = \mathbf{F} \cdot \mathbf{v}_C. \tag{D.60}$$

D.3.5 Huygens' Theorem

Consider the angular momentum $\mathbf{H}$ with respect to the center of mass of a rigid body in planar motion, given by Eq. (D.48). Knowledge of the moment of inertia I_{zz} and the products of inertia I_{xz} and I_{yz} is required to compute $\mathbf{H}$.

However, in many cases, one already possesses information about the inertia components I_{zz}^A, I_{xz}^A, and I_{yz}^A, pertaining to the axes $\mathbf{t}_j = \mathbf{e}_j$ $(j = 1, 2, 3)$ passing through a point A located by the position vector $\boldsymbol{\pi}_A = X_A\mathbf{e}_1 + Y_A\mathbf{e}_2 + Z_A\mathbf{e}_3$ relative to the center of mass, as shown in Fig. D.6. To obtain the desired components I_{zz},

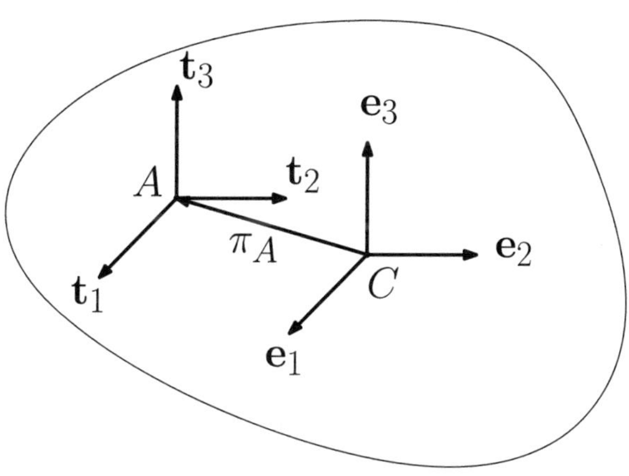

Fig. D.6 A rigid body with axes $\{\mathbf{e}_j\}$ passing through its mass center C. The parallel axes $\{\mathbf{t}_j = \mathbf{e}_j\}$ pass through point A, which is located by the position vector $\boldsymbol{\pi}_A$ relative to C

I_{xz}, and I_{yz}, Huygens' theorem (or the parallel axis theorem) can be applied:

$$I_{zz}^{A} = I_{zz} + m(X_A^2 + Y_A^2), \tag{D.61}$$

$$I_{xz}^{A} = I_{xz} + mX_A Z_A, \tag{D.62}$$

$$I_{yz}^{A} = I_{yz} + mY_A Z_A. \tag{D.63}$$

Appendix E
Solving Initial Value Problems with MATLAB

In this appendix, the fundamental aspects of solving initial value problems with MATLAB are presented.

E.1 Initial Value Problems

The vibrations of a mechanical system are governed by a system of n second-order ODEs for the generalized coordinates q_1, q_2, ..., q_n. With the change of variables provided by

$$y_j = q_j, \quad y_{n+j} = \dot{q}_j, \quad (j = 1, \ldots, n) \tag{E.1}$$

for example, these equations can be recast as a system of $N = 2n$ first-order ODEs

$$\dot{\mathbf{y}} = \mathbf{h}(t, \mathbf{y}), \tag{E.2}$$

where $\mathbf{y} = [y_1 \quad y_2 \quad \ldots \quad y_N]^T \in \mathbb{R}^N$ and $\mathbf{h}$ is a nonlinear $N \times 1$ vector function. The objective of an initial value problem (IVP) is to determine the solution $\mathbf{y}(t)$ of (E.2) for given a IC vector

$$\mathbf{y}(0) = \mathbf{y}_0. \tag{E.3}$$

If t does not appear explicitly in the equation of motion, the system is called autonomous and the associated system of first-order ODEs becomes

$$\dot{\mathbf{y}} = \mathbf{h}(\mathbf{y}). \tag{E.4}$$

R. Gonçalves Salsa Junior, *Fundamentals of the Theory of Mechanical Vibrations*,
https://doi.org/10.1007/978-3-031-83995-5

In an autonomous system, there are no time-dependent applied forces, although there may be constant forces and forces depending on $\mathbf{y}$. A linear system of ODEs can be expressed as

$$\dot{\mathbf{y}} = \mathbf{A}(t)\mathbf{y} + \mathbf{b}(t), \tag{E.5}$$

where $\mathbf{A}(t)$ is an $N \times N$ matrix function, and $\mathbf{b}(t)$ is an $N \times 1$ vector. If $\mathbf{A}$ does not depend on t, then this is a constant coefficient linear system:

$$\dot{\mathbf{y}} = \mathbf{A}\mathbf{y} + \mathbf{b}(t). \tag{E.6}$$

Conditions for the existence of a unique solution for (E.2) are well-known. Let $h_1, h_2, \ldots, h_n$ be continuous functions having continuous partial derivatives $\partial h_l / \partial y_j$ $(l,\ j = 1, 2, \ldots, n)$ in some domain R of $(\mathbf{y}, t)$-space containing the point $(\mathbf{y}_0, 0)$. Then (E.2) has a unique solution on some interval $|t - 0| < \delta$ satisfying the IC $\mathbf{y}(0) = \mathbf{y}_0$.

Instead of the continuity of partial derivatives, let $\mathbf{h}(t, \mathbf{y})$ be continuous and satisfy the Lipschitz condition

$$\|\mathbf{h}(t,\ \mathbf{y}_1) - \mathbf{h}(t,\ \mathbf{y}_2)\| \le M\|\mathbf{y}_1 - \mathbf{y}_2\| \tag{E.7}$$

in some domain R of $(\mathbf{y}, t)$-space containing the point $\mathbf{y}(0) = \mathbf{y}_0$. Then there exists a unique solution for (E.2).

E.2 Some Basic Numerical Methods

One of the simplest approaches to build numerical methods to solve Eq. (E.2) is by using the Taylor series expansion for $\mathbf{y}(t + \Delta t)$ about $\mathbf{y}(t)$:

$$\mathbf{y}(t + \Delta t) = \mathbf{y}(t) + \dot{\mathbf{y}}(t)\Delta t + \frac{1}{2}\ddot{\mathbf{y}}(t)(\Delta t)^2 + \ldots, \tag{E.8}$$

where Δt is called time step. In the *forward Euler* method, terms up to the first power of Δt are used to advance the solution from instant t to $t + \Delta t$ with the following formula:

$$\mathbf{y}(t + \Delta t) = \mathbf{y}(t) + \mathbf{h}(t,\ \mathbf{y}(t))\Delta t, \tag{E.9}$$

where $\dot{\mathbf{y}}$ has been replaced by the ODE (E.2). This numerical scheme is considered to be first-order convergent. Convergence is a fundamental criterion that any reliable numerical method should fulfill. It means that as the time step Δt decreases, the numerical solution approaches the exact solution more closely. In essence, by decreasing Δt sufficiently, the numerical method can yield arbitrarily accurate solutions. The forward Euler method exhibits first-order convergence, indicating that when the time step is halved, the error between the true solution and the numerical approximation is expected to decrease by approximately half for small Δt.

By considering higher-order terms in the Taylor expansion, it is possible to derive numerical methods with improved convergence properties. For instance, the second-order *Taylor series method* includes terms up to $(\Delta t)^2$ and can be expressed as follows:

$$\begin{aligned} \mathbf{y}(t+\Delta t) = \mathbf{y}(t) + \mathbf{h}(t,\ \mathbf{y}(t))\Delta t + \\ \frac{(\Delta t)^2}{2}\left[\frac{\partial \mathbf{h}}{\partial t}(t,\ \mathbf{y}(t)) + \frac{\partial \mathbf{h}}{\partial \mathbf{y}}(t,\ \mathbf{y}(t))\mathbf{h}(t,\ \mathbf{y}(t))\right], \end{aligned} \tag{E.10}$$

where $\partial \mathbf{h}/\partial \mathbf{y}$ denotes the Jacobian matrix of $\mathbf{h}$. This method incorporates additional terms to enhance the accuracy of the numerical approximation, but with the obvious drawback that it requires evaluation of the derivatives of $\mathbf{h}$.

The idea of *Runge–Kutta methods* is to obtain higher-order methods by evaluating $\mathbf{h}(t,\ \mathbf{y})$ at several points for each time step, instead of computing its derivatives. A general Runge–Kutta method has s stages:

$$\begin{aligned} \mathbf{k}_1 &= \mathbf{h}\left(t + c_1\Delta t,\ \mathbf{y}(t) + \Delta t(a_{11}\mathbf{k}_1 + a_{12}\mathbf{k}_2 + \ldots + a_{1s}\mathbf{k}_s)\right), \\ \mathbf{k}_2 &= \mathbf{h}\left(t + c_2\Delta t,\ \mathbf{y}(t) + \Delta t(a_{21}\mathbf{k}_1 + a_{22}\mathbf{k}_2 + \ldots + a_{2s}\mathbf{k}_s)\right), \\ &\vdots \\ \mathbf{k}_s &= \mathbf{h}\left(t + c_s\Delta t,\ \mathbf{y}(t) + \Delta t(a_{s1}\mathbf{k}_1 + a_{s2}\mathbf{k}_2 + \ldots + a_{ss}\mathbf{k}_s)\right), \\ \mathbf{y}(t+\Delta t) &= \mathbf{y}(t) + \Delta t\left(b_1\mathbf{k}_1 + b_2\mathbf{k}_2 + \ldots + b_s\mathbf{k}_s\right), \end{aligned} \tag{E.11}$$

where the scalars a_{lj}, c_j, and b_j $(l,\ j = 1,\ 2,\ \ldots,\ s)$ are chosen so that the method satisfies certain order conditions. For example, the two-stage *Heun's method*

$$\begin{aligned} \mathbf{k}_1 &= \mathbf{h}\left(t,\ \mathbf{y}(t)\right), \\ \mathbf{k}_2 &= \mathbf{h}\left(t + \frac{2}{3}\Delta t,\ \mathbf{y}(t) + \frac{2}{3}\Delta t\mathbf{k}_1\right), \\ \mathbf{y}(t+\Delta t) &= \mathbf{y}(t) + \Delta t\left(\frac{1}{4}\mathbf{k}_1 + \frac{3}{4}\mathbf{k}_2\right) \end{aligned} \tag{E.12}$$

is second-order convergent.

E.3 Basic MATLAB Commands

The MATLAB function `ode45` is designed to solve systems of ordinary differential equations in the form (E.2). It employs the Dormand–Prince scheme, which combines fourth and fifth-order Runge–Kutta methods. The objective of this scheme is to estimate the error at each step by comparing the solution from the lower-order method to the one of the higher-order method. This dynamic step size adjustment ensures that the errors remain within a specified tolerance.

To use the `ode45` function, the syntax is as follows:

```
[t,y] = ode45(dydt,tspan,y0)
```

The 2×1 vector `tspan = [t0 tf]` specifies that the system of differential equations (E.2) will be integrated from instant `t0` to `tf`. The $N \times 1$ vector `y0` provides ICs $\mathbf{y}(0) = [y_1(0) \quad \dots \quad y_N(0)]^T$ for the system of ODEs. The function `dydt` is a user-defined function that takes two arguments: a scalar `t` and an $N \times 1$ column vector `y`. This function returns a column vector of data type single or double, representing the derivative $\mathbf{h}(t, \mathbf{y})$, i.e., the right-hand side of the ODE (E.2). On the other hand, the output `t` is a column vector of time values, and the solution array `y` contains the corresponding solution values. Each row in `y` corresponds to a value returned in the column vector `t`, and the N columns refer to solutions for y_1 up to y_N.

For demonstration purposes, consider the IVP

$$\dot{y} = -y, \tag{E.13}$$

with IC $y(0) = 1$. Suppose one wants to solve this IVP for $t \in [0, 5]$. Define the time span and IC in the MATLAB command window as follows:

```
>> tspan = [0,5];
>> y0 = 1;
```

The ODE function `dydt` may be created in a separate function file. But it is more didactic and easier to visualize if anonymous functions are used instead. An anonymous function is not stored in a program file but is associated with a variable, which is of function handle data type:

```
>> dydt = @(t,y) -y;
```

Finally, the IVP can be solved using the following code:

```
>> [t,y] = ode45(dydt,tspan,y0);
>> plot(t,y)
```

The solution is plotted in Fig. E.1. The data points represent the computed values of y at different time instants.

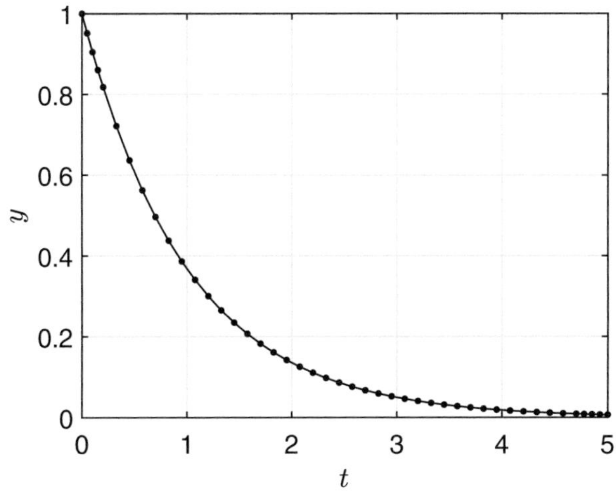

Fig. E.1 Numerical solution of IVP (E.13)

The process of solving a system of ODEs is similar to that of a scalar equation. Consider the van der Pol equation as an example:

$$\ddot{x} - \mu(1 - x^2)\dot{x} + x = f_0 \cos(\omega t), \tag{E.14}$$

where $\mu = 0.6$, $f_0 = 1.2$, $\omega = \pi/5$, and the ICs are $x(0) = 0.5$ and $\dot{x}(0) = 0$. To use the `ode45` function, convert this second-order ODE into a first-order system of equations by introducing the change of variables $y_1 = x$ and $y_2 = \dot{x}$:

$$\frac{d}{dt}\begin{bmatrix} y_1 \\ y_2 \end{bmatrix} = \begin{bmatrix} y_2 \\ \mu(1 - y_1^2)y_2 - y_1 + f_0 \cos(\omega t) \end{bmatrix}. \tag{E.15}$$

To solve this equation over the interval [0, 60], open MATLAB command window and enter the following:

```
>> tspan = [0,60];
>> mu = 6;
>> f0 = 1.2;
>> w = pi/5;
```

Since this is a system of two equations, it is necessary to provide a vector of ICs $\mathbf{y}(0) = [y_1(0) \quad y_2(0)]^T$:

```
>> y0 = [0.5; 0];
```

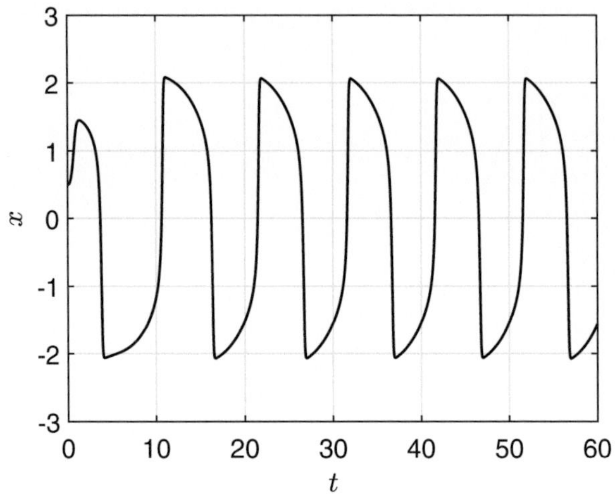

Fig. E.2 Numerical solution of the van der Pol equation

The next line defines a function for the ODE and integrates it using `ode45` to produce the plot shown in Fig. E.2:

```
>> dydt = @(t,y) [y(2);...
        mu*(1 - y(1)^2)*y(2) - y(1) + f0*cos(w*t)];
>> [t,y] = ode45(dydt, tspan, y0);
>> plot(t, y(:,1))
```

The input `dydt` is now a 2×1 vector function, and the order of the equations in each line should correspond to the order in `y0`. Additionally, the output `y` now has two columns: the first column represents the solution vector for $y_1 = x$, and the second column represents the solution vector for $y_2 = \dot{x}$.

E.4 Remarks on Stiff Problems

The methods discussed in Sects. E.2 and E.3 are known as *explicit* one-step methods because they only rely on the system's state at time t to advance the solution to the next time step $t + \Delta t$. In contrast, implicit methods involve solving nonlinear equations that include both the current state and the future state of the system in order to advance the solution. An example of an implicit method is the *backward Euler method*:

$$\mathbf{y}(t + \Delta t) = \mathbf{y}(t) + \mathbf{h}(t + \Delta t,\ \mathbf{y}(t + \Delta t))\Delta t. \tag{E.16}$$

Explicit Runge–Kutta methods are characterized by having $a_{lj} = 0$ for $l \leq j$, and all others are implicit.

The extra computational cost associated with implicit methods arises from solving a nonlinear system of equations at each time step. However, this cost is justified when dealing with *stiff* problems. Stiff problems occur in physical situations where there are significant variations in time scales. Explicit methods require excessively small time steps to yield accurate results for stiff problems, whereas implicit methods provide accurate results at a reasonable computational cost. To illustrate this, consider the IVP

$$\ddot{x} + 101\dot{x} + 100x = 0. \tag{E.17}$$

This ODE is stiff because it exhibits two distinct timescales, as evidenced by its general solution:

$$x(t) = Ae^{-t} + Be^{-100t}. \tag{E.18}$$

The following lines of code can be used in MATLAB to solve this IVP with `ode45`:

```
>> A = [0,1;-100,-101];
>> dydt = @(t,y) A*y;
>> tspan = [0,10];
>> y0 = [1;-1];
>> [t,y] = ode45(dydt,tspan,y0);
```

The number of time steps taken can be checked by examining the number of entries in the `t` array:

```
>> length(t)
ans =
        1093
```

MATLAB took a large number of steps to solve this simple linear IVP because `ode45` uses explicit methods. The implicit solver `ode15s` can be used instead to solve this problem with way less effort:

```
>> A = [0,1;-100,-101];
>> dydt = @(t,y) A*y;
>> tspan = [0,10];
>> y0 = [1;-1];
>> [t,y] = ode15s(dydt,tspan,y0);
>> length(t)
ans =
    43
```

There are plenty of ODE solvers available in MATLAB, and they all share a similar syntax. The `ode45` solver is versatile and should be the first choice for most problems. However, if `ode45` fails to produce accurate results or is inefficient, and if it is suspected that the problem is stiff, then `ode15s` can be employed. This function uses a variable order, variable step-size backward differentiation formula method combined with a nonstiff Adams–Moulton method. This approach allows `ode15s` to handle stiff problems more efficiently than explicit methods. Additionally, `ode15s` is suitable for solving differential algebraic equations, which are systems of differential equations where some variables appear in the equations without their derivatives.

Bibliography

1. Andronov, A.A., Vitt, A.A.F., Khaikin, S.E.: Theory of Oscillators, Adiwes International Series in Physics, vol. 4. Pergamon Press, Oxford (1966)
2. Baker, G.L., Blackburn, J.E.: The Pendulum: a Case Study in Physics. Oxford University Press, Oxford (2005)
3. Balachandran, B., Magrab, E.B.: Vibrations, 3rd edn. Cambridge University, Cambridge (2018)
4. Balakrishnan, V.: Mathematical Physics: Applications and Problems. Springer, Cham (2020). https://doi.org/10.1007/978-3-030-39680-0
5. Baruh, H.: Analytical Dynamics. McGraw-Hill, New York (1999)
6. Benaroya, H., Nagurka, M., Han, S.M.: Mechanical Vibration: Theory and Application, 5th edn. Rutgers University Press, New Brunswick (2022)
7. Betty Jr., M.F.: Principles of Engineering Mechanics: Kinematics—the Geometry of Motion, vol. 1. Springer, Berlin (2006)
8. Betty Jr., M.F.: Principles of Engineering Mechanics: Dynamics—the Analysis of Motion, vol. 2. Springer, Berlin (2006)
9. Billingsley, J.: Essentials of Dynamics and Vibrations. Springer, Cham (2017). https://doi.org/10.1007/978-3-319-56517-0
10. Brogan, W.L.: Modern Control Theory, 3rd edn. Prentice Hall, Upper Saddle River (1991)
11. Burden, R.L., Faires, J.D.: Numerical Analysis, 9th edn. Brooks/Cole, Pacific Grove (2011)
12. Casey, J.: Elements of Dynamics. Unpublished notes
13. Casey, J.: The bobbing buoy. Phys. Teach. **26**, 33 (1988). https://doi.org/10.1119/1.2342414
14. Chapra, S.C.: Applied Numerical Methods with Matlab for Engineers and Scientists, 3rd edn. McGraw Hill, New York (2012)
15. Childs, D.W., Conkey, A.P.: Dynamics in Engineering Practice, 11 edn. CRC Press, New York (2015)
16. Chopra, A.K.: Dynamics of Structures: Theory and Applications to Earthquake Engineering, 5th edn. Pearson, Hoboken (2017)
17. Churchill, R.V.: Operational Mathematics, 3rd edn. McGraw-Hill, New York (1958)
18. Clough, E.W., Penzien, J.: Dynamics of Structures, 2nd edn. McGraw-Hill, New York (1993)
19. Craig Jr., R.R., Kurdila, A.J.: Fundamentals of Structural Dynamics, 2nd edn. Wiley, New York (2006)
20. Den Hartog, J.P.: Mechanical Vibrations, reprint of 1956 edn. Dover, New York (1985)
21. Ewins, D.J.: Modal Testing: Theory and Practice, 2nd edn. Wiley, Philadelphia (2000)
22. Fabien, B.: Analytical System Dynamics: Modeling and Simulation. Springer, New York (2009). https://doi.org/10.1007/978-0-387-85605-6

R. Gonçalves Salsa Junior, *Fundamentals of the Theory of Mechanical Vibrations*,
https://doi.org/10.1007/978-3-031-83995-5

23. Gander, W., Gander, M.J., Kwok, F.: Scientific Computing—an Introduction Using Mapple and Matlab. Springer, Berlin (2014). https://doi.org/10.1007/978-3-319-04325-8
24. Gantmacher, F.R.: The Theory of Matrices, vol. 1. Chelsea Publishing Company, New York (1960)
25. Gawronski, W.K.: Dynamics and Control of Structures: A Modal Approach. Springer, New York (1998)
26. Genta, G.: Vibration Dynamics and Control. Springer, New York (2009)
27. Ginsberg, J.H.: Advanced Engineering Dynamics, 2nd edn. Cambridge University, Cambridge (1995)
28. Gohberg, I., Lancaster, P., Rodman, L.: Matrix Polynomials. Academic Press, New York (1982)
29. Golan, J.S.: The Linear Algebra A Beginning Graduate Student Ought To Know, 3rd edn. Springer, New York (2012)
30. Greenwood, D.T.: Principles of Dynamics, 2nd edn., pp. 450-455. Prentice Hall, Englewood Cliffs (1988)
31. Grigoriu, M.D.: Linear Dynamical Systems. Springer, Cham (2021). https://doi.org/10.1007/978-3-030-64552-6
32. Hagedorn, P.: Nonlinear Oscillations, 2nd edn. Oxford University, Oxford (1988)
33. He, J., Fu, Z.F.: Modal Analysis. Butterworth-Heinemann, Oxford (2001)
34. Horn, R.A., Johnson, C.R.: Matrix Analysis, 2nd edn. Cambridge University, New York (2013)
35. Humar, J.L.: Dynamics of Structures, 3rd edn. CRC Press, Boca Raton (2012)
36. Hurty, W.C., Rubinstein, M.F.: Dynamics of Structures. Prentice-Hall, Englewood Cliffs (1964)
37. Ince, L.E.: Ordinary Differential Equations. Dover, New York (1956)
38. Inman, D.J.: Vibration with Control, Measurement, and Stability. Prentice Hall, New Jersey (1989)
39. Inman, D.J.: Engineering Vibration, 4th edn. Prentice Hall, New Jersey (2013)
40. Jardin, M.R., Mueller, E.R.: Optimized measurements of unmanned-air-vehicle mass moment of inertia with a Bifilar Pendulum. J. Aircr. **46**, 3 (2009). https://doi.org/10.2514/1.34015
41. Jordan, D.W., Smith, P.: Nonlinear Ordinary Differential Equations, 4th edn. Oxford University, Oxford (2007)
42. Kelly, S.G.: Mechanical Vibrations: Theory and Applications, 2nd edn. Cognella Academic Publishing, San Diego (2022)
43. Khalil, H.K.: Nonlinear Systems, 3rd edn. Prentice Hall, Upper Saddle River (2002)
44. Kreyszig, E.: Advanced Engineering Mathematics, 10th edn. Wiley, Hoboken (2011)
45. Lancaster, P., Tismenetsky, M.: The Theory of Matrices, 2nd edn. Academic Press, San Diego (1985)
46. Lancaster, P.: Lambda-Matrices and Vibrating Systems, reprint of the 1966 edn. Dover, New York (2002)
47. Lanczos, C.: The Variational Principles of Mechanics, 4th edn. University of Toronto Press, Toronto (1970)
48. Lathi, B.P., Green, R.: Linear Signals and Systems, 3rd edn. Oxford University, Oxford (2018)
49. LeVeque, R.J.: Finite Difference Methods for Ordinary and Partial Differential Equations: Steady-State and Time-Dependent Problems. SIAM, Philadelphia (2007)
50. Ma, F.: Theory of Vibration. Unpublished notes.
51. McKenna, P.J., Tuama, C.O.: Large torsional oscillations in suspension bridges visited again: vertical forcing creates torsional response. Am. Math. Mon. **108**, 8 (2001). https://doi.org/10.1080/00029890.2001.11919805
52. Mathworks: Choose an ODE Solver. https://www.mathworks.com/help/matlab/math/choose-an-ode-solver.html
53. Mazzoleni, M.J., Krone, M.B., Mann, B.P.: Dynamics of rocking semicircular, parabolic, and semi-elliptical disks: equilibria, stability, and natural frequencies. J. Vib. Acoust. **137**, 4 (2015). https://doi.org/10.1115/1.4030169
54. Meirovitch, L.: Analytical Methods in Vibrations. Macmillan, New York (1967)
55. Meirovitch, L.: Fundamentals of Vibrations. McGraw-Hill, New York (2001)

56. Moler, C.B.: Numerical Computing with Matlab. SIAM, Philadelphia (2004)
57. Müller, P.C., Schiehlen, W.O.: Forced Linear Vibrations. Springer, Berlin (1977)
58. Müller, P.C., Schiehlen, W.O.: Linear Vibrations. Martinus Nijhoff Publishers, Dordrecht (1985)
59. Nakayama, T., Shima, H.: Higher Mathematics for Physics and Engineering. Springer, Berlin (2010). https://doi.org/10.1007/b138494
60. Newland, D.E.: Mechanical Vibration Analysis and Computation. Longman, Essex (1989)
61. Nayfeh, A.H., Mook, D.T.: Nonlinear Oscillations. Wiley, New York (1979)
62. Nayfeh, A.H., Pai, P.F.: Linear and Nonlinear Structural Mechanics. Wiley-Interscience, Hoboken (2004)
63. O'Reilly, O.M.: Engineering Dynamics: A Primer, 3rd edn. Springer, Cham (2019). https://doi.org/10.1007/978-3-030-11745-0
64. O'Reilly, O.M.: Intermediate Dynamics for Engineers: Newton-Euler and Lagrangian Mechanics, 2nd edn. Cambridge University, Cambridge (2020)
65. Pain, H.J.: The Physics of Vibrations and Waves, 6th edn. Wiley, New York (2005)
66. Palm, W.J.: Mechanical Vibration. Wiley, New York (2007)
67. Pars, L.A.: A Treatise on Analytical Dynamics. Wiley, New York (1965)
68. Preumont, A.: Twelve Lectures on Structural Dynamics. Springer, Dordrecht (2015). https://doi.org/10.1007/978-94-007-6383-8
69. Rade, D.A., Steffen Jr., V.: Absorbers, vibration. In: Encyclopedia of Vibration, vol. 1. Academic Press, New York (2001). https://doi.org/10.1006/rwvb.2001.0176
70. Rao, S.S.: Mechanical Vibrations, 6th edn. Pearson, London (2016)
71. Rayleigh, J.W.S.: The Theory of Sound, vol. 1, reprint of the 1894 edn. Dover, New York (1945)
72. Roseau, M.: Vibrations in Mechanical Systems: Analytical Methods and Applications. Springer, New York (1987)
73. Rosenberg, R.M.: Analytical Dynamics of Discrete Systems. Plenum Press, New York (1977)
74. Rossing, T.D., Fletcher, N.H.: Principles of Vibration and Sound, 2nd edn. Springer, New York (2004). https://doi.org/10.1007/978-1-4757-3822-3
75. Salsa, R.G., Ma, F.: Advances in the Theory of System Decoupling. Springer, Cham (2020). https://doi.org/10.1007/978-3-030-60846-0
76. Shampine, L.F., Reichelt, M.W.: The matlab ode suite. SIAM J. Sci. Comput. **18**, 1 (1997). https://doi.org/10.1137/S106482759427642
77. Schiehlen, W., Eberhard, W.: Applied Dynamics. Springer, Cham (2014). https://doi.org/10.1007/978-3-319-07335-4
78. Schmerr, L.W.: Engineering Dynamics 2.0: Fundamentals and Numerical Solutions. Springer, Cham (2019). https://doi.org/10.1007/978-3-319-98470-4
79. Schmitz, T.L., Smith, K.S.: Mechanical Vibrations: Modeling and Measurement. Springer, New York (2014). https://doi.org/10.1007/978-1-4614-0460-6
80. Shabana, A.: Vibration of Discrete and Continuous Systems, 3rd edn. Springer, Cham (2019). https://doi.org/10.1007/978-3-030-04348-3
81. Spivak, M.: Physics for Mathematicians, Mechanics I. Publish or Perish (2010)
82. Srinivasan, P.: Nonlinear Mechanical Vibrations. Wiley, New York (1996)
83. Strogatz, S.H.: Nonlinear Dynamics and Chaos. Perseus Books, Reading (1994)
84. Synge, J.L., Griffith, B.A.: Principles of Mechanics, 2nd edn. McGraw-Hill, New York (1949)
85. Szeidl, G., Kiss, L.P.: Mechanical Vibrations: An Introduction. Springer, Cham (2020). https://doi.org/10.1007/978-3-030-45074-8
86. Tang, K.: Mathematical Methods for Engineers and Scientists 2: Vector Analysis, Ordinary Differential Equations and Laplace Transforms. Springer, Berlin (2006). https://doi.org/10.1007/978-3-540-30270-4
87. Tang, K.: Mathematical Methods for Engineers and Scientists 3: Fourier Analysis, Partial Differential Equations and Variational Methods. Springer, Berlin (2006). https://doi.org/10.1007/978-3-540-44697-2

88. Thomsen, J.J.: Vibrations and Stability: Advanced Theory, Analysis, and Tools, 2nd edn. Springer, Berlin (2013). https://doi.org/10.1007/978-3-662-10793-5
89. Thomson, W.T.: Theory of Vibration with Applications, 4th edn. CRC Press, New York (1996)
90. Tongue, B.H.: Principles of Vibration, 2nd edn. Oxford University, Oxford (2001)
91. Weaver Jr., W., Timoshenko, S.P., Young, D.H.: Vibration Problems in Engineering, 5th edn. Wiley-Interscience, New York (1990)
92. Wiggins, S.: Introduction to Applied Nonlinear Dynamical Systems and Chaos, 2nd edn. Springer, New York (2003)

Index

R. Gonçalves Salsa Junior, *Fundamentals of the Theory of Mechanical Vibrations*,
https://doi.org/10.1007/978-3-031-83995-5

MIX
Papier aus verantwortungsvollen Quellen
Paper from responsible sources
FSC® C105338

If you have any concerns about our products,
you can contact us on
ProductSafety@springernature.com

In case Publisher is established outside the EU,
the EU authorized representative is:
Springer Nature Customer Service Center GmbH
Europaplatz 3, 69115 Heidelberg, Germany

Printed by Libri Plureos GmbH
in Hamburg, Germany